About Island Press

Island Press is the only nonprofit organization in the United States whose principal purpose is the publication of books on environmental issues and natural resource management. We provide solutions-oriented information to professionals, public officials, business and community leaders, and concerned citizens who are shaping responses to environmental problems.

In 1994, Island Press celebrates its tenth anniversary as the leading provider of timely and practical books that take a multidisciplinary approach to critical environmental concerns. Our growing list of titles reflects our commitment to bringing the best of an expanding body of literature to the environmental community throughout North America and the world.

Support for Island Press is provided by The Geraldine R. Dodge Foundation, The Energy Foundation, The Ford Foundation, The George Gund Foundation, William and Flora Hewlett Foundation, The James Irvine Foundation, The John D. and Catherine T. MacArthur Foundation, The Andrew W. Mellon Foundation, The Joyce Mertz-Gilmore Foundation, The New-Land Foundation, The Pew Charitable Trusts, The Rockefeller Brothers Fund, The Tides Foundation, Turner Foundation, Inc., The Rockefeller Philanthropic Collaborative, Inc., and individual donors.

THE OCEANS AND ENVIRONMENTAL SECURITY

IN MEMORY OF OUR FRIEND AND COLLEAGUE
JAMES M. BROADUS
FEBRUARY 24, 1947 – SEPTEMBER 28, 1994

THE OCEANS AND ENVIRONMENTAL SECURITY

Shared U.S. and Russian Perspectives

Edited by
James M. Broadus and
Raphael V. Vartanov

Contributing Authors
Yuri G. Barsegov, Anna Bistrova, Lawson W. Brigham, James M. Broadus, Jonathan I. Charney, Suzanne M. Demisch, Mark Eiswerth, Arthur G. Gaines, Jr., Kristina Gjerde, Peter M. Haas, J. Christopher Haney, Yoshiaki Kaoru, Anna Korolenko, Vladimir Korzun, Matthew J. LaMourie, Philip A. McGillivary, Elena N. Nikitina, M. J. Peterson, Giulio Pontecorvo, Alexei Yu. Roginko, Artemy A. Saguirian, Mary Schumacher, Michael N. Soloviev, Tom Tietenberg, Sergey N. Tikhomirov, Raphael V. Vartanov, Michael Vexler, Miranda Wecker

ISLAND PRESS

Washington, D.C. □ Covelo, California

Copyright © 1994 by Island Press

All rights reserved under International and Pan-American Copyright Conventions. No part of this book may be reproduced in any form or by any means without permission in writing from the publisher: Island Press, 1718 Connecticut Avenue N.W., Suite 300, Washington, DC 20009.

Island Press is a trademark of the Center for Resource Economics.

Versions of some of the material in chapters 3 and 6 appeared previously in *Marine Policy*, volume 16, number 4 (Mirovitskaya and Haney 1992; Roginko and LaMourie 1992).

Library of Congress Cataloging-In-Publication Data
The Oceans and environmental security : shared US and Russian perspectives / James M. Broadus and Raphael V. Vartanov, editors (Marine Policy Center, Woods Hole Oceanographic Institution and Institute of World Economy and International Relations, Russian Academy of Sciences).
    p. cm.
    Includes bibliographical references and index.
    ISBN 1-55963-235-6 (cloth).—ISBN 1-55963-236-4 (paper)
    1. Marine resources conservation—Law and legislation.   2. Marine pollution—Law and legislation.   3. Environmental law, International.   4. Hazardous substances—Transportation—Law and legislation.   5. Radioactive pollution of the sea—Law and legislation.   6. Maritime law—Russia (Federation)   7. Maritime law—United States.   I. Broadus, James M.   II. Vartanov, R. V. (Rafael Vramovich)   III. Marine Policy Center (Woods Hole Oceanographic Institution)   IV. Institut mirovoĭ ekonomiki i mezhdunarodnykh otnosheniĭ (Rossiĭskaia akademiia nauk)
K3485.6.027 1994
333.91—dc20   93-48894
                                                                                    CIP

Printed on recycled, acid-free paper.

Manufactured in the United States of America
10 9 8 7 6 5 4 3 2 1

Contents

Preface xi
Acknowledgments xv
About the Contributors xvii

CHAPTER 1. *Introduction* 3

    Evolving Notions of Security 6
    The Emergence of the Concept of Environmental Security 7
    Threats to Environmental Security 13

CHAPTER 2. *Land-based Marine Pollution: The Gulf of Mexico and the Black Sea* 17

    Regional Cooperation 21
    Comparative Cases 24
    The Gulf of Mexico 24
    The Black Sea 39
    Conclusions 48

CHAPTER 3. *Living Resource Problems: The North Pacific* 50

    The Ecosystems of the Bering Sea and the Adjacent North Pacific 51
    Fisheries Depletion 53
    Ecosystem Perturbations from Fisheries Depletion 58
    The Doughnut Hole and the Problem of Straddling Stocks 60
    Economic Considerations for Straddling Stocks 61
    Legal Options for Straddling Stocks 63
    Driftnet Fisheries and Incidental Take 67
    Policy Implications of Incidental Take 70
    National Legal Initiatives on Driftnets 72
    International Legal Responses to Living Resource Problems 76

Gaps in Scientific Information and Future Research Priorities   81
A Regional Regime for Environmental Security   82

CHAPTER 4. *Hazardous Materials Transport*   86

Problems of Definition, Classification, and Information   89
The Existing Legal and Regulatory Structure   92
Regulatory Inadequacies and Their Implications for Marine Environmental Protection   99
Economic Perspectives on Seaborne Hazardous Materials   111
Technology for Hazardous Materials Transport   117
Conclusions   120

CHAPTER 5. *Radioactivity in the Oceans*   122

Potential Sources of Radioactivity in the Oceans   123
Projections of Future Sources of Contamination   131
Nuclear Proliferation and the Threat to the Oceans   140
Public Perceptions and Resulting Countermeasures   142
Existing Legal Rights, Obligations, and Controls   152
Conclusions   157

CHAPTER 6. *Environmental Protection for the Arctic Ocean*   163

Environmental Vulnerabilities in the Arctic Basin   163
Current Economic Development Activities in the Arctic   166
Cooperative Agreements and the Emerging Multilateral Regime   172
The Role of Political Developments in Russia   185
Prospects and Prerequisites for Success   187

CHAPTER 7. *The Southern Ocean*   190

The Existing Legal Rules   190
Competing Management Principles   195
Environmental Hazards in the Southern Ocean   201
Which Regime?   216

CHAPTER 8. *The Law of the Sea*   223

The Sources of International Environmental Law   224
An Overview of the Law of the Sea Convention, Part XII   226
Pollution Control and Ecosystem Protection   228

The Law of the Sea Convention as a Framework for
   Assurance Building   235
Conclusions   239

CHAPTER 9. *Conclusions*   245

Critical Problems of International Environmental
   Security   247
Cross-Cutting Themes   249

Notes   256
References   279
List of Abbreviations   311
Index   315

Preface

THIS BOOK BEGAN as a multidisciplinary, collaborative research project among scholars from our two institutes: the Institute for World Economy and International Relations (IMEMO) of the Russian Academy of Sciences and the Woods Hole Oceanographic Institution (WHOI) in the United States. The project was suggested initially by Raphael Vartanov and his IMEMO colleagues during a visit to Moscow by WHOI's James Broadus. IMEMO's Department of Oceans and Environment is virtually an exact counterpart to WHOI's Marine Policy Center in terms of size (15 to 20 scholars), disciplinary orientation (economics and law), research emphasis (oceans, environment, and international relations), and location within a larger research organization. The idea was to compare and combine the thinking of the two groups on the concept of environmental security as it applies to the world's oceans, to define the concept precisely, and to identify and analyze environmental security problems of high mutual interest to our two countries.

Our work together began in 1990, when Mikhail Gorbachev's Soviet reforms were just taking hold and Russia was still the USSR. Much of the material in the book was completed along the way, as the Cold War ended, the Soviet Union disintegrated, and Russia was cast into deepening economic and political peril. These massive and continuing changes, along with more routine developments in ocean environmental affairs, have challenged the editors to keep the book up to date. We have tried our best to do so and believe most of the material included is current through early 1994, with pending or expected future changes noted in the text.

We hope this book will be interesting and useful to a broad spectrum of readers. It offers an introduction to the concept of environmental security and an overview of several of the most prominent ocean environmental problems. Our treatment of the subject is intended to be thorough enough to serve as a reference for ocean policy professionals and accessible enough to provide an introduction for students and other newcomers to the subject.

This is a multiauthored book, with twenty-nine American and Russian scholars contributing. Working together on this project, we discovered some apparent Russian–American differences in outlook and approach. The Russian scholars seemed to favor more holistic and comprehensive treatments than did many of the Americans, who tended toward more reductionist, analytical approaches. As a group, the Russian scholars also seemed somewhat more internationalist in orientation than did the Americans. The Russians sometimes pushed harder for consideration of multilateral and global approaches to problems, whereas the Americans tended to look first to bilateral bargains or regional arrangements. The Russians seemed to exhibit more faith in international institutions than did the Americans, and they appeared less comfortable than the Americans in accepting the inevitability of uncertainties and probabilities, tending instead to seek institutional arrangements that would eliminate risk and provide security guarantees. Of course, these differences are only subtle tendencies, and they probably depend more on the individuals involved than on their nationalities. Despite group tendencies, the most insistently holistic and enthusiastically internationalist individuals were American; some of the Russian scholars, on the other hand, were as analytically reductionist as any of the Americans.

The book's many co-authors teamed up in different combinations for the different chapters. They also contributed in varying degrees and left the editors with drafts of the chapters in varying states of completion. Some chapters were written entirely by just two or three co-authors, others were written largely by a few co-authors with some contributions by several colleagues, and yet others were a collection of materials from many hands. It was the editors' job to join these contributions into a consistent flow, to keep the material more or less up to date, to referee and resolve internal inconsistencies and remaining differences of opinion, and generally to tie up loose ends.

Because each chapter is intended as a largely freestanding treatment of its topic, some repetition was inevitable, and for this we beg the reader's patience. Although general agreement was achieved among most co-authors on most issues, no effort was made to reach consensus. The co-authors left completion and resolution of unsettled or evolving questions in the hands of the editors. Each chapter, therefore, may not fully represent the views of all of its contributors. Indeed, some co-authors may well disagree with some points made in chapters to which they contributed or made elsewhere in the book. For this reason and to reflect the fact that this was truly a joint effort in which all contributed to the whole, we grouped all co-authors together in the list of contributors (which indicates the chapters to which they contributed most directly). The editors salute their collegiality, hard work, honest inquiry, and commitment to an improved ocean environment.

James M. Broadus
Raphael V. Vartanov
Woods Hole, Massachusetts

Acknowledgments

WE ARE GRATEFUL FOR the support provided by the John D. and Catherine T. MacArthur Foundation, the Peace Research Institute of the Russian Academy of Sciences, the Woods Hole Oceanographic Institution and its Marine Policy Center, IMEMO and its Department of Oceans and Environment, and Island Press. Partial support for work on one of the case studies was provided by the National Sea Grant College Program Office of the National Oceanic and Atmospheric Administration, US Department of Commerce, and occasions for collaborative work by the co-editors were made possible in part by a grant from IREX, the International Research and Exchange Board. The book's many authors and two editors also appreciate the guidance, input, and assistance provided by several interested and knowledgeable individuals: Ruth Adams, Igor M. Averin, Kennette Benedict, Craig Dorman, Scott Farrow, Sarah Gille, Abraham Hallenstvedt, Charles Hollister, Michael Horgan, Levan B. Imnadze, Lee Kimball, Alexander K. Kislov, Hauke Kite-Powell, Anatoliy L. Kolodkin, Ivan S. Korolev, Hugh Livingston, Vitaliy N. Lystsov, Vladlen A. Martinov, Ilya M. Mogilyovkin, Rem A. Novikov, Horace B. Robertson, Fred Sayles, Tucker Scully, Sid Smith, Tim D. Smith, Catherine Tinker, Marilyn Tischbin, Nikolai N. Vorontsov, Dolores Wesson (chapter 4), and Alexei V. Yablokov. Any faults remaining in the book are, of course, our own.

Special credit is due Mary Schumacher for her superb editing, book project management, research assistance, and co-authorship. Expert administrative, research, and editing suport were provided in Woods Hole by Suzanne Demisch, Kim Fahnley, Ellen Gately, and Matthew LaMourie and in Moscow by Karen V. Antonyan, Elena N. Nikolskaya, Marina A. Svanidze, and Vera I. Torchilina. The encouragement, guidance, and patience received from Joe Ingram and Christine McGowan at Island Press are gratefully acknowledged.

About the Contributors

Yuri G. Barsegov (chapters 7 and 8) is head of the Department of Oceans and Environment at the Institute of World Economy and International Relations (IMEMO) of the Russian Academy of Sciences, where he specializes in scientific research on international relations and international law. Barsegov was a member of the Soviet delegation to the Preparatory Commission for the International Sea Bed Authority and the International Tribunal for the Law of the Sea in 1986–87 and a member of the UN International Law Commission from 1987 to 1989. In addition to three books, he has published more than 200 articles on various problems of international law and international relations in Soviet/Russian and foreign journals and newspapers.

Anna Bistrova (chapter 2) is a senior research fellow in the Department of Oceans and Environment, IMEMO. Her research concentration is in international environmental regimes, with particular emphasis on problems of environmental security in the Baltic and Black Sea regions.

Lawson W. Brigham (chapter 6) is a captain in the US Coast Guard, currently serving as commanding officer of the icebreaker *Polar Sea*. A career seagoing officer, Captain Brigham has also served as chief of the Coast Guard's Office of Strategic Planning. He has published extensively on the Soviet and Russian Arctic, polar ship technology, and US Arctic policy. During 1989–90 he was a marine policy research fellow at the Woods Hole Oceanographic Institution (WHOI), where he edited the volume *The Soviet Maritime Arctic*.

James M. Broadus (volume editor and chapters 1, 2, 4, 5, and 9) is director of the Marine Policy Center, WHOI, and an economist with interests in marine resources, the organization and regulation of marine industries, environmental economics, and the economic effects of environmental change. Dr. Broadus has served on a number of national and international advisory bodies, including GESAMP, IPCC, UNEP Regional Working Groups, the Man and the Biosphere Program, and the National Research Council's Marine Board.

Jonathan I. Charney (chapter 1) is a professor of law at Vanderbilt University School of Law. He currently serves as chair of the Senior Advisors Committee of the Marine Policy Center, WHOI, and as vice president of the American Society of International Law. Professor Charney is also a member of the editorial boards of the *American Journal of International Law* and *Ocean Development and International Law*. As project director of the Maritime Boundary Project of the American Society of International Law since 1988, he edited the two-volume treatise *International Maritime Boundaries*, which was published by the society in 1993 and is now being supplemented.

Suzanne M. Demisch (chapters 2 and 5) is a research assistant at the Marine Policy Center, WHOI, with an academic background in political science and international relations, concentrating on international ocean environmental issues. She served previously on the staff of the Office of International Activities, US Environmental Protection Agency, specializing in US-Soviet environmental cooperation.

Mark Eiswerth (chapters 2 and 6) is a senior associate with the international consulting firm of RCG/Hagler, Bailly, Inc. His areas of specialization are environmental and energy economics. Dr. Eiswerth's recent publications appeared in *Ecological Economics, Land Economics*, and *Theory, Modeling and Experience in the Management of Nonpoint-Source Pollution*.

Arthur G. Gaines, Jr. (chapter 2), a research specialist at the Marine Policy Center, WHOI, comes to coastal and ocean policy studies from a background in the earth sciences. Dr. Gaines' research interest is in applications of marine science and technology toward improved understanding of the use and protection of resources at the ocean's margin. His recent publications have focused on technological applications for improved maritime transportation safety and protection of the marine environment from lost cargoes.

Kristina Gjerde (chapters 2, 3, 4, and 8) is a research fellow at the Law School of the University of Hull in England, specializing in international marine environmental law. Her current research interests include regional regimes for the prevention of land-based pollution, evolution of the concept of particularly sensitive sea areas at the International Maritime Organization, and national approaches to marine biodiversity conservation.

Peter M. Haas (chapter 1) is associate professor of political science at the University of Massachusetts at Amherst and a former research fellow in WHOI's Marine Policy Center. He has published widely on international environmental subjects, including pollution control in the Mediterranean and in the Baltic and North seas, UNEP's regional seas programs, stratospheric ozone protection, and international environmental institutions. He is the author of *Saving the Mediterranean: The Politics of International Environmental Cooperation* and a co-editor of *Institutions for the Earth: Sources of Effective International Environmental Protection*.

J. Christopher Haney (chapters 3 and 6) is assistant professor of wildlife technology at the Pennsylvania State University at Dubois, where his research emphasis is in ecology and conservation of marine vertebrates. He is currently working with the P.P. Shirshov Institute of Oceanology and the Russian Institute of Nature Protection and Reserves on an ecosystem project in the Bering Sea sponsored by the US Department of State.

Yoshiaki Kaoru (chapters 2, 3, and 4) is an associate scientist at the Marine Policy Center, WHOI. A resource and environmental economist, his research interests and recent publications focus on valuation of the benefits of resource and environmental quality improvement, development and assessment of benefit valuation techniques, and economic policy analysis of marine resource management.

Anna Korolenko (chapter 2) is an assistant scientist at IMEMO who has published several articles in the field of the economic geography of the former Soviet Union.

Vladimir Korzun (chapter 7) is lead researcher at IMEMO, where he specializes in the integration of economics, ecology, and law. The author or co-author of more than 100 works, Dr. Korzun's current concentration is in the intersections between marine and fisheries research and environmental security.

Matthew J. LaMourie (chapters 4, 5, and 6), until recently a research assistant at the Marine Policy Center, WHOI, is presently completing a J.D. at Northeastern University School of Law. He has also worked at the US Maritime Administration and the US Department of Justice. His research interests include international environmental law and marine environmental management regimes.

Philip A. McGillivary (chapter 3), research associate, Center for Ocean Analysis and Prediction, NOAA, in Monterey, California, is an oceanographer with research interests in advanced technologies and their applications in fisheries and coastal resources management.

Natalia S. Mirovitskaya (chapter 3) is a senior researcher at IMEMO who has published numerous works on the biological and environmental security problems of the world ocean.

Elena N. Nikitina (chapters 1 and 2) is a senior researcher at IMEMO. Her current research interests include the problems of international environmental security, international environmental regimes and their national implementation, and Russian environmental policy. Among her many publications on environmental issues in both Russian and English is the book *World Meterological Organization and the World Ocean*.

M.J. Peterson (chapters 1 and 7) is associate professor of political science at the University of Massachusetts, Amherst, and a former WHOI marine policy research fellow. Her research focuses on international institutions. She is the author of *Managing the Frozen South: The Creation and Evolution of the Antarctic Treaty System*.

Giulio Pontecorvo (chapter 3) is a professor of economics and banking at the Columbia University Graduate School of Business. A member and past chairman of the Senior Advisors Committee of the Marine Policy Center, WHOI, Dr. Pontecorvo has also served on the Advisory Group to the Economics Division of the National Marine Fisheries Service, US Department of Commerce; the International Marine Science Affairs Policy Committee of the National Academy of Sciences; and the Executive Committee of the Law of the Sea Institute.

Alexei Yu. Roginko (chapters 4, 6, and 8) is a senior researcher at IMEMO. A graduate of the Geographical Faculty of Moscow State University, he has published numerous works on international marine and Arctic environmental protection issues. His current research examines the effectiveness of international environmental regimes.

Artemy A. Saguirian (chapters 1 and 4) is a senior research fellow at IMEMO, currently specializing in the theoretical and practical problems of marine policy, with particular emphasis on the formulation of international regimes.

Mary Schumacher (chapters 2, 5, and 6) is a research assistant at the Marine Policy Center, WHOI. Formerly a research fellow in national security at Harvard University, she has written a number of case studies on US national security policy and weapons acquisition programs, including an extended series on the Trident strategic system.

Michael N. Soloviev (chapters 4 and 6) is a research fellow at IMEMO and a retired first captain of the Soviet Navy. His current research interests include the environmental problems confronting the Russian Navy and merchant fleets and the main straits and channels of the world ocean.

Tom Tietenberg (chapters 2, 4, and 5) is Christian A. Johnson Distinguished Teaching Professor at Colby College, where he teaches in the economics department. He is the author or editor of seven books and more than 50 articles and essays on environmental and resource economics, including the best-selling textbook *Environmental and Natural Resource Economics*. As a senior research fellow at the Marine Policy Center, WHOI, in 1990–91, he edited the volume *Innovation in Environmental Policy: Economic Aspects of Recent Developments in Environmental Enforcement and Liability*. Dr. Tietenberg has served as a consultant to the World Bank, the US Agency for International Development, the US Environmental Protection Agency, and the United Nations.

Sergey N. Tikhomirov (chapter 5) is an assistant scientist at IMEMO. His research interests include environmental security for the Baltic and Black seas and international environmental regimes.

Raphael V. Vartanov (volume editor and chapters 1, 2, 5, 6, and 9) is head of the Section on Ocean Development and Environment, IMEMO, and from December 1991 through June 1994 was concurrently a senior fellow of the Marine Policy Center, WHOI. He is also a senior environmental advisor at the Education Development Center in Newton, MA; president of EcoDevelopment, a nongovernmental international environmental research organization; and a member of the editorial board of *Ocean and Coastal Management*. An economist and coauthor of some 50 articles, monographs, and books, Dr. Vartanov's research interests and publications focus on environmental problems and marine science policy, ocean development and management, and issues of international environmental security.

Michael Vexler (chapter 5) is a research fellow at IMEMO. His primary research interest is the problem of radioactive contamination of the world ocean.

Miranda Wecker (chapter 4) is an international legal scholar and consultant in ocean and coastal management. From 1985 to 1992 she served as associate director of the Council on Ocean Law (COL) in Washington, DC, and was also the editor of the COL periodical *Oceans Policy News*. She served as a US delegate to the Bureau of the Cartagena Convention for the Protection and Development of the Marine Environment of the Wider Caribbean Region and has worked in support of the Panel on the Law of Ocean Uses.

THE OCEANS AND ENVIRONMENTAL SECURITY

CHAPTER 1

Introduction

IN THE WAKE OF the Cold War, the global security agenda is shifting from a predominantly military conception of threat to a growing sense of urgency about economic, social, and environmental challenges. Despite the many signs of improving trust and cooperation between the United States and the former Soviet republics, there has been little apparent easing of our collective anxiety about the world's habitability for future generations. As the complexity and interdependence of the many factors involved become more apparent, so, too, does the inadequacy of our knowledge and of our current legal and institutional framework to support the sustainable global solutions that are required.

The need for greater openness, awareness, and cooperation is particularly pressing with respect to the natural environment, where the processes at work cannot be confined within national borders or readily reversed merely by a change of national policy. Recognition of these qualities and the potential for their exploitation give environmental affairs new geopolitical significance.

This reorientation is seen throughout the traditional defense and security establishments as they seek to define new roles for themselves in the post–Cold War world (Weston 1990; Oswald 1992). New norms of "military humanitarianism" are emerging, and American troops have been involved in an unprecedented series of relief efforts for both man-made and natural disasters (Weiss and Campbell 1991). Alliances devoted to military and economic security arrangements, such as the North Atlantic Treaty Organization (NATO), the Conference on Security and Cooperation in Europe (CSCE), and the Association of South East Asian Nations (ASEAN), are launching a variety of environmental affairs programs.

In the United States, a 1991 White House memorandum signed by President Bush referred to "dramatic changes in US defense planning" and the growing interest "in our intelligence services tackling new issues and problems," headed by "environment, natural resource scarcities" (White House Memorandum, cited in Funke 1992). Legislation for a Strategic Environmental Research and Development Program to convert strategic defense industries to work on environmental threats was introduced in the US Senate by Senate Armed Services Chairman Sam Nunn (D-GA), who spoke of "a new and different threat to our national security emerging—the destruction of the environment" (Nunn, cited in Funke 1992). The US military services, meanwhile, are devoting much greater attention to their own environmental practices (Kelso 1991; Broadus 1991; Shepherd 1993; Kraska 1993), and such military intelligence assets as subsea hydrophone acoustic data and nuclear submarine operations under the polar ice are for the first time being made available for civilian science (*Ocean Science News* 1993; *Washington Letter of Oceanography* 1993; *Marine Technology Society Journal* 1994).

In Russia, changes of this nature have been somewhat slower in coming, but the Russian military is nonetheless acutely aware of the country's environmental problems and of its own responsibility to contribute to their solution by bringing to bear the scientific and technological capabilities and assets that it controlled for so long. One example of recent initiatives in this direction is the State Committee for Special Underwater Operations Dealing with Nuclear Contamination, which is chaired by Tengiz Borisov, a senior officer in the Russian Navy ("Delayed-Action Mines in the Seas" 1993; WHOI 1993).

Moreover, increased attention is being given to armed conflict itself as an environmental threat (Schachte 1991; Oxman 1991; Caggiano 1993). An ominous example was Iraq's intentional release of Kuwaiti oil into the Persian Gulf in 1991 in an apparent attempt to impede a military assault and to contaminate supplies of Saudi Arabian drinking water (Readman et al. 1992; Plante 1992). Issues of responsibility, liability, and compensation for the resulting environmental damages were simplified somewhat in that case because a vanquished Iraq was forced militarily to accept terms of disengagement (see UN Sec. Res. 687 and 692 in *International Legal Materials* 1991a, 1991b).

Similar issues arise in other cases involving the marine environment, outside the context of armed conflict, and resolution of these issues cannot depend on military victories. Quite the opposite; an important goal of their resolution is to avoid or contain international conflict and to prevent or minimize environmental damage, to enhance *environmental security*.

Imagine the complications if resources and livelihoods in the Soviet Union had borne the brunt of damages from the 1989 *Exxon Valdez* oil spill. How would international institutions have assured the proper assignment of responsibility and a fair compensation for losses? Experience with transboundary effects from the 1986 Chernobyl disaster suggests that international mechanisms for dealing with accidental environmental damage are neither well developed nor effective. In fact, widely acknowledged mechanisms to govern oil transportation practices, to assign responsibility for accidental oil spill damages, to guide the resolution of disputes, and to counsel the allocation of compensation have evolved since World War II. Comparable mechanisms for the much more complex, potentially more dangerous problem of other hazardous seaborne cargoes or for transboundary toxic releases have yet to be worked out.

A similar problem involves contamination from nuclear devices. Published reports estimate that fifty nuclear weapons and twenty-three reactors have been lost at sea, all by the US and Soviet navies. In 1989, for instance, a Soviet *Mike*-class nuclear-powered submarine carrying two plutonium-tipped torpedoes sank in the Norwegian Sea. Iceland worried that mere suspicion of radioactive contamination of its fish products could damage its export markets. More recently, Russian authorities disclosed that a heavily armed *Yankee*-class nuclear submarine, which sank 500 miles east of Bermuda in 1986, is broken up on the sea floor and is leaking radioactivity into strong ocean currents (Broad 1994). Revelations of three decades of secret Soviet dumping of radioactive materials in the arctic seas have occasioned alarm among the region's indigenous populations and environmentalists everywhere, who have scarcely been reassured by preliminary scientific assessments finding no evidence of any regional-scale threat to human health or the environment (WHOI Conference Statement 1993). In October 1993, a dispute broke out between Japan and Russia over continued Russian dumping of low-level radioactive wastes in the Sea of Japan (Sanger 1993).

The concept of environmental security helps us think clearly about such problems. It draws on the widely understood ideas of international, strategic interdependence (in facing threats of nuclear war or economic collapse) to focus attention on the similarly shared exposure to threats from global environmental degradation. Implicit in the concept is a direct link to conventional understanding of international security arising from the potential for conflict over resource use and environmental practices.

Early in our collaboration we formulated a working definition:

Environmental security is the reasonable assurance of protection against threats to national well-being or the common interests of the international community associated with environmental damage.

Critical problems of international environmental security are those that are likely to destabilize normal relations between nations and to provoke international countermeasures.

Evolving Notions of Security

Security is a central concept of all politics. Individuals, groups, and political communities alike prefer safety to danger, predictability to unpredictability. Traditionally, security has involved assuring the physical survival of a political unit and of the individuals or groups populating it. Security has also included maintaining (and often expanding) the territorial domain of the unit, protecting its basic institutional structure against change imposed from outside, and assuring that domestic systems of command, control, production, and distribution are not disrupted by external forces.

A critical component here is not just undesirability itself but the fact of imposition by others. States wish to avoid any undesirable outcome, but those stemming from internal choice or error do not raise security concerns in quite the same way. Through conciliation, deterrence, and defense, each political community attempts to protect human life, territorial possessions, institutional continuity, economic viability, internal order, and a distinctive way of life.

Traditional writings on world politics, whether "realist" or "idealist," tend to define security in terms of *national security*—that is, from the point of view of the individual political unit. In the twentieth century, an emerging alternative conception treats security as an international system-level objective. Proponents of collective security, followers of the Grotian tradition of a "society of states" bound by common rules and adherents of the postwar "interdependence school," all believe that it is possible to combine nation-state autonomy with common rules to avoid or lessen armed conflict.

The national and international approaches are not irreconcilable but can reinforce each other. It has long been understood that unilateral actions intended to enhance security might in fact reduce it by provoking countermeasures. The dangers of the arms race were widely understood before 1914, and notions of a "security dilemma," in which actions viewed as defensive by the doer are perceived as aggressive by others, began to be articulated after World War II (Richardson 1938; Wolfers 1952; Herz 1959; Jervis 1978). The resulting notion of "strategic interdependence" describes situations where outcomes are determined not by the choice of any one player (be it an individual, a group, or a political community) but by the combination of choices made by all (e.g., Von Neumann and Morgenstern 1944, 1953; Shelling 1963; Shubik 1975). The focus on mutual dependence in avoiding the common aversion that underlay the postwar concept of a balance of terror was elaborated, when both superpowers developed second-strike capability, into the aptly named condition of mutually assured destruction.

More recently, stress on common aversion and the system level have converged toward the concept of environmental security, founded on a belief that individuals, groups, and political communities cannot be secure if they fail to take account of adverse consequences that may result from human behavior affecting the environment. Precursor notions may be traced back into the late eighteenth century, when Malthus (1798) provided early visions of disasters due to human populations' outgrowing of available food supplies. Late nineteenth and early twentieth century writers on geopolitics stressed the importance of climate, topography, and location for the organization, expansion, and perpetuation of political units (e.g., Mahan 1897; Mackinder 1904; Huntington 1924; Haushofer 1932a, 1932b). Though all these writers focused on constraints imposed by nature, they usually regarded these constraints as manageable through technology.

THE EMERGENCE OF THE CONCEPT OF ENVIRONMENTAL SECURITY

Only in the 1960s did concern about the natural environment congeal in the form of respect for the limits imposed by nature. The first wave of writings warned of the dangers of environmental degradation and advocated policies to avoid catastrophe, but they did not explicitly link environmental with security concerns (e.g., Carson 1962; Kennan 1970; Falk 1971; Meadows et al. 1972). Yet certain commonalities existed and could be used to bring the two concepts together. In particular, these writings and the responses they evoked all suggested a situation with some features reminiscent of the balance of terror. There was an obvious common aversion and enough analysis to show that the undesired outcome could not be avoided unilaterally, but only through compatible choices by many or all.

These similarities made it possible to merge security and environmental concerns in two ways. The first was a fairly simple stretch of traditional security notions to include maintenance of ecological balance within countries. This encouraged a new look at population pressures, soil erosion, deforestation, overexploitation of renewable resources, desertification, drought, and climate change as possible sources of international tension as people migrated or governments sought new territory in search of resources (e.g., Choucri and North 1975; Ullman 1990; Myers 1986, 1987a, 1987b, 1989; Homer-Dixon, Boutwell, and Rathjens 1993; Suhrke 1993). Environmental disruptions might even create new or exacerbate existing national social pressures sufficiently to inspire a challenge to the legitimacy of a government unable to provide sufficient resources to a needy population.

Yet the global or regional nature of many environmental problems has encouraged a broader conception, a notion of international environmental security requiring mutual self-restraint and cooperation among states (e.g., WCED 1987; Mathews 1989; Westing 1989). Just as the maintenance of international security requires acknowledgment of other countries' legitimate concerns for their own integrity and continuity, the maintenance of international environmental security requires acknowledgment that all humans exist on one planet and that all must help preserve its ability to sustain life. Just as avoiding an all-out nuclear exchange becomes a first priority of superpowers, so avoiding general environmental collapse becomes a first priority of all responsible states.

The most prominent efforts at elaborating the concept of international environmental security were made by the Socialist International, whose 18th Congress in 1989 produced the document "Toward Environmental Security: A Strategy for Long-Term Survival," and by the Soviet government under Mikhail Gorbachev. The Gorbachev team, eager to gain the trust of the United States and other democratic societies, seized upon environmental affairs as an arena in which they could demonstrate the Soviet Union's good faith and even stake out a position of international leadership. In a speech at Murmansk in October 1987, Gorbachev called for the establishment of an "arctic zone of peace," which provided the impetus for regional scientific and environmental cooperation among the Arctic and Nordic states. In turn, international attention to these and other environmental initiatives strengthened the Soviet leaders' hand as they challenged a massive network of entrenched interests to reverse many decades of systematic environmental neglect and abuse at home. *Glasnost* having begun to take root, the 1986 Chernobyl accident propelled environmental issues to the top ranks of the Soviet domestic agenda. They have remained there since, through the tumultuous transition to the successor government of Russian President Boris Yeltsin, and have also continued to be a centerpiece of Russia's international policies.

When the Yeltsin government confirmed in late 1992 the thirty-year history of extensive nuclear dumping by the Soviets in the seas and shallow coastal waters of the Arctic and when continued Russian dumping was observed in the Sea of Japan in late 1993, the revelations sparked serious studies and sober debate about the extent of the radioactive contamination and the degree of risk posed to human health and the ocean environment. Continuing cooperative efforts and dialogue provide encouraging signs of increasing candor and cooperation in the handling of international environmental affairs. At the same time, they exemplify how far we remain from achieving a level of scientific certainty, political consensus, and shared legal and regulatory norms that is adequate to address issues of global environmental security effectively.

Nowhere on the global environmental agenda are these shortfalls more apparent or more consequential than in our understanding and use of the shared resource of ocean space (Westing 1992; Johnston 1993). The world's ocean, as a single, uninterrupted medium, exemplifies the continuity and interconnectedness of environmental process (see map 1.1). The marine environment encompasses a wealth of resources and complex ecosystems, and its highly dynamic processes exert a tremendous influence over global climate and other conditions. Yet our basic approach to the wise use and protection of the oceans remains severely fragmented, with most initiatives focused narrowly on a targeted region (e.g., the Baltic or North Sea) or a contentious issue (e.g., fishing rights or technologies in specific fisheries or navigational practices in specific settings). Although such regional and issue-oriented approaches have proved valuable in alleviating some acute problems in the short term, their ultimate effectiveness depends on their consistency with national economic objectives, implementation through institutions of widely recognized legitimacy, and firm foundation in international law.

Self-interested voluntary actions and agreements by sovereign states must provide the basis for environmental security, as they do for international security more generally, but there is also an important place for constructive collective action and the development of binding norms of general international law. The United Nations system, with its relevant specialized organs—International Maritime

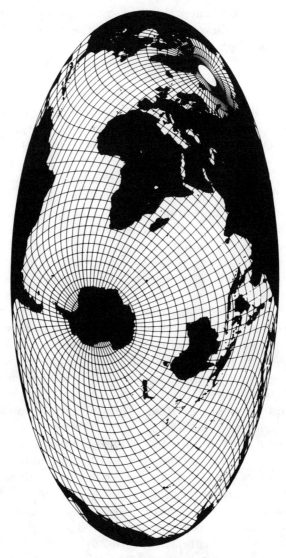

Map 1.1. The global ocean: oblique Hammer equal area map with dissected continents. (From Spilhaus 1991, by permission)

Organization (IMO), World Meteorological Organization (WMO), Intergovernmental Oceanographic Commission (IOC) of UNESCO, Food and Agriculture Organization (FAO), International Atomic Energy Agency (IAEA), UN Environment Programme (UNEP), and others—obviously has a role to play. To be most useful, however, the United Nations and its organs must maximize their credibility, competence, and operating efficiency (Lefever 1993). Moreover, just because environmental affairs are recognized as security issues, one should not conclude that environmental security is a matter for the UN Security Council under the special powers granted it by the UN Charter (Tinker 1992), as these powers are limited to threats to or breaches of the peace. On the other hand, environmental security fits well within the broad authority of the UN General Assembly. By its resolutions, declarations, and endorsements of draft international agreements, and especially by virtue of its democratic character, the General Assembly can facilitate the evolution of new international norms binding on states. Evolution along this route would eliminate any question of interference with a state's domestic jurisdiction and would not rely on mandatory decisions of the limited group of states represented on the Security Council. This evolution could be augmented by dispute settlement procedures before the International Court of Justice (ICJ) and other third-party fora (Tinker 1992).

The United Nations has been instrumental in organizing broad agreement on a general set of principles that can help provide a normative foundation for enhanced environmental security. Among the most important principles is Principle 21 of the 1972 Stockholm Declaration on the Human Environment, which is universally recognized as customary international law (Tinker 1992).

> States have, in accordance with the Charter of the United Nations and the principles of international law, the sovereign right to exploit their own resources pursuant to their own environmental policies, and the responsibility to ensure that activities within their jurisdiction or control do not cause damage to the environment of other States or of areas beyond the limits of national jurisdiction.

This principle was reaffirmed as Principle 2 in the Rio Declaration of the 1992 UN Conference on Environment and Development (UNCED), which, among its total of twenty-seven principles, also stated several others that speak most directly to environmental security concerns:

States shall develop national law regarding liability and compensation for the victims of pollution and other environmental damage. States shall also cooperate in an expeditious and more determined manner to develop further international law regarding liability and compensation for adverse effects of environmental damage caused by activities within their jurisdiction or control to areas beyond their jurisdiction. (Principle 13, an elaboration of Principle 22 of the 1972 Stockholm Declaration)

States shall provide prior and timely notification and relevant information to potentially affected States on activities that may have a significant adverse transboundary environmental effect and shall consult with those States at an early stage and in good faith. (Principle 19)

Warfare is inherently destructive of sustainable development. States shall therefore respect international law providing protection for the environment in times of armed conflict and cooperate in its further development, as necessary. (Principle 24)

States shall resolve all their environmental disputes peacefully and by appropriate means in accordance with the Charter of the United Nations. (Principle 26)

One of the principles of the 1992 Rio Declaration not found in the 1972 Stockholm Declaration deserves particular attention because it is subject to widely varying interpretations and misunderstanding and it is found in other international resolutions and instruments. This is the so-called precautionary principle:

In order to protect the environment, the precautionary approach shall be widely applied by States according to their capabilities. Where there are threats of serious or irreversible damage, lack of full scientific certainty shall not be used as a reason for postponing cost-effective measures to prevent environmental degradation. (Principle 15)

This approach is intended to deal with those situations, so common in environmental affairs, where there is great scientific uncertainty about the risks or long-term consequences of certain practices (such as chemical emissions). The precautionary approach instructs regulators to err on the side of safety and to take preventive or remedial action even if the certainty of harm has not been scientifically established. This is simply good sense, but it gives no guidance about just how serious a supposed threat must be, how uncertain the science must be, or what actions are most appropriate (Bodansky 1991;

Broadus 1992). The precautionary approach seems to assume that the choice is between risk and caution, but real choices are most often among different risks (Bodansky 1991). This shortcoming becomes especially clear in those interpretations that would demand proof of safety before a suspect activity is permitted, regardless of the benefits the activity might otherwise produce.

Perhaps the most important institutional basis for environmental security is clear and universally recognized international law. In that regard, the 1982 UN Law of the Sea Convention promises to be the single most influential and effective instrument of environmental security for ocean areas and activities. That promise will be fully realized, however, only with a universally recognized Law of the Sea Treaty in force. The failure of the international community to achieve that status for the Law of the Sea thus must be included among the threats to ocean environmental security.

THREATS TO ENVIRONMENTAL SECURITY

Using the working definition and criteria discussed above for guidance, our collaborative Russian-American research team identified seven post–Cold War problems of ocean environmental security that are of high mutual interest to Russia and the United States. The case study chapters in this volume explore the threats to and opportunities for environmental security that arise with these problems.

Land-based Marine Pollution. Land-based sources are estimated to account for more than 75 percent of total marine pollution, yet international control efforts concentrate almost exclusively on vessel-source pollution. Although specific instances of land-based marine pollution are mostly confined to individual nations, the problem is widespread globally and is one of the most urgent priorities for marine environmental protection. There are several other ways in which land-based marine pollution may constitute a problem of international environmental security. The pollution may cross national boundaries or affect shared resources. More localized pollution may cross borders by such indirect means as tourism or contaminated seafood exports. Even foreigners who never visit the polluted area or consume its exports may suffer second-order damages associated with their concern for the well-being of others and for such ethical questions as intergenerational equity. In chapter 2 we explore and compare the problem of land-based marine pollution in two regional seas of special interest to the United States and Russia, the Gulf of Mexico and the Black Sea.

Living Resource Problems: The North Pacific. Restrictions imposed by the United States and Russia in their 200-mile exclusive economic zones (EEZs) in the North Pacific–Bering Sea region have driven international fleets, especially those of Japan, Korea, and Taiwan, onto the high seas beyond. The central Bering Sea, an anomalous area of high seas that is completely surrounded by the Russian and US EEZs, constitutes an enclave where putatively straddling fish stocks may have been fished close to depletion. The use of very long driftnets in the high seas has been implicated in the "incidental" losses of nontargeted fish species, seabirds, and protected marine turtles and mammals. There have also been charges of illegal taking at sea of Russian, US, and Canadian salmon. Some of these problems have been addressed by a multilateral agreement that temporarily suspends fishing for pollock in the Central Bering Sea and by a nonbinding UN General Assembly resolution that effectively has placed a global moratorium of indefinite duration on the use of driftnets on the high seas. As discussed in chapter 3, however, the potential for destabilized relations between nations remains. An environmentally secure solution to these problems will most likely require a multilateral scheme for cooperative management of North Pacific fishery resources on an ecosystems basis.

Hazardous Materials Transport. Well over 90 percent of all goods carried in foreign trade move by marine transportation, including a very substantial amount of hazardous cargoes. Unlike petroleum, whose properties are well known and for which a comprehensive network of international institutions and pollution-control regulations has been developed, a confusing array of hazardous materials continue to enter the market by the hundreds each year, defying international efforts at definition, classification, and reporting. The unfinished agenda of international regulation of hazardous materials transport constitutes a pressing issue of environmental security for coastal states, whose interest in protecting their marine environments may conflict with freedom of navigation by maritime states and thus with one of the most fundamental principles of international ocean law. Chapter 4 examines the inadequacies of the existing legal and regulatory framework and of current liability and insurance regimes and considers some potential remedies in the areas of advanced technologies and techniques of risk estimation and damage assessment.

Radioactivity in the Oceans. In the post–Cold War era, fears of a cataclysmic nuclear exchange have given way to anxiety about the more localized consequences of longstanding, routine nuclear activities around the globe. Although radioactive waste management practices have thus far not had catastrophic consequences for human health or the marine environment, nuclear activities of all types have been officially shrouded in secrecy by virtually all participating nations. Accounts of incidents such as the Chernobyl accident feed the public's fears about nuclear contamination, which stem from a lack of certainty as to the nature, extent, and long-term environmental consequences of radioactivity in the oceans. Chapter 5 provides an overview of the various sources of natural and artificial radioactivity in the oceans and explores the relationship between scientific uncertainty and public misinterpretation of the environmental security threats associated with marine radioactivity. Several environmental security flashpoints are reviewed, including US and Russian plans to decommission hundreds of nuclear-powered submarines and the increased threat of nuclear weapons proliferation in the wake of the Cold War.

Environmental Protection for the Arctic. North of the Bering Sea, the Arctic Ocean remains one of the planet's least polluted environments. It is also unusually sensitive to human disturbance. Harsh conditions, short food chains, and a low level of species richness render Arctic Ocean ecosystems highly vulnerable, while low temperatures slow the breakdown of many pollutants and extend both their residence time and their geographic range within the environment. The notion of environmental security is particularly relevant to the Arctic because of its profound influence upon global (or at least hemispheric) environmental processes and also because of longstanding strategic objectives in the region. Chapter 6 explores the interplay of these factors as the military objectives of the former superpowers increasingly give way to problematic mining, oil and gas production, and shipping initiatives.

The Southern Ocean. Another cold, harsh, pristine, and vulnerable marine environment surrounds Antarctica. Economically, the Southern Ocean supports limited fishing, minimal navigation, and no minerals development. Since 1959, many activities in this ocean have been governed de facto by a limited club of nations participating in the Antarctic Treaty regime. Most of the Southern Ocean has thus been demilitarized, and national claims to maritime jurisdiction have been placed in abeyance. Future management of the Southern Ocean and its resources, however, is a matter of great controversy. Many nations not involved in the Antarctic Treaty system wish to substitute a UN-based regime of universal participation. Chapter 7 analyzes three different possibilities for future management of Antarctica and the Southern Ocean: (1) a strictly preservationist regime, which would set the entire region aside as a world park or nature preserve; (2) a multilateral conservation system that would allow controlled multiple uses and limited resource exploitation; and (3) a territorial division that would partition the region into separate national jurisdictions.

The Law of the Sea. The United Nations Convention on the Law of the Sea contains the most comprehensive statement of international law ever negotiated and shows great potential as a constitutional instrument of ocean environmental security. However, neither the United States nor Russia has ratified the treaty (in large part because of objections to its seabed mining provisions), and on November 16, 1993, the sixtieth instrument of ratification was deposited, thus fixing November 16, 1994, as the date on which it will enter into force. The convention's ultimate fate remains uncertain, as does the legal force of its environmental provisions. Ongoing efforts at the United Nations have been seeking to resolve questions surrounding the deep seabed regime, giving hope that the convention will enter into force with widespread participation, including that of the United States, Russia, and other developed states of the Northern Hemisphere. In any event, recognition by the convention that countries have an obligation to "protect and preserve the marine environment" supports the view that this obligation has become incorporated into customary international law. Chapter 8 explores the practical implications of this legal obligation and the convention's potential as a framework for assurance building in the context of environmental security.

After this review of cases, chapter 9 revisits the concept of environmental security, considers the status or potential of critical problems of international environmental security, and identifies for all of the cases important cross-cutting themes: the pervasive issue of responsibility, liability, and compensation; the common global perspective; the implications of population growth; criteria and standards; monitoring, data, and information; international cooperation and coordination; the role of international organizations; the application of technology and technology transfer; and problems and opportunities arising with military conversion.

CHAPTER 2

Land-based Marine Pollution: The Gulf of Mexico and the Black Sea

LAND-BASED MARINE POLLUTION is the world's most serious marine pollution problem. The United Nations Joint Group of Experts on the Scientific Aspects of Marine Pollution (GESAMP) has estimated that land-based sources contribute more than 75 percent of the pollutants entering the sea (although the unit of measure is left undefined) (GESAMP 1990). Land-based marine pollution—from agricultural runoff, sewage discharges, industrial emissions, and atmospheric deposition—presents a much more complex problem than do the relatively minor sources of pollution from vessels. The fundamental obstacle is that land-based marine pollution arises throughout the very fabric of daily life and from virtually all economic activity. Measures to address land-based marine pollution must reach practically all polluting aspects of society's activities. The complexity of the problem thus arises from its broad range of sources, large variety of pollutants, constant and daily discharge modes, cumulative effects, and far-reaching influences on national economic and social developments (Meng 1987).

Unlike such widely dispersing global phenomena as global warming or hazardous materials transport, land-based marine pollution may seem to be primarily a domestic problem, not an issue of international environmental security. Indeed, regarding land-based pollution, GESAMP (1990) stated, "Only a small part of those contaminants has spread beyond the limits of the continental shelf. The bulk remains in coastal waters and, in places, particularly in poorly flushed areas, has built up to significant levels." The GESAMP report also noted, however, that increasing signs of pollution in semi-enclosed areas "rais[e] the question of whether all coastal zones, including well-flushed areas, will become damaged to the same degree, and whether eventually even the open ocean will be affected."

Despite widespread problems of land-based marine pollution, instances of significant transboundary damages currently seem to be rare exceptions. Yet there are several ways, some indirect or subtle, in which land-based marine pollution does express itself as an issue of international environmental security.

Transboundary Pollution. This is the most obvious mechanism. Though uncommon, it is not unheard of. Europe's second largest river, the Danube, for example, rises in southwestern Germany and traverses or forms the boundaries of eight other countries on its course to the Black Sea, where it introduces 75 percent of all pollutants (Keondzhyan, Kudina, and Terehena 1990). International institutions to address the problem of transboundary marine pollution are just emerging.

Shared Resource Stocks. Although pollution may remain within one country's jurisdiction, shared resource stocks may become contaminated or damaged there and thus harm a neighboring country's well-being. The United States and Mexico, for instance, share roughly equally in the shrimp stocks in the Gulf of Mexico (Nance, pers. comm., 1993), some of which migrate across national jurisdictions; thus, shrimp contaminated in one country's waters could very well be landed and consumed in the neighboring country.

Damage to Export Goods. Similarly, export goods, such as seafood products, may be damaged by domestic pollution but consumed in another country, whose well-being would be reduced.

Pollution and Tourism. Consumers from one country who travel to a polluting country can suffer directly from the marine pollution there. For example, land-based pollution of the Black Sea affects not only the twenty million people in the former Soviet republics who vacation on the Black Sea coast each year, but also an additional ten million visitors from Bulgaria, Romania, and Turkey (Mitin 1990). The large number of visitors not only means potentially more suffering from marine pollution, but also may itself contribute to the amount of pollution.

Pollution and Non-use Values. More subtly, citizens in a foreign country may feel a real loss because of their concern about damage to human populations, wildlife, or other natural resources (coral reefs, endemic species, marine mammals, etc.) in the polluting country. This is a genuine loss in well-being even if there is no intention ever to visit the polluting country or to use its natural resources first-hand. Such "second-order damages" may reflect a concern not so much over "pollution or environmental degradation per se, . . . [but] over the welfare of current and future generations of people as it is impacted by broad concepts of ethics and equity. Thus, the concern is really over the standard of living in all its facets" (Emerson and Collinge 1993).

Emissions "Export" through Capital Mobility. Perhaps the most subtle international connection occurs when investors from one country locate their polluting facilities in another country. By this means they indirectly export the polluting emissions and threaten the environmental well-being (although not necessarily the net economic well-being) of the country whose waters they pollute. The potential for this form of transboundary pollution between the United States and Mexico figured prominently in the stormy US domestic debate over the North American Free Trade Agreement (NAFTA). The argument most frequently advanced against NAFTA has been that the lowering of trade barriers will result in a loss of US jobs and an increase in pollution for both countries by making it easier for US companies to exploit Mexico's lower labor costs and less stringent pollution-control enforcement practices. It has also been suggested that, in industries like agriculture and meat products, US domestic manufacturers will be forced by the competitive pressures of lower trade barriers to intensify production, with potentially damaging consequences for the environment (Yancey 1993).

In contrast to vessel-source pollution and ocean dumping, there is no comprehensive global legal regime for land-based marine pollution. Within the global legal structure, provisions for land-based pollution are made only through very general obligations or guidelines (UN Convention on the Law of the Sea, Montreal Guidelines on Land-based Pollution, and International Law Commission [ILC] draft Articles on International Watercourses). Regional regimes for the control of land-based marine pollution, on the other hand, typically are more detailed and inclusive by design (i.e., 1992 Convention on the Protection of the Marine Environment of the Baltic Sea Area, 1992 Convention for the Protection of the Marine Environment of the North-East Atlantic, 1976 Convention for the Protection of the Mediterranean Sea against Pollution).

The 1985 Montreal Guidelines for the Protection of the Marine Environment against Pollution from Land-based Sources constitute perhaps the first globally agreed legal instrument devoted to land-based marine pollution control (Meng 1987). Based on principles and common elements from existing agreements, the Montreal guidelines "are presented as a checklist of basic provisions rather than a model agreement, from which governments may select, adopt or elaborate, as appropriate, to meet the needs of special regions" (Meng 1987). The guidelines, which are recommendatory in nature and thus of limited legal force, are intended to serve as a basic framework for the "development of regional agreements in those areas where such agreements are called for" and to provide "guidance for states in areas which may not be covered by regional agreements." They also call for the preparation of a global convention at some future time (Meng 1987).

Since the guidelines were adopted in 1985, however, the goal of a global convention has given way to what is generally seen as a more realistic international effort to devise some kind of agreement on what can and should be done, primarily at the regional level but within a global context, to control and prevent land-based marine pollution. At present, the precise nature of such an agreement remains a somewhat delicate question. As one close observer explained, the shift in goals reflects widespread skepticism about the value of binding legal obligations with respect to a problem that is universally recognized and deplored but that developing nations have neither the institutional nor the financial means to address.[1]

In May 1993, the UNEP Governing Council held the first in a series of intergovernmental meetings to develop an agreement on land-based marine pollution, as called for by the 1992 UN Conference on Environment and Development (UNCED). The next meeting was planned for June 1994 in Montreal, where the main business would be to review and revise the Montreal guidelines, organize an intersession drafting group, and outline preparations for the final, two-week meeting, scheduled to be held in Washington, DC, in late 1995. If work on the as-yet-undefined agreement proceeds as planned, most of the details will have been worked out by the close of a March 1995 meeting to be hosted by Chile or Iceland.

REGIONAL COOPERATION

A regional framework for controlling land-based marine pollution may be the most effective mechanism because it can accommodate the need to resolve potentially conflicting national and international interests. Moreover, most problems of land-based marine pollution are felt regionally. States sharing a regional sea discharge pollutants into it as a common pool, and they share in the resulting damages (or benefits of remediation) on a common basis, if in varying degree. Moreover, states in the same region often share similarities in culture, economic geography, public preferences, and trading relationships.

Still, there seem to be shortcomings in attempts to address marine pollution at the regional level. In several regions, organizational efforts "have not yet led to concrete results" (UN Secretary General 1989). Also, support at the domestic level for regional programs is sometimes quite limited. If a state perceives that its costs of participating in collective action exceed its benefits, it has a strong incentive to avoid regional cooperation even if region-wide benefits are greater than costs.

In view of the widespread perception and complaint that regional agreements on land-based marine pollution have been notably ineffective, the WHOI Marine Policy Center intensively examined and compared three pioneering regional programs to control land-based marine pollution. Sponsored by the US Environmental Protection Agency, this research identified the efforts made in the Baltic, North Sea, and Mediterranean regions and sought to show what has worked best, what has not worked, and why (Broadus et al. 1993). Relevant lessons learned from the experience in those three regions were reported (table 2.1).

Several noteworthy observations were made about all three European programs. In many instances it was hard to show that active investment or strict compliance was in the self-interest of each of the parties. None of the programs included explicit mechanisms for trade among parties (of emission quotas, quid pro quos, know-how, etc.), with the exception of the Med Plan's foreign assistance and technology transfer activities. All of the programs have been hindered by inadequate compliance reporting and lack of transparency, although these factors seem to be improving markedly for the Baltic and the North Sea. All three programs have been subject to and have apparently benefited from the superimposition of high-level political oversight structures.

Table 2.1
LESSONS DERIVED FROM THE EXPERIENCE IN THE BALTIC, NORTH SEA, AND MEDITERRANEAN REGIONAL PROGRAMS

Clarity Costly misunderstandings, false starts, and wasted effort can be avoided by seeking maximum clarity in defining program goals, the relations among parties, measures and recommendations, and the role of the secretariat. Particular effort should be made to excise hidden agendas (e.g., if income transfer or lobbying of governments is to be a program objective, this should be set out clearly and incorporated explicitly into program design).

Capacity to evolve The ability to adjust the program to changed circumstances and improved knowledge is vital to its effectiveness over time. The method of keeping the framework convention quite spare and leaving program elaboration to subsequent protocols, directives, recommendations, or flexible actions plans is useful in this regard. So, too, is the periodic oversight and intervention of high-level political authority from among the parties.

Political commitment Program effectiveness will depend inevitably on the political commitment of the parties. Examples of movement toward enhanced effectiveness were observed in all three regions as a result of collective political intervention by parties at a ministerial level. The device of an overarching high-level political forum above the regional program, as in the North Sea Conferences, appears to have been useful in this regard. Such a forum also provides a highly visible focal point for public pressure, and it contributes to program transparency, which enhances the relevance and effectiveness of public pressure.

Specificity of objectives Demonstrating effectiveness and clarity of program measures is important both in sustaining program support and in operational program implementation. For this it is necessary to specify objectives whose accomplishment can be measured and demonstrated. Targets and timetables are useful, especially if they are sensitive to differences in the stakes and economic capabilities among the parties. Black lists and gray lists are useful in specifying the scope of concern, but experience has pointed to the need to narrow program focus in practice to priority targets (sometimes called "red lists"). The persuasive establishment of baseline conditions, in terms of both ambient measures and emissions, must be a priority, as demonstrated by the hindrances created by shortcomings in doing so in all the cases examined. It should be recognized that the baseline profiles can be useful even if somewhat crude and that assembling credible baselines will be expensive and time consuming.

Scientific involvement A mechanism for expert scientific input and advice is essential, to clarify the nature and magnitude of problems and to monitor changing conditions, but it is important to assure that scientific research interests do not altogether run away with program resources. The persuasive establishment of baseline conditions, in terms of both ambient measures and emissions, must be a priority, as demonstrated by the hindrances created by shortcomings in these areas in all of the cases examined. It should be recognized that the base-

Table 2.1 continued

line profiles can be useful even if somewhat crude, and that assembling credible baselines will be expensive and time consuming.

Self-interest test States can be expected to serve their own self-interests, but program effectiveness will be improved to the extent that program design takes account of this and explicitly seeks to accommodate the genuine (rather than presumed) self-interest of its parties. Program structures that encourage or accommodate agreements and actions by subgroups of parties, that allow for flexible financial or technical assistance, or that provide explicitly for variable schedules or an "opt out" option on some measures, may also be helpful in this regard.

Mechanisms for trade Because the self-interests of states will differ and because one of the principal rationales for collective action is to failitate sharing and exchange, program effectiveness will be enhanced by provision of mechanisms for explicit "trade" (quid pro quos, specialization) among participants. Similarly, efficiency is served by implementation measures that take into account differences among states and that facilitate trading to accomplish mutual objectives (such as tradable quota schemes). Implicit trades or explicit quid pro quos may be achieved through program funding schemes.

Reporting and compliance Assessing the degree of state compliance with program recommendations has been precluded in all three cases by the poor performance of state reporting. Special emphasis in program design should be placed on specifying expectations for compliance, means for monitoring compliance, and, particularly, procedures that encourage accurate and timely state reporting. In this regard, keeping the reporting burden simple and at a minimum is important. It may also be useful to assure that the results of reporting are useful to all parties and to suggest a reporting system that meshes routinely and automatically into the states' own practices. Some provision for nonintrusive, third-party inspection may also be useful, though this is only implicit in the cases examined in their lack of success with more conventional, passive means of collecting state reports.

Transparency Most of the lessons proposed above speak to the value of transparency in program design and execution, but transparency nonetheless warrants explicit inclusion. Program effectiveness is enhanced (in terms of both party state support over time and verifiability by other interests) by the maximum transparency compatible with the protection of proprietary interests and the rights of national sovereignty. Specific mechanisms to enhance transparency are identified below.

Realistic expectations Regional programs in the real world can at best be catalytic. Measures to control land-based marine pollution will necessarily occur almost exclusively through domestic actions. A program may succeed as a medium for exchange and consensus among members and as a promoter of external funding support, which can be used to reinforce internationally agreed program objectives and reward compliance. The effectiveness of the regional trust fund device is not clear from the cases examined and must be carefully questioned. Again, clarity of purpose and transparency cannot be overemphasized.

SOURCE: Broadus et al. 1993.

Comparative Cases

Two of the regional seas of greatest interest to the United States and Russia have only recently been included within regional international cooperative programs. Land-based marine pollution in the Gulf of Mexico and the Black Sea remains almost entirely a domestic concern, although in each case it clearly exhibits many of the aspects of international environmental security cited above. Each of these large, semi-enclosed seas is economically important to its region's major economic and political presence, which shares the sea with neighboring states and is a significant contributor to coastal pollution.

The Gulf of Mexico

The Gulf of Mexico, with a surface area of 1.6 million square kilometers (617,600 square miles) and a volume of 2.3 million cubic kilometers (table 2.2), is the ninth largest body of water in the world. The maximum depth is estimated at 3,750 meters, but the average depth is less than 2,000 meters. The Gulf of Mexico spans three national jurisdictions: the United States, Mexico, and Cuba. The Gulf coastline extends for 4,800 kilometers, bordering five US states (Florida, Alabama, Mississippi, Louisiana, and Texas) and Mexico. The Gulf itself contains 33 major river systems and 207 estuaries (AMS and Webster 1991). By far the largest carrier of water and associated materials into the Gulf of Mexico is the Mississippi River, whose principal tributaries are the Missouri River from the west and the Ohio River from the east. Other significant US river systems that drain into the Gulf of Mexico include the Pecos–Rio Grande, the Suwannee, the Chatahoochee-Apalachicola, the Tombigbee-Alabama, and the Brazos rivers.

Two-thirds of the contiguous United States, an area of 4.4 million square kilometers (1.7 million square miles), lies in the drainage basin of the Gulf of Mexico (map 2.1). The population of the drainage area is about 87.3 million.

The water exchange between the Gulf of Mexico and the Atlantic is limited, and pollutants tend to accumulate and concentrate in the Gulf. There are two loop currents in the Gulf of Mexico. The Yucatan Current carries some thirty million cubic meters of water per second northward, loops to the north, and exits as the Florida Current, which then turns into the Gulf Stream. The five million acres of Gulf coastal wetlands comprise about half of the national total and, together with

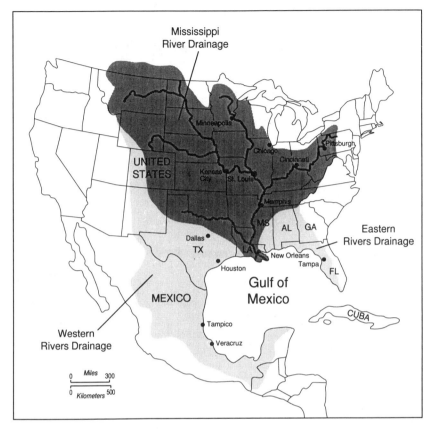

Map 2.1. The Gulf of Mexico and its drainage area.

the extensive barrier islands, provide critical habitat for 75 percent of the migratory waterfowl traversing the continent.

The US natural resources and commercial and recreational activities supported by the Gulf of Mexico are of vital economic significance to the nation, especially to its five bordering US states (NOAA 1985). The Gulf region, with its warm waters and extensive beaches, supports an important tourist industry for each of the Gulf coastal US states (fig. 2.1). Florida alone hosted some fifty million tourists in 1990, for example, and tourism in the state has grown at an annual rate of nearly 9 percent for more than the past decade (Puffer, pers. comm., 1993). Together the five Gulf coastal states contribute roughly twenty billion tourism-related dollars to the national economy (AMS and Webster 1991).

Table 2.2
SELECTED FEATURES OF THE GULF OF MEXICO AND THE BLACK SEA

Feature	Gulf of Mexico	Black Sea
Size		
Area	1,600,000 km^2	423,000 km^2
Maximum depth	3750 m	2200 m
Average depth	2000 m	1197 m
Maximum length × width	1770 km × 1300 km	608 km × 1204 km
Shoreline	4800 km	4090 km
Volume	2,332,000 km^3	534,000 km^3
Properties		
Salinity	34.86–36.7‰	17.598–22.507‰
Surface temperature	18°–24°C	−0.5°–26°C
Dissolved oxygen	4.8–5.2 mg/L	absent below 200 m
Watershed		
Drainage area	4,441,300 km^2	1,864,000 km^2
Population of drainage area	87 million	130 million
Population density	20/km^2 (52/mi^2)	70/km^2 (181/mi^2)
Water discharge	>779 km^3/yr (US)	374 km^3/yr
Sediment discharge	230 million tons/yr	150 million tons/yr
Nitrogen input	883,940 tons/yr	>2800 tons/yr (USSR sewage only)
Phosphorus input	215,240 tons/yr	>700 tons/yr (USSR sewage only)

Boundaries
National jurisdictions — US, Mexico, Cuba — Russia, Ukraine, Georgia, Bulgaria, Romania, Turkey

Principal coastal lagoons and estuaries (US and former USSR)
- Galveston Bay — Karkynit Bay
- Mobile Bay — Danube estuary
- Tampa Bay — Dniester estuary
- Laguna Madre — Tendzov Bay
- Pensacola Bay — Kalamit Bay
- Lake Ponchartrain — Dnieper estuary
- Atchafalaya Bay
- Apalachee Bay

Major straits
- Florida Strait — Bosporus
- Yucatan Channel — Kerch Strait

Major deltas and rivers
- Mississippi Delta and River System (Missouri and Ohio) (US) — Danube (Germany, Austria, former Czechoslovakia, Hungary, former Yugoslavia, Romania, Bulgaria, Moldova, Ukraine)
- Pecos-Rio Grande (US/Mexico) — Don (Russia)
- Suwannee (US) — Dnepr (Ukraine, Byelorussia, Russia)
- Chatahoochee-Apalachicola (US) — Dnestr (Moldova)
- Tombigbee-Alabama (US) — South Bug (Ukraine)
- Brazos (US)

Fig. 2.1. A popular vacation spot on the Gulf of Mexico coast. According to the EPA (1989b), nearly 13 million adults visited the Florida Gulf coast alone in 1984, accounting for an estimated $4.5 billion in related sales. Virtually identical scenes are commonplace at many sites along the Black Sea coast. (Courtesy of the Gulf of Mexico Program)

The Gulf of Mexico also accounts for more than two billion pounds of commercial fish and shellfish landings annually. Fetching more than $780 million at dockside, this catch represents 40 percent of the total annual US harvest of commercial fish (AMS and Webster 1991).

Gulf oil and gas production are likewise valuable to the region's economy and represent an important part of the nation's total energy supply. In 1990, more than sixteen hundred outer continental shelf (OCS) leases were in production, yielding revenues to the US Federal Treasury of about three billion dollars (AMS and Webster 1991).

The Effects of Coastal Development

Coastal development and the subsequent rise in population density are commonly associated with habitat loss and increased sewage and toxic runoff entering coastal waters. In 1988, fourteen million people lived in the ninety-nine US coastal counties of the Gulf of Mexico. The region currently ranks fourth in total population among the five US coastal regions, accounting for 13 percent of the nation's total coastal population. The Gulf region is not as densely settled as other

US coastal regions, but it is experiencing the second-fastest rate of growth. The Gulf-area population increased by 33 percent between 1970 and 1980, and it is expected to increase another 22 percent, to eighteen million, by 2010 (NOAA 1990). Western Florida is expected to remain the fastest-growing area on the Gulf, with a population increase of more than 1.5 million anticipated over the next twenty years. Texas is the second most rapidly growing Gulf state; its coastal population is expected to increase by more than 1.1 million over the same period (NOAA 1990).

Rapid population growth in Mexico, especially in industrialized areas of the otherwise sparsely populated US-Mexico border, may be contributing to increases in sewage discharge into the Gulf of Mexico. The population of the border area (which the two countries have defined for administrative purposes as extending 100 kilometers on each side of the international boundary[2]) has doubled over the last decade, to more than six million, and by the year 2000 it is projected to increase by an additional 50 to 230 percent (Emerson and Bourbon 1991).

The Effects of Nutrient Enrichment

Human sources of aquatic nutrients, such as nitrogen and phosphorus, include both *point sources* (10–30 percent)—on-site domestic septic systems, discharge pipes from municipal sewage treatment plants, and industrial sources—and *nonpoint sources* (70–90 percent)—agricultural runoff (fertilizer and animal wastes), automotive exhaust, and stack emissions. The combined drainage of the Mississippi-Atchafalaya River system contributes 70–75 percent of the total nitrogen and phosphorus concentrations entering the Gulf from all US sources. An overabundance of nutrients like that in some parts of the Gulf of Mexico can give rise to oxygen depletion (hypoxia), fish kills, and noxious algal blooms, resulting in diminished fishery stocks and reduced recreational and aesthetic values.

Hypoxia is typically caused by nutrient enrichment acting in conjunction with the temperature and salinity differences of a stratified water column to prevent an adequate rate of reaeration of bottom waters. In hypoxic waters most life disappears, except for anaerobic bacteria, fungi and yeasts, and some protozoa (Clark 1989). For the temperature conditions that prevail in the Gulf of Mexico and its surrounding bays and estuaries, a dissolved oxygen concentration at or below 2 mg/L or 2 ppm (the level at which trawling yields no catch) is considered hypoxic (Rabalais, pers. comm., 1993). Hypoxic conditions on the Louisiana continental shelf were first monitored in 1986, but it is not known when they first developed or whether they predate human intervention in the nutrient cycle. As much as 4,000 square miles (10,360 square kilometers) of hypoxic bottom waters, known as the *dead zone*, has been documented off the Louisiana and Texas coasts (Rabalais et al. 1991). This condition occurs at the terminus of the Mississippi River and amidst the nation's second richest fishing grounds (after Alaska). In addition, hypoxic conditions have been intermittently detected in Mobile, Pensacola, and Tampa bays, among others, when an event such as dead fish washing ashore has prompted investigation (Rabalais, pers. comm., 1993).

The Effects on Public Health

Red tides, the most common form of toxic algal bloom, are often cited as one consequence of nutrient overenrichment, but the relationship has not been sufficiently validated. It is likewise unclear whether red tides have become more frequent or intensive than in the past. Red tides are known to have occurred in the Gulf for centuries, but the danger to human health seems to be minimal. Between 1960 and 1990 there were only forty reported cases of shellfish-related disease outbreaks in the Gulf coastal states; still, it is estimated that only 5–10 percent of all such cases in the United States are actually reported (Abdelghani et al. 1993). Closures of inshore waters due to the presence of red tides, however, can have significant economic effects on tourism, recreation, and commercial fishing. Of the shellfish-producing areas along the Gulf coast, 57 percent, or 3.4 million acres, were permanently or conditionally closed as of 1991 because of the threat of red tides (AMS and Webster 1991, 1992).

To date there are no records pertaining to human illness occurring in Alabama, Mississippi, Louisiana, or Texas as a result of paralytic shellfish poisoning, ciguatera, neurotoxic shellfish poisoning, or diarrhetic shellfish poisoning (Abdelghani et al. 1993). Records for Florida include five cases of neurotoxic shellfish poisoning (four in 1973 and one in 1974), plus one 1988 case of "shellfish-associated poisoning" by an undetermined toxic agent following the consumption of oysters from Louisiana (Abdelghani et al. 1993).

The risks of pathogen-related diseases have diminished greatly since the advent of enforced sanitary standards for on-site domestic septic systems, the installation of municipal sewage treatment, laws prohibiting discharge of raw sewage into US waters, and the use of on-board disinfectant equipment. Nonetheless, in 1991, high levels of fecal coliform resulted in 299 beach closures/advisories in Florida alone, primarily in the St. Petersburg, Pasco County, and Dade County areas.[3] Of the five Gulf coastal states, only Florida and Texas currently perform regular microbial monitoring of marine beaches; Mississippi discontinued its summer-only monitoring in 1989, and Alabama and Louisiana have never monitored beaches on a regular basis (Abdelghani et al. 1993).

Human exposure to toxic chemicals released into the environment is a matter of widespread public concern, and contamination of fish, shellfish, and other aquatic resources by toxic chemicals is well documented by regulatory agencies (EPA 1989a) and international agencies. Twenty-five percent of monitored US coastal waters register elevated levels of substances like pesticides, polychlorinated biphenyls (PCBs), and heavy metals (EPA 1990b).

Industrial and Municipal Sources

Three hundred forty-seven US industrial facilities discharge through pipelines directly into the Gulf of Mexico or its estuaries. Most of these facilities are petroleum, forest, or fish-processing plants. Still others discharge into municipal treatment facilities.

One hundred thirteen US cities and towns discharge wastes directly into Gulf estuaries and coastal waters. Urban runoff puts oil and grease, lead, chromium, and many other contaminants into marine waters. Municipal septic systems are also significant contributors of nutrients and human pathogens into the Gulf marine environment. US municipalities surrounding the Gulf dump more than a billion gallons of sewage effluent into Gulf waters each day (Puffer, pers. comm., 1993).

Providing proper sewage treatment is especially difficult in Mexico, where the problem is compounded by a lack of enforceable environmental laws and a shortage of funds for wastewater treatment facilities. In Nuevo Laredo, on the Texas border, twenty-seven million gallons of untreated sewage are being discharged each day into the Rio Grande (Emerson and Bourbon 1991). Tests by the Texas Water Commission show that at least as far as twenty-five miles downstream, and only forty miles from Laredo, the contamination level of the river greatly exceeds the applicable health and safety standards (Texas Water Commission 1990).

The US Management Approach

US mitigation efforts in response to land-based marine pollution in the Gulf of Mexico are embodied in local, state, and national efforts that reflect the nation's federalist form of government. These efforts include both information gathering on pollution and the legislation and implementation of regulations. Federal legislation establishes broad goals, and typically regulations and standards as well, that must be satisfied (and often are exceeded) in programs administered at the state and local levels, often with federally appropriated funds.

In recent years, private citizens groups and nongovernmental organizations (NGOs) have come to play an increasingly important role in helping to shape and advance the nation's environmental agenda. In the arena of domestic politics, NGOs like the Health Effects Institute, with strong scientific credentials and a reputation for working from neutral ground, have helped to overcome legislative gridlock by serving as "a bridge between government regulators and the affected sectors of industry" (Carnegie Commission 1993). The influence of NGOs is likewise growing where environmental concerns transcend national boundaries, as in the case of the Environmental Protection Agency's (EPA's) heavy reliance on NGOs for analysis and other assistance in improving the environmental aspects of the North American Free Trade Agreement (Carnegie Commission 1993).

Information Gathering. Adding to some longstanding data systems, such as fisheries and water resources statistics, are new environmental information sources being developed at the federal level for use as resource management tools.

The National Oceanic and Atmospheric Administration's National Status and Trends Program (NS&T) seeks "to determine the current status of, and to detect changes in, the environmental quality of US estuarine and coastal waters related to toxic contaminants" (Intergovernmental Task Force on Monitoring Water Quality et al. 1992). The NS&T Mussel Watch Project and the NS&T Benthic Surveillance Project are the two components of this program that monitor coastal environmental conditions. Altogether, the NS&T monitors the concentrations of toxic chemicals and trace elements in bottom-feeding fish, shellfish, and sediments at almost three hundred coastal and estuarine locations, fully a third of which are in the Gulf coast region (EPA and NOAA 1992). Among its many databases, the National Coastal Pollutant Discharge Inventory Program (NCPDIP) documents pesticide transfer from streams and rivers into marine systems and measures toxicity, contamination levels, and bioconcentration factors. This represents the first use of a uniform set of techniques to measure coastal and estuarine environmental quality. The National Institute of Standards and Technology maintains a specimen bank of samples taken annually at about 10 percent of the NCPDIP sites for "future, retrospective analyses" (NOAA 1992).

In 1990, the EPA launched an Environmental Monitoring and Assessment Program (EMAP) to respond to the growing demand for information characterizing the condition of the environment and the types and locations of changes in the environment. When fully implemented in 1995, EMAP will provide the EPA administrator, Congress, and the public with statistical data summaries and periodic interpretive reports on ecological status and trends for the entire country. These reports are intended to include information on the current status, extent, and geographical distribution of ecological resources; the extent to which these resources are degrading or improving and the likely causes of these changes; and how these ecosystems are responding to control and mitigation programs.

EMAP is implemented through seven resource groups (Agroecosystems, Arid Ecosystems, Forests, Near Coastal, Great Lakes, Surface Waters, and Wetlands). Sampling has been conducted by the Near Coastal component for estuaries in the Virginian Province and the Louisianan Province, which includes the Gulf of Mexico (EPA 1992a). According to a recent report of the first EMAP demonstration project, which was conducted in the estuaries of the mid-Atlantic states (the Virginian Province) in 1990, an early "major accomplishment" of the program has been the development and application of a method for calibrating and validating biological indicators of estuarine condition (EPA 1993).

Legislation and Regulation. One of the most comprehensive bodies of federal marine legislation is the Clean Water Act (CWA). Originally enacted by Congress in 1972 as the Federal Water Pollution Control Act, the CWA's major objective "is to restore and maintain the 'chemical, physical, and biological integrity of the Nation's waters' by regulating the discharge of pollutants into 'navigable waters'" (Imperial, Robadue, and Hennessey 1992). The CWA contains three primary components: the National Pollution Discharge Elimination System (NPDES), which prohibits the discharge of pollutants into navigable waters without a permit; the construction grants program, which provides financing for the construction of municipal waste treatment facilities; and four regional planning programs (Imperial, Robadue, and Hennessey 1992).

The NPDES (Section 403(c) of the CWA) sets out criteria applicable to discharges into US territorial seas and contiguous zones and into the ocean based on whether a discharge will cause degradation of those waters. Permits to discharge legally accepted material into navigable US waters are issued by the EPA or its authorized agencies in certain states. Section 403(c) also provides broad authority for regions and states to impose controls on discharges, requires the evaluation of alternatives to discharging, and may require process changes, if necessary, to ensure that no unreasonable degradation of the marine environment occurs (EPA 1990e).

Of the five Gulf coastal states, only Florida, Texas, and Louisiana have discharges that are subject to 403(c) regulation. Most of the major 403(c) dischargers in Florida are large publicly owned treatment works (POTW). More than $100 billion has been allocated to construct POTW through the Section 208 construction grants program (Goldman 1990), but it is estimated that an additional $88 billion is needed to meet current needs (EPA 1990a). There are ten small dischargers in the Florida Keys. Texas has two permitted dischargers, and Louisiana has fifteen. There is also one general permit for offshore oil and gas activities in the Gulf of Mexico (EPA 1990e).

The 1987 amendments to the CWA direct the states to identify water bodies where toxic pollutants prevent the attainment of water quality standards or of designated beneficial uses. For each identified water body, individual toxic control strategies to reduce point-source discharges must be developed for implementation within the NPDES permitting process. Another program under the 1987 CWA amendments requires each state to assess its water and to develop a management program to control nonpoint pollution sources, identifying actions needed to mitigate the problems caused by each source category.

In the Gulf Coast region, responses by individual states reflect the predominance of agriculture, construction, and forestry as important nonpoint sources of pollution. Alabama's management program has emphasized the application of best management practices (BMP) to control sediment and properly manage animal waste, together with education on the proper use and application of agricultural chemicals. Florida's efforts have centered on enforcement of regulations requiring BMP for erosion and sediment control during and after construction; enforcement of wetland protection regulations; BMP evaluations; and public education, including distribution of the *Florida Developmental Manual: A Guide to Sound Land and Water Management*. In Mississippi, the state Bureau of Pollution Control has drafted a statewide Erosion and Sediment Control Law intended to reduce the nonpoint source effects of construction activities. In Louisiana, some state and federal agencies cooperate with the Louisiana Department of Environmental Quality through an interagency coordinating committee that has targeted two river basins, the Tangipahao and the Mermentau, for agricultural controls to prevent groundwater contamination. The Texas Nonpoint Source Management Program includes a voluntary agriculture/silviculture component that targets dairy wastes in the North Bosque River and arsenic in groundwater in the Southern High Plains for technical assistance and intensive educational efforts (EPA 1992b).

The construction grants program authorized under Section 208 of the 1972 CWA provides federal funds for the construction of municipal water treatment facilities that meet a uniform standard of secondary treatment. The program's management plan identified treatment works needed to meet municipal and industrial needs for the next twenty years and established construction priorities and a regulatory program to control point and nonpoint sources of pollution, including those resulting from land-use activities. The plan specified which agencies should construct, operate, and maintain the treatment facilities; identified appropriate implementation measures and the role of existing state and regulatory programs in carrying them out; and projected the economic, environmental, and social influences of the implementation measures (Imperial, Robadue, and Hennessey 1992). In the 1987 reauthorization of the CWA, the construction grants program was redefined as a revolving land fund for states, financed by direct federal appropriations that are scheduled to end after fiscal year 1994 (Knopman and Smith 1993).

Despite the apparent thoroughness of the original Section 208 management plan, implementation of the program has generally been weak. Among its chief shortcomings have been inadequate federal funding and guidance during the formative years and inconsistency with other water quality planning initiatives (Imperial, Robadue, and Hennessey 1992). Moreover, scientists have no reliable method for measuring its environmental effects (Comptroller General 1981; Knopman and Smith 1993).

The National Estuary Program (NEP), established in 1987 under CWA authority, was developed to identify nationally significant estuaries, protect and improve their water quality, and enhance their living resources. To meet these goals, representatives of citizen and user groups, scientific and technical institutions, and all relevant government agencies and resource managers at the state, local, and federal levels work together in a collaborative decision-making process known as *management conferences*. Management conferences are convened for the purpose of producing a comprehensive conservation and management plan (CCMP) for a given estuary within a five-year time frame. A CCMP should assess the environmental condition of the estuary and develop plans that make use of existing management and regulatory systems for the coordinated implementation of priority mitigation actions.

NEP status is strictly voluntary; states nominate their estuaries to be considered for participation in the program. The NEP currently includes twenty-one estuary projects, of which five are in the Gulf area. Estuaries are selected on the basis of the state's potential to address issues of significant national concern and its demonstrated commitment to taking protective measures. Mitigation efforts often differ from region to region and even from state to state because of the specificity of problems in a given area and the regulatory structure in place there (EPA 1990c, 1992b).

The NEP has been appropriated roughly $15 million in federal funds each year for the past several years. Large estuaries, such as Galveston Bay, receive a maximum of $1 million, and smaller estuaries receive a maximum of $600,000. A new program, such as the one in Corpus Christi, may receive about $150,000 for its first year (Nance, pers. comm., 1993).

The 1972 Coastal Zone Management Act (CZMA) emerged from a national reevaluation of the effectiveness of US environmental protection and land-use planning. The CZMA established a cooperative state-federal program for comprehensive coastal zone management, launched in 1974 and administered by NOAA, with an emphasis on encouraging voluntary and flexible responses from states to wide-ranging and sometimes conflicting national policy objectives (Owens 1992). These are broadly summed up in the legislation's statement of purpose: "to preserve, protect, develop, and where possible, to restore or enhance the resources of the Nation's coastal zone for this and succeeding generations" (Section 303(1)). More specifically, coastal zone management program objectives include (1) protecting natural resources; (2) managing coastal development; (3) giving development priority to uses that are coastal dependent; (4) concentrating new development in existing developed areas; (5) participating in the planning of ports and urban waterfront redevelopment; (6) providing coastal access for recreational purposes and protecting historic, cultural, and aesthetic resources; and (7) improving governmental decision making and coordination (Imperial, Rodabue, and Hennessey 1992).

Three major sections of the CZMA lay out the primary techniques and requirements for its implementation: initial grants to develop management programs for federal approval (Section 305), follow-up grants to implement these programs (Section 306), and a requirement that federal actions (i.e., licenses, permits, and financial assistance) be consistent with approved state coastal management programs (Section 307) (Godschalk 1992). These federal funding and consistency provisions also constitute the primary incentives for states to join the coastal zone management program (Imperial, Rodabue, and Hennessey 1992).

Through 1988, the federal investment in the coastal program was less than $696 million, yet this was sufficient to "motivat[e] 29 states to design and carry out far-ranging coastal management efforts," in most cases contributing matching funds on at least a dollar-for-dollar basis (Godschalk 1992). For the period 1982–87, more of total available program funds were spent to improve government decision making (39 percent) than for any other purpose. Such expenditures declined over the six-year period, however, from 46 to 36 percent, while expenditures for natural resource protection, the second-largest expenditure category, rose from 23 to 31 percent (Owens 1992).

Of the Gulf coast states, Alabama, Florida, Mississippi, and Louisiana are long-time Coastal Zone Management (CZM) participants, while Texas is a newcomer. Alabama and Louisiana spent somewhat more than half their grant funds from 1982 to 1987 on improving government decision making (53 percent and 62 percent, respectively), whereas Florida placed unusually heavy emphasis on natural resource protection (70 percent). In Mississippi, natural resource protection was the top spending priority as well (at 29 percent), with considerable investments also made in improved government decision making (23 percent), increased public access (17 percent), and urban waterfront development (15 percent) (Owens 1992).

The CZMA was reauthorized in 1980, 1986, and 1990. In 1990, the consistency provision was strengthened and extended to apply to oil and gas leases and other activities "directly affecting" the coastal zone (Godschalk 1992).

In other areas, state and federal environmental legislation is frequently inconsistent, and gaps in authority often translate into little action. In the Gulf region, coastal states sometimes respond to this problem by pushing for measures that allow for more comprehensive state jurisdiction over sensitive coastal areas. Louisiana, for example, which contains nearly half of all US wetlands, passed the Louisiana Wetland Conservation and Restoration Act of 1990 to enhance protection of the fragile and endangered wetlands resources that were ignored by the Clean Water Act. Similarly, gaps in legislative/regulatory programs and sluggish administrative mechanisms have led to the recent growth of citizen advisory committees in each of the Gulf states, local volunteer clean-up programs, and an increase in voluntary waste reductions by industrial companies.

The Gulf of Mexico (GOM) Program, developed by EPA Regions IV and VI and NOAA's Sea Grant programs in the five Gulf coastal states, is an example of an intraregional initiative launched to address comprehensively the problems of the marine environment. Initiated in 1988, the GOM Program was formed to pioneer a broad, geographic focus to address the major environmental issues in the Gulf. The program's overall goals are to provide (1) an infrastructure capable of addressing interjurisdictional and multijurisdictional environmental problems; (2) better coordination among federal, state, and local programs; (3) a regional perspective to research; (4) a forum in which affected Gulf user groups, public and private educational institutions, and the general public can participate in the solution process; and (5) a strategy and framework for action.

The Gulf of Mexico Program is a cooperative effort administered by federal, state, and local agencies and managed by three committees: a policy review board, a technical steering committee, and a citizens advisory committee representing environmental, agricultural, business/industry, development/tourism, and fisheries interests in each state. Specific environmental priorities in the Gulf (e.g., habitat degradation, freshwater inflow, nutrient enrichment, and public health threats) are addressed by technical subcommittees.

The GOM Program has succeeded in defining the problems facing the Gulf marine environment. The next steps are to develop options for action and to implement them. That is a much greater challenge.

International Regional Cooperation: The Wider Caribbean

At the international level, the Action Plan for the Wider Caribbean, coordinated through the Regional Seas Program of the United Nations Environment Programme (UNEP), was adopted by representatives of twenty-three states in April 1981 at Montego Bay, Jamaica, in response to the growing concern for conservation, protection, and development of the marine and coastal resources of the region, including the Gulf of Mexico. The action plan called for a Convention for the Protection and Development of the Marine Environment of the Wider Caribbean Region, which was signed by representatives of sixteen states on March 24, 1983, at a conference in Cartagena de Indias, Colombia. The implementation strategy includes the development of several regionally coordinated comprehensive programs between 1990 and 1995.

As part of the Program for Assessment and Control of Marine Pollution (CEPPOL), a regional protocol for land-based sources of marine pollution is being developed (UNEP 1992). The United States is a leading participant in these negotiations. As the region's dominant economic power and one of its most progressive environmental managers, the United States has a strong interest in assuring that the negotiations result in a protocol that is both realistic (in matching the priorities and capabilities of the region's states) and effective (in achieving its marine environmental objectives).

The development of the Land-based Marine Pollution Sources Protocol involves consideration of both the Montreal Guidelines and the provisions of the Rio Declaration and Agenda 21 (UNEP 1992). The draft protocol relies more heavily than do earlier agreements on an approach based on water quality and effluent standards. Like most of the existing regional agreements, it refers to black and gray list substances; unlike other agreements, however, it links these lists to water quality standards that are to be jointly developed by the parties.

With negotiations toward a final version of the protocol continuing, the Conference of Plenipotentiaries, where final adoption of the protocol may take place, has been deferred to 1995 so that additional UNEP Meetings of Experts on the technical and legal requirements of the protocol can be held first (UNEP 1992).

Although there is tremendous economic and cultural diversity among the countries in the region, there is also reason to believe that an effective protocol can be devised based in part on US experience in the Gulf of Mexico.

THE BLACK SEA

The Black Sea, with a surface area of 423,000 square kilometers (163,000 square miles) and a volume of 534,000 cubic kilometers, is somewhat more than one-third the size of the Gulf of Mexico (table 2.2). The maximum depth is approximately 2,200 meters, and the average depth is 1,197 meters. The Black Sea is almost completely anoxic, with oxygen being absent below 200 meters, primarily because of a permanent halocline between 100 and 200 meters. The 4,090-kilometer Black Sea shoreline borders the former Soviet republics of Russia, the Ukraine, and Georgia, as well as Bulgaria, Romania, and

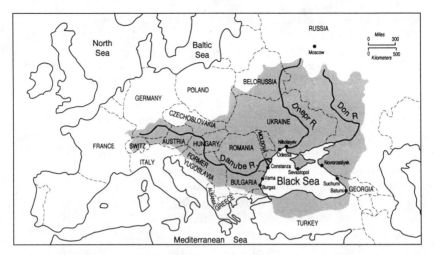

Map 2.2. The Black Sea and its drainage area.

Turkey. The major rivers draining into the Black Sea are the Danube, Dnepr, Don, Rioni, Dnestr, Kuban, and Bug.

The Danube is Europe's second largest river and carries about 75 percent of the pollutants entering the Black Sea (Keondzhyan, Kudina, and Terehena 1990). The Danube's riparian states are Germany, Austria, the former Czechoslovakia, Hungary, the former Yugoslavia, Romania, Bulgaria, Moldova, and the Ukraine, and it transfers water from the nonriparian countries of France, Switzerland, Italy, Albania, and Poland. Like the Mississippi River, the Danube is an important natural resource for its riparian states. It is a navigable waterway, a natural wildlife habitat, and a source of electric power, water supply, and recreation (Linnerooth 1990).

The Black Sea's watershed (map 2.2), which is somewhat less than one-half the size of the Gulf of Mexico's, is 4.4 times as large as its water surface, encompassing parts of fifteen European states and Turkey, or more than 1.86 million square kilometers (700,000 square miles), and a population of 130 million.

The Black Sea flushes out to the Mediterranean through the Strait Bosporus to the Sea of Marmara. The very narrow straits virtually shut off the Black Sea from the Mediterranean, allowing a complete water exchange only once in 2,250 years. In addition, because of persistent stratification of water layers, complete vertical mixing requires several decades.

The Black Sea is an important industrial and recreational center. It is also rich in marine resources; total annual commercial fishery landings in the Black Sea amounted by the end of the 1970s to about 880 million pounds, of which the former USSR's share was about 550 million. The only warm water seacoast within the territory of the former Soviet Union, the Black Sea's is the most popular coastline with vacationers and tourists. Nonetheless, it is seriously threatened by land-based pollution.

Since the collapse of the Soviet Union, the Black Sea has become increasingly important to each of the former Soviet coastal republics. Russia, having lost direct access to several European ports, now has only the Black Sea port city of Novorossiysk, on the northeast coast, and has been forced to intensify its usage of the Black Sea within a smaller jurisdictional area to satisfy its transportation and recreational needs. Ukraine and Georgia, meanwhile, are interested in capitalizing on their newfound control of important Black Sea coastal areas and ports. In addition, the Ukraine is pursuing new opportunities in oil and gas exploration in the northwestern section of the Black Sea, which is the area most vulnerable to the harmful effects of pollution. In early 1994, a survey of 9,300 miles of the Ukrainian offshore sector—the largest speculative survey ever undertaken in the Black Sea—was launched under an exclusive agreement among the Ukraine government, Western Atlas Inc., and EasternOil Services Ltd. of Bulgaria ("Spec Seismic Survey Planned in Black Sea" 1994).

The current environmental situation in the Black Sea is rapidly deteriorating. Some shelf areas have already suffered irreparable changes, especially in the northwestern Black Sea and part of the Bulgarian coastal area (Keondzhyan, Kudina, and Terehena 1990). Increased exploitation of the Black Sea is likely to aggravate the situation.

The Effects of Coastal Development

The population of the Black Sea drainage area is roughly 130 million, including 68 million inhabitants in the Russian and Ukrainian portions of the Black Sea basin. More than 8 percent of world industrial output is produced here, and Black Sea ports serve as connections to international trade and interstate transport facilities for all the republics of the former USSR. Growing populations and coastal development have had major adverse effects on the coastal marine environment.

The Effects of Nutrient Enrichment

Although the lack of vertical mixing is the main cause of the anoxia that characterizes some 90 percent of the Black Sea basin, contamination via rivers, such as the uncontrolled input of nitrogenic and phosphoric compounds, plays an important secondary role (UNEP 1987). In the agricultural sector, widespread and uncontrolled use of chemical fertilizers steadily increases. The use of mineral fertilizers in Georgia, for example, increased fourfold during the last thirty years. Animal wastes continue to go unmanaged. Purification installations are available at all fifty-five stock-breeding plants in Georgia, but at least twenty-one of them do not work. (These conditions prevailed even before the outbreak of civil war in Georgia, where the Black Sea coast has been a major area of confrontation.) Red tides in the northwestern part of the Black Sea shelf are common, but a direct relationship with nutrient loading has not been documented. In general, there has been mounting anxiety as the hypoxic, lifeless water layers have accelerated their rise into the upper layers.

The Effects on Public Health

Communal sewage from Black Sea cities typically is purified only slightly or not at all before discharge. In resort areas, the number of peak-season vacationers surpasses the capacity of purification facilities by a factor of five to seven. Increased rates of illness are associated with elevated pollution in recreational coastal areas and with unsafe drinking water.

These public health effects entail other economic and social consequences as well. Many Black Sea sanatoria resorts, for example, have lost business and revenue. In 1987, some 300,000 reservations for accommodations in the seaport resort of Batumi were cancelled because of threats to public health, reducing tourism income by 30 percent. In 1990, outbreaks of intestinal bacillus on Black Sea beaches resulted in numerous temporary beach closures.

Industrial and Municipal Sources

One-quarter of all household and municipal sewage discharged into the Black Sea is utterly unpurified. So are livestock and railway sewage, and the latter contains high concentrations of strong disinfectants. The waters around the naval fuel-storage facilities in Sevastopol have for many years contained levels of petroleum-related substances that far exceed, sometimes by 100-fold, the maximum permissible concentration. Airborne pollutants reach the Black Sea via automotive and other exhausts generated in connection with industrial activities. Petrochemical facilities and steel plants are heavily concentrated in the coastal region (fig. 2.2). Along the banks of the Prypjat River, not

Fig. 2.2. Recently a major polluter, the Novorossiysk oil farm on the Black Sea coast in Russia has built major failsafe purification facilities. Comparable installations are a part of the landscape along several stretches of the Gulf of Mexico coast as well. (Novosti from Sovfoto)

far from the Black Sea itself, stands the Chernobyl nuclear power plant, which spread radioactive pollution to the Black Sea and to parts of Poland, the Baltic states, Finland, Norway, and Sweden from its infamous reactor disaster in April 1986 (Hanemann et al. 1992). Perhaps even more damaging has been the extensive dumping of nuclear waste for many years at numerous sites throughout the former Soviet Union, including locations near the Black Sea port of Odessa in the Ukraine.

Studies are still being done to determine the likely future effects of the Chernobyl event, but scientists predict that it will be years before the problems for public health and agriculture begin to surface (Shlyakhter and Wilson 1992). Much depends on the long-term behavior of radionuclides, about which relatively little is known (see chapter 5). Some scientists believe, however, that the harmful effects from Chernobyl will eventually be felt all over Europe (Kwaczek, Kerr, and Mooney 1990).

The Management Approach of the Former Soviet Union

The management approach of the former Soviet Union reflects its centrally planned form of government, and land-based marine pollution is among the notorious environmental abuses engendered by that system (Vartanov 1991). With its tight, centralized control, the political structure of the former USSR provided no basis for the kinds of checks and balances among different interest groups and levels of government that have characterized the environmental decision-making process in the United States. Pollution and resource degradation attributable to "Soviet mold" environmental practices are estimated to have cost the former USSR 17 percent of GNP annually (25–30 billion rubles in 1990, according to the Soviet Ministry of Foreign Affairs) and are generally accepted to be the cause of chronic illness among one-quarter of the adult population and one-sixth of all children under the age of fifteen (Canfield 1993a and references therein). Despite sustained popular dissent and the growing influence of social organizations in the political system, however, today the Soviet mold remains largely intact. Even the State Committee for Environmental Protection of the USSR, established by decree in 1988 in response to a burgeoning green movement, was almost totally ineffectual (as its successor the Russian Ministry of the Environment has been) because it was completely dependent for its operation on existing agencies of the government bureaucracy.

It would be unfair to say, however, that the Soviet Union did nothing to protect the Black Sea from land-based pollutants. An extensive monitoring system has long been in place, and much legislation concerning industrial practices, recreational uses, and other land-based sources of pollution in the Black Sea has been introduced and enacted over the past twenty years. Sadly, implementation and enforcement of these measures left much to be desired.

Nor has the breakdown of the Soviet Union brought much positive change in the area of environmental protection. Newly developed environmental activities and legislation are still subject to the limitations of the old system. Meanwhile, local authority over environmental activities has increased, but local knowledge of environmental technologies and practices, as well as appropriate legal instruments, is distinctly lacking. Of course, developing an effective, integrated environmental protection regime takes time under the best of circumstances. For countries like Russia and the other newly independent republics, which are confronted by a tremendous amount of instability at the local level, the obstacles are daunting indeed.

Information Gathering. In the former Soviet Union, the State Oceanographic Institute (SOI) monitored Black Sea pollution levels for many years, as part of the "Seas of the USSR" project, in cooperation with the All-Union Service of Monitoring and Environmental Pollution Control under the auspices of the USSR State Committee on Hydrometeorology (*Goskomgidromet*). The SOI succeeded in obtaining more than three thousand water quality analyses annually from 242 monitoring stations. Criteria included levels of oil and chlorinated hydrocarbons, detergents, mercury, and phenols. Sanitary and Epidemiological Stations (SESs) were responsible for water quality in recreational areas. The sampling frequency for each index was determined by the local SES but was not to be less than twice a month during the swimming season for any one index.

The system suffered from organizational and financial fragmentation, however, which resulted in information gaps between national and local services. Moreover, *Goskomgidromet* did not always fulfill its responsibilities with respect to the marine environment. As a result, its regional branches often did not receive data from the enterprises discharging wastes into the Black Sea, and thus they were unable to pass that information to the services responsible for water-quality mitigation efforts. Monitoring activities have become even more fragmented since the dissolution of the Soviet Union, and it will take much time and effort to establish cooperative and routine interactions among the cognizant authorities in the newly independent states.

Turkey and Romania have been independently monitoring the concentrations of various pollutants in the Black Sea for roughly the past decade. Similar efforts in Bulgaria have been only sporadic, however, and have frequently been undermined by "incompetent" application of sampling and storage techniques (UNEP 1987). Overall, regional efforts have suffered from the same kind of institutional fragmentation that plagues the former Soviet Union. Perhaps most debilitating of all has been the lack, until very recently, of any regional intercalibration exercises on instrumentation and analytic techniques, making it "practically impossible to compare what is available as data on marine contaminants in [the] Black Sea" (UNEP 1987).

Legislation and Regulation. The December 1972 Decision of the Central Committee of the Soviet Communist Party and the USSR Council of Ministers restricted the use of valuable recreational areas for agricultural and industrial purposes, but with little effect on Black Sea coastal areas (or the coastal areas of any other major seas in the former Soviet Union). The 1976 Decision of the Central Committee and the USSR Council of Ministers, "On Measures of Pollution Prevention in the Black and Azov Seas' Basin," charged the Councils of the Ukraine, Byelorussia, Georgia, and Moldova, along with the interested Soviet ministries and agencies, with developing measures to reduce, by 1985, input into the Black Sea watershed of untreated wastewaters from the Soviet Union. Decisions concerning the Black Sea were adopted by the republics as well. Unfortunately, all of these measures were ineffective. For example, 80 percent of the enterprises affected by the 1976 decision were unable to make the planned investments, and those that were financially able could not find contractors to build the necessary installations.

Shortly before the Soviet collapse, the USSR and the republics attempted to create an infrastructure of environmental quality-control and management branches in the Black Sea region, in effect planting the first seeds of transfer to independent republics. This network included the State Committee on Hydrometeorology (*Goskomgidromet*); republican representation from the Ukraine, Moldova, and Georgia; the USSR Ministry of the Environment (formerly *Goskompriroda*); republican environmental protection committees; and numerous corresponding local branches.

Agency fragmentation, however, was and continues to be a characteristic problem of management structures in the Black Sea region, with the separation between regional and central management offices causing severe disruption in the flow of information. This became especially clear during the Chernobyl disaster, when *Goskomgidromet* did not pass vital operational information to *Goskompriroda* in time for action. Soviet scientists did present a detailed report on the aftermath of Chernobyl to the International Atomic Energy Agency in 1986, but in 1989 it was revealed that important information about the deposition of large quantities of radionuclides northeast of Chernobyl and in Russia had been intentionally omitted. Moreover, the 1986 report to the International Atomic Energy Agency was classified as secret. Not only is this type of government secrecy inconsistent with a responsibility to protect the public health and provide international notice, but also it is indicative of "a breakdown of trust between the Soviet citizens and the central government and its experts" (Shlyakhter and Wilson 1992).

In short, lack of effective implementation was and remains the pervasive problem. Legally fixed permissible concentrations of pollutants are routinely exceeded by tens to hundreds of times in the Black Sea. No integrated coastal zone management program has been established, and local authorities have lacked the authority and the technical skills to regulate or otherwise influence the activities of huge, state-sponsored industrial and agricultural enterprises.

Citizen organizations in the Black Sea region have had little influence over state environmental control structures, much less over polluters themselves. To a large extent, state bureaucracies continue to be driven by the old governing structure and the entrenched industrial complex.

International Regional Cooperation

Preparatory work began in 1969 on the International Convention on Protection and Preservation of the Marine Environment of the Black Sea, with the support of the Soviet Union, Bulgaria, Romania, and Turkey. After the withdrawal of Turkey, however, cooperation was suspended until 1988. The breakup of the Soviet Union further postponed development of the convention until April 1992, when the Bucharest Convention on the Protection of the Black Sea Against Pollution was adopted in Bucharest by Bulgaria, Georgia, Romania, Russia, Turkey, and the Ukraine (*International Legal Materials* 1993). It calls for the establishment of a Black Sea Commission in Istanbul. Of its three legal protocols, that dealing with land-based marine pollution receives the most attention.

International attention to the Black Sea marine environment was stimulated in part by the United Nations Environment Programme. In 1987 UNEP expressed to Bulgaria, Romania, Turkey, and the Soviet Union an interest in the development of a Black Sea Action Plan, under the aegis of UNEP's Regional Seas Programme. UNEP hoped to begin the development of a Black Sea program in 1990–91. Although some of the involved nations favored a UNEP approach, others preferred the development of a Black Sea marine protection program through a direct, multilateral agreement among the interested states.

UNEP continues to assist efforts toward a Black Sea action plan. In an effort that paralleled the 1992 Bucharest meeting and with initial funding of $9.3 million from the Global Environment Facility (GEF) and over $15 million from other donors, the riparian countries commenced a Project for Environmental Management and Protection of the Black Sea. The GEF project will be executed by the UN Development Programme (in partnership with UNEP and the World Bank), and it is generally considered to be the first step, or perhaps the reinitiation, of the UNEP Action Plan for the Black Sea (Ozhan 1993).

After the collapse of the Soviet Union, Russia, Georgia, and the Ukraine made a number of very general declarations about the need to establish strong collaborative measures to preserve their common ecology and traditions in the Black Sea region. Although cooperation among their governments has nonetheless deteriorated since then, the scientists and technicians who have long worked together at the local level continue to enjoy good communication and cooperation.

This working relationship can serve as a foundation for a newly conceived program of regional cooperation in the management of Black Sea resources and the protection of the Black Sea ecosystem. In addition to providing sorely needed funds, international support for such a program can help the newly independent states fill important gaps in technical skills and establish the kinds of private companies and nonprofit organizations that have played such a vital role in environmental protection efforts elsewhere around the world.

CONCLUSIONS

Although the widespread problem of land-based marine pollution is largely confined to coastal waters within national jurisdictions, it does raise issues of international environmental security. This occurs not only where there is direct transboundary pollution, but also more indirectly—where shared resource stocks are affected and through trade, tourism, and capital mobility.

The United States and Russia each borders a major, semi-enclosed regional sea that it shares with neighboring states. The Gulf of Mexico is economically and culturally important to the United States, which is a dominant economic and political presence in the region. Russia, as a successor state to the former Soviet Union, has a similar relationship to the Black Sea, with a few important distinctions. Since the collapse of the Soviet Union, Russia has lost its claim to vast areas of the Black Sea, including the traditionally vital ports of Sevastopol and Nikolayev, and it now ranks third in terms of Black Sea coastline (after Turkey and the Ukraine). Russia is therefore acutely concerned with maintaining social and political stability in the region, since the conduct of its marine and coastal activities depends upon international cooperation.

Both the US coast of the Gulf of Mexico and the Black Sea coast of the former Soviet Union are subject to serious land-based marine pollution, although perhaps more is known about the US case. Management approaches in the United States reflect the federal form of government and involve a complex of interactions among local, state, and national programs. Coordination is at a premium, and efforts to enhance coordination are still largely nascent. Citizen action is influential. In the Black Sea region, on the other hand, coastal zone management approaches and skills remain highly fragmented.

In both cases, new efforts are under way to coordinate regional control of land-based marine pollution at an international level. Constructive experience is available from efforts in other regions, but it remains to be seen how effectively those lessons will be transferred to the wider Caribbean and Black Sea regions.

Perhaps the most striking conclusion to be drawn from these case studies is not a surprising one. It merely notes one symptom of the disintegration of the previous centralized state and economy of the former Soviet Union. The severe and very complex political and socioeconomic problems that now confront the post-Soviet republics mean that no significant environmental activity is likely to be undertaken in the region in the near future. Most republics simply cannot afford to participate in such efforts. In the Black Sea region, moreover, Georgia has been plagued by civil unrest and, like the Ukraine, has a relationship with Russia that is heavily weighted with rivalry and mistrust. Currently, the Commonwealth of Independent States (CIS) represents the best hope for a framework for inter-republic cooperation to replace the hierarchical Soviet structure in which regional cooperation was previously managed, however inadequately; of the three former Soviet Black Sea coastal states, however, only Russia is a CIS member.

CHAPTER 3

Living Resource Problems: The North Pacific

THE NORTH PACIFIC IS a resource-rich region used intensively for commercial purposes. Three principal issues pertaining to the use of living resources have threatened international environmental security in the Bering Sea and adjacent North Pacific: depletion of straddling fish stocks, particularly the unregulated fishing of walleye pollock (*Theragra chalcogramma*) from the Central Bering Sea; illegal and preemptive take of anadromous fish species on the North Pacific high seas; and incidental take of such nontarget protected species as marine mammals and seabirds by drift gillnet fisheries operating within and immediately adjacent to the Bering Sea.

Besides the United States and Russia, the states most involved with current management practices in the Bering Sea include Canada, Poland, South Korea, North Korea, Japan, Taiwan, and the Peoples Republic of China (Zilanov et al. 1989). Illegal fishing, unregulated fishing, and driftnetting have all resulted in bitter disputes, impaired relationships between "victim" and "culprit" states, and international countermeasures (Johnston 1990). Recently, however, developments in the region's complicated legal status have somewhat lessened the potential for international resource conflicts.

In this chapter, we describe the affected ecosystem, explore the kinds and magnitude of transboundary environmental problems affecting relationships among the involved states, and identify potential legal, economic, and management solutions for resolving the interconnected environmental security issues in the region.

The Ecosystems of the Bering Sea and the Adjacent North Pacific

The Bering Sea–North Pacific region exhibits several biological and geological features with important implications for the use of its resources.[1] The shallowness and width of continental shelves, large-scale upwellings, the hydrochemical structure, seasonal sea ice, and other factors result in extremely high productivity (see Favorite, Dodimead, and Nasu 1976). The North Pacific is characterized by a variety of current systems, an inverse correlation between productivity and the depth of the sea floor, a strong connection between ocean climate and fisheries, and, due to its cold temperatures and seasonal sea ice, sensitivity to anthropogenic disturbances. The average productivity of the North Pacific is 1.3 to 4.5 times the average background values for the oceans in general. The Bering Sea proper has high concentrations of nutrients, which in turn support large growths of phytoplankton and zooplankton. This ecosystem includes 450 species of fishes, crustaceans, and mollusks, of which 50 species are commercially important. Fisheries have targeted various species of anadromous, pelagic, and demersal fish, as well as several sedentary species such as king and tanner crabs. Important pelagic fish species include Japanese sardine, Pacific herring, Japanese chub, jack mackerel, Pacific saury, and Pacific anchovy. Walleye pollock, Pacific cod, hake, and sole are important demersal fishes. The Bering Sea has some of the highest concentrations of apex predators (including at least 50 species of marine mammals) at upper trophic levels in any ocean, including the Antarctic-Southern Ocean region. Seabird populations in the region, including breeders and nonbreeders, exceed eleven million individuals.

The Bering Sea itself is divided geologically into two major portions: a very large and wide (>200 km) continental shelf, the edge of which runs mostly northwest to southeast past the Pribilof Islands, and a deeper Aleutian Basin lying just north of the western Aleutian and Commander Islands. As the water connection between two large continental shelves (the Bering and Siberian), the Bering Sea is the pathway for water exchange between the Pacific and Arctic Ocean basins (Coachman, Aagard, and Tripp 1975). Water flows in from the North Pacific Ocean through the straits and passes of the Aleutian Arc (especially at Near and Kamchatka Straits) and out to the Chukchi Sea and Arctic Ocean through the Bering Strait (Reed 1990). On the eastern side of the Bering Sea, fresh and relatively warm Alaskan coastal water flows northward along the Alaska coastline. On the western side, colder and more saline Anadyr and Bering Shelf water moves northward out of Anadyr Gulf and along the shelf break through Anadyr Strait and into the Chirikov Basin and Chukchi Sea.

The Bering Sea is among the most productive of all subpolar marine ecosystems (Khen 1989; Highsmith and Coyle 1990; McRoy, Goering, and Shields 1972; Sambrotto, Goering, and McRoy 1984; Hansell et al. 1989; Conference on Shared Living Resources of the Bering Sea 1990). The variety and magnitude of its production sources (including ice algae and extended seasonal blooms) are the bases for exceptional concentrations of both pelagic and benthic consumers in the Bering Sea. Productivity originating in the pelagic water column in both the Bering Shelf and Anadyr waters assimilates into pelagic food webs (Springer et al. 1987; Springer, McRoy, and Turco 1989). These support the millions of both planktivorous and piscivorous seabirds (Shuntov 1972; Sowls, Hatch, and Lensink 1978; Springer et al. 1987). Portions of primary productivity ungrazed by pelagic zooplankton are exported to rich benthic communities on the shallow shelf (Grebmeier, McRoy, and Feder 1988; Grebmeier, Feder, and McRoy 1989; Grebmeier and McRoy 1989). These benthic communities are dominated by bivalves and amphipods (Highsmith and Coyle 1990), whose high standing stocks in turn support large populations of walrus (*Odobenus rosmarus*) and migratory gray whales (*Eschrichtius robustus*), respectively.

The northern Bering Sea is typically ice covered for five to six months each year, with sea ice occasionally extending as far south as the Pribilof Islands and eastern Aleutian Islands. Seasonal sea ice adds more structural variety for marine animals, with several species limited to, or preferring, pack ice, polynyas, and open leads (Fay and Cade 1959; Portenko 1972, 1973; Bogoslovskaya and Votrogov 1981). Some 90 percent of the world's stocks of anadromous fish species are concentrated in the North Pacific Ocean, including multitudes of pink, chinook, chum, sockeye, and coho salmon stocks. The various salmon species all coexist during their growth periods, inhabiting the North Pacific between 35 degrees and 50 degrees north latitude.[2] These anadromous fish originate in Russia, the United States, Canada, and Japan. Only a relatively small portion of the anadromous stocks originates in Korea and China. The fact that anadromous species transit the Bering Sea yet originate far up the Yukon, Mackenzie, and other Canadian rivers illustrates the transboundary nature of fishery resources and how a state outside an immediate geographic area can be affected by management practices far removed from its borders.

Although most seals, walruses, and seabirds are associated with specific island and coastal sites during reproduction and haul-out, foraging and migration patterns are such that populations of most marine vertebrates use very large portions of the Bering Sea and regularly cross political and management units within it. Several species also move to adjacent areas of the North Pacific Ocean, particularly to ice-free waters during winter (Trukhin and Kosygin 1987). The North Pacific–Bering Sea ecosystem is thus a fluid one, with considerable exchange and transport of both biotic and abiotic materials into, through, and out of various sectors. On the bases of geological, physical, and biological parameters, we have therefore chosen to define our study region as reaching from the North Temperate Zone of the Pacific Ocean at about 35 degrees north latitude to the northern extremity of the Bering Sea at Bering Strait.

FISHERIES DEPLETION

The North Pacific Ocean accounted for one-third of the world's marine fish landings during the 1980s (FAO fishery statistics). Commercial take of walleye pollock constituted the largest single species fishery in the world, in excess of three million tons for the entire North Pacific Ocean (FAO 1989). Since the 1930s, this fishery has evolved from (1) primarily a Japanese far-seas fishery to (2) an international fishery with vessels from Japan, the Soviet Union, Korea, and Taiwan to (3) a US-Japanese joint venture and, finally, to (4) primarily a US and USSR (now Russian) domestic fishery (see fig. 3.1). This trend illustrates a form of policy problem displacement: to a large extent, management of declining or depleted stocks was merely transferred from the international to the national level. Like vessel and effort substitution from other depleted fisheries (another form of policy problem displacement), this practice does not constitute a long-term solution for environmental security.

Pollock represents such a large fraction of ocean-based protein in the world market that a significant increase in world fish prices is likely to occur if overfishing continues. Despite remaining uncertainties about biological management units for fish stocks, interrelationships among stocks, and environmental influences independent of fishing operations, there is substantial evidence that stocks in the Bering Sea are under increasing stress (table 3.1). The effort by all nations to harvest pollock, for example, increased fourfold between 1968 and 1973, and during the same period the catch per unit of fishing effort fell by 50 percent. Stocks of other groundfish, including yellowfin sole, Pacific perch, and sablefish, have shown similar evidence of stress.

Movement toward extended ocean jurisdiction by coastal states has progressed rapidly. In 1976, the United States enacted the Magnuson Fishery Conservation and Management Act (MFCMA), which declared exclusive management rights over all fishery resources except tuna within 200 nautical miles. Under the MFCMA, which became effective in 1977, "maximum sustainable" yields of fishery resources in the 200-mile zone are determined, and "optimum" yields are defined after the maximum sustainable yields are adjusted according to economic, social, and ecological conditions. Fishing by foreign nations is allowed only for the part of optimum yields not harvested by domestic fishers.[3] For the eastern Bering Sea and Aleutian Islands region, four categories of finfish and invertebrates have been established for US management purposes: prohibited species, target species, other species, and nonspecified species.. Prohibited species are highly regulated and must be returned to the sea when caught

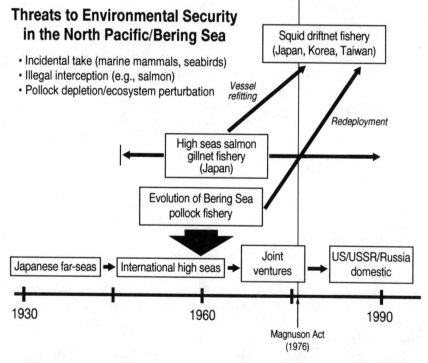

Fig. 3.1. The evolution of threats to environmental security in the North Pacific–Bering Sea, 1930–1990.

Table 3.1
MAJOR FISHERY STOCKS IN THE NORTH PACIFIC OCEAN, 1970–1987

Stock	Estimated Potential ($\times 10^3$ tons)	Catches ($\times 10^3$ tons)				State of Exploitation (1987)	Main Fishing Countries (1987)
		1970–74	1975–79	1980–84	1985–87		
Salmon	550–600	385	449	593	691	Asian stock depleted but recovering; American stock fully exploited	Japan, USSR, Canada, US
Alaska pollock	4,150–6,600	4,077	4,452	4,704	6,548	Fully exploited	USSR, Japan, Poland, Korea, US
Other demersal	3,550–3,850	2,015	2,332	2,165	2,632	Most stocks fully exploited or overfished	Japan, China, USSR, Korea, US, Canada
Japanese pilchard	5,800–6,500	157	1,433	3,959	5,028	Close to maximum	Japan, Russia, Korea
Chub mackerel	1,000	1,554	1,808	1,354	1,344	Overfished	Japan, Russia, China, Korea

(continues)

Table 3.1 continued

Stock	Estimated Potential (× 10³ tons)	Catches (× 10³ tons)				State of Exploitation (1987)	Main Fishing Countries (1987)
		1970–74	1975–79	1980–84	1985–87		
Pacific herring	1,000(?)	579	349	223	323	Fully exploited	USSR, Japan, US, Canada
Tunas, billfishes	280–380	399	430	467	483	Fully exploited or overfished	Japan, China, Taiwan
Other pelagic fish	?	2,301	2,099	2,038	2,178	Moderately to fully exploited	Japan, Taiwan, US, Korea, Canada
Squids	300–400	425	285	267	157	Moderately exploited	Japan, Taiwan, Korea
King, snow crabs[a]	80–130	71	104	71	57	Fully exploited; king crabs depleted	US, Japan
TOTALS							
Northeast Pacific	19,800–21,600	14,570	17,962	20,742	24,993		
Northwest Pacific	?	2,392	2,059	2,322	3,211		

source: Mirovitskaya and Haney 1992.

[a]Fisheries only in the northeast Pacific Ocean.

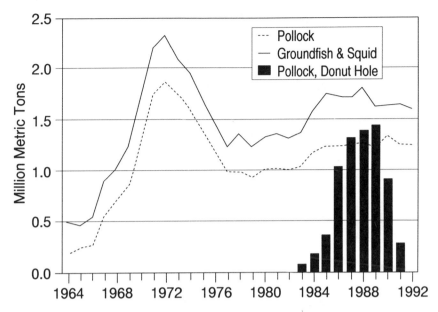

Fig. 3.2. Groundfish and squid catches in the eastern Bering Sea and Aleutian Islands and pollock catches in the Central Bering Sea or Doughnut Hole by all nations, 1964–1992. (Data sources: Plan Team for Groundfish Fisheries 1989, 1992)

without permission. Allowable catch quotas are established for each target species, and aggregate total allowable catch is established for the group of other species. In 1993, the total allowable catch (TAC) for all groundfish in the Gulf of Alaska was a recommended 306,651 metric tons, and the corresponding TAC for the Bering Sea/Aleutian Islands was 1.998 million metric tons. These quotas represent 7 percent and 45 percent, respectively, of total US commercial fish/shellfish landings for 1992 (NMFS 1993a).

Groundfish and squid catches of target and other species within the eastern Bering Sea and Aleutian region are shown in figure 3.2 for the period 1964–92; pollock clearly dominated total catches, accounting for 71 percent on average during the twenty-five-year period through 1988 and for 78 percent on average from 1989 through 1992. Before the enactment of the MFCMA, aggressive fishing effort by several nations led to the peak total catch of 2.3 million metric tons in 1972. Of this total, pollock accounted for 1.9 million metric tons. Catch restrictions and other management measures had lowered the total catch to 1.2 million metric tons by 1977.

Fish stocks within the Russian exclusive economic zone now account for 1.6–1.7 million metric tons of potential annual yield, including 1.2 million metric tons of pollock. The total potential yield of Bering Sea stocks is estimated at 3.8–4.5 million metric tons, including 2.6–2.9 million metric tons of pollock. In recent years the actual catch of pollock has been as high as 3.3–3.8 million metric tons, but stock depletion has been so severe that one 1993 estimate put the actual catch at about 300 metric tons (Niemeier, pers. comm., 1993).

Ecosystem Perturbations from Fisheries Depletion

Depletion of fish stocks or removal of other food web components can cause dramatic changes in entire marine ecosystems (e.g., see Hamre 1988; Estes, Duggins, and Rathbun 1989; Røttingen 1990; Vader et al. 1990). Apex predators are particularly sensitive to perturbations of food webs and direct exploitation. One species of marine mammals (the Stellar's sea cow) has already been driven to extinction in the North Pacific–Bering Sea (see Domning 1978). Several coincidental effects from pollock overfishing have been hypothesized or manifested within the Bering Sea ecosystem, including changes in the size classes of pollock taken by their predators, precipitous declines in pinniped populations that rely on pollock, and increases in populations of apex predators that feed primarily on the prey (i.e., zooplankton) no longer consumed by adult pollock that have been removed by the fishery. The possibility that distinct parts of the Bering Sea food web respond in a cascade fashion to changes in pollock abundance has prompted calls to manage decisions for the commercial fishery on a multispecies basis (Springer et al. 1987). Detrimental effects of overfishing on some components of the Bering Sea food web are compounded when the same fisheries also remove these predators via incidental take and intentional killing. Competition between marine mammals and fisheries is intense, with the amount of finfish alone consumed by pinnipeds and cetaceans estimated as high as 2.4 million tons annually (Laevastu and Favorite 1977).

Pollock are important food species for northern fur seals (*Callorhinus ursinus*) in the North Pacific Ocean (Lander and Kajimura 1982). Swartzman and Haar (1984) found that, during the period of fishery expansion for pollock, northern fur seal diets included progressively more pollock, despite a drop in the size of the fish eaten. Commercial fisheries take larger fish (which are cannibalistic on smaller fish), and the fact that there are fewer large pollock enables smaller size classes and younger cohorts of pollock to survive in greater numbers. Assessing the relative degree of competition between the northern fur seal and commercial fishing operations has been complicated, however, because survival rates and population sizes of fur seals have not changed in response to the cessation in 1968 of direct harvesting of females (Northridge 1984).

Ecosystem-wide stress is also thought to be contributing to the precipitous decline in numbers of Steller's sea lion (*Eumatopias jubatus*) in the Bering Sea region (Environmental Defense Fund 1990). Based on its feeding habits and demography, Lowry and Frost (1984) found this species to be one of three marine mammals most likely to be involved directly in interactions with commercial fisheries. Although Steller's sea lions have declined over the past four decades, a joint US-USSR survey conducted in 1989 revealed a recent acceleration in the rate of this decline. The population had fallen 63 percent since 1985 (Loughlin, Pervlov, and Vladimirov 1990). The decline was most pronounced in the eastern Aleutian Islands (93 percent since 1960), but it eventually extended from the Gulf of Alaska to the western Aleutians. Projections have indicated that the species could become extinct in little more than twenty years if this rate of decline continues (Conference on Shared Living Resources of the Bering Sea 1990). The Steller's sea lion is listed as threatened under the US Endangered Species Act.

Different trophic groups of marine animals may be responding differently to the changes in pollock abundance within the Bering Sea. The reproductive success of seabirds feeding primarily on pollock has been linked to the age-class strength of juvenile fish (Springer et al. 1987). Springer and colleagues proposed that poor pollock recruitment in recent years led to persistent seabird breeding failures in 1981–84, as well as to declines in population sizes of black-legged kittiwakes (*Rissa tridactyla*) and other piscivorous seabirds at island colonies in the Bering Sea (see also Craighead and Oppenheim 1985). In contrast, planktivorous seabirds, which possibly compete with juvenile pollock for zooplankton, may be benefiting from lower abundances of older pollock. Plankton-feeding auklets have doubled in number at some island colonies in the Bering Sea during the last two decades (Piatt, Roberts, and Hatch 1990). Although this finding is in agreement with the hypothesis that reductions in pollock favor other (i.e., planktivorous) components of the food web, alternative explanations for the increase in auklet populations have precluded definitive conclusions. Other environmental factors that favor increases in auklet populations (e.g., removal of predators) occurred during the same period as the expansion in the pollock fishery (see Piatt, Roberts, and Hatch 1990).

THE DOUGHNUT HOLE AND THE PROBLEM OF STRADDLING STOCKS

Conflicts over harvesting and depletion rates will typically arise in any open-access fishery, but they can become especially heated when the stocks being fished are suspected to straddle or move back and forth across the open area and an area of regulated national jurisdiction. Anadromous fish (such as salmon) cross these boundaries as they alternate between spawning in continental streams (mainly in North America and Russia) and ranging widely in the North Pacific high seas. The region has witnessed a painful history of international conflict over fishing rights for anadromous species, but customary international law, as reflected in Article 66 of the Law of the Sea Convention, has evolved clearly to assign primary responsibility to the state of origin (Burke 1991; Kwiatkowska 1993; Kaitala and Munro 1993). Thus, directed anadromous fisheries on the high seas may be deemed illicit, although the status of nondirected by-catch of these species remains less clear (Burke 1991). Even less clear, and a continuing source of international conflict, is the conservation and management status of straddling stocks in enclaves of high seas surrounded by areas of exclusive national jurisdiction.

This is exactly the problem that has arisen in an anomalously isolated patch of high seas in the Central Bering Sea. Since the extension of ocean jurisdiction to 200 nautical miles by both the United States and the former Soviet Union (now Russia), the Central Bering Sea, an area of approximately 55,000 square miles, is completely surrounded by the two nations' EEZs. This area thus remains high seas and is frequently termed the *Doughnut Hole*.[4]

Although there are serious scientific uncertainties about how much stocks in the Doughnut Hole actually straddle their national jurisdictions, both the United States and the former Soviet Union became increasingly concerned in the late 1980s about uncontrolled fishing there by Japan, South Korea, China, Poland, and possibly Taiwan and North Korea (Miovski 1989). Large-scale fishing by these nations within the Central Bering Sea was a relatively recent phenomenon and was targeted mostly on pollock. Rapid development of this fishery brought a total pollock catch of only a few thousand metric tons before 1984 up to nearly 1.5 million metric tons in 1989, an amount that exceeded the total pollock catch from the eastern Bering Sea and Aleutian Island region (see fig. 3.2). Japan alone took 750,000 tons from this area in 1988. Indeed, some in the US fishing industry voiced suspicion that the foreign operations in the Doughnut Hole were being used as a cover for illicit forays into the richer resources of the

EEZ (Canfield 1993b). This international management anomaly of the Doughnut Hole threatened to create a truly critical problem of environmental security by destabilizing the normal relations between states and provoking international countermeasures.

By August 1992, however, pollock stocks had declined so drastically that China, Japan, South Korea, and Poland agreed to join the United States and Russia in a voluntary suspension of fishing for pollock in the Central Bering Sea for all of 1993 and 1994. The United States and Russia also agreed to suspend pollock fishing in their Bering Sea EEZs (Canfield 1993b). The two-year multilateral agreement also calls for a scientific survey by research vessels, trial fishing for pollock by a limited number of fishing vessels, scientific assessments of stock structure, and mutual effort toward a long-term arrangement before the beginning of 1995 (Joint Resolution of the Fifth Bering Sea Conference 1992).

The Central Bering Sea international fishery developed, in large part, as a result of phasing out foreign fishing allocations within the EEZs of the United States and the former Soviet Union (Miles and Fluharty 1991). As domestic fishers became capable of harvesting the entire EEZ optimum yield, foreign fishing activities were cut back. Under joint venture agreements, pollock landings by US domestic vessels were delivered to non-US processors. During several years immediately after US enactment of MFCMA, pollock catches had been almost exclusively allocated to foreign fisheries. However, foreign allocations decreased markedly during the second half of the 1980s, and no foreign allocations were made in 1987 and 1988. The domestic pollock fishery and joint ventures were developed rapidly during this phase-out period of foreign fishing activities (with joint ventures phased out as well by 1991; see fig. 3.3). Foreign efforts were thus pushed farther seaward to concentrate on the Doughnut Hole. This trend illustrates another element of policy problem displacement, that is, the geographical transfer of fisheries exploitation from one location to another. Rarely a satisfactory solution to stock depletion, this practice is particularly unhelpful when stocks are transboundary and unlikely to remain within the limited portions of the ecosystem being managed.

ECONOMIC CONSIDERATIONS FOR STRADDLING STOCKS

Overfishing problems in the Central Bering Sea typify the problems of fishery resources with unregulated open access where multiple na-

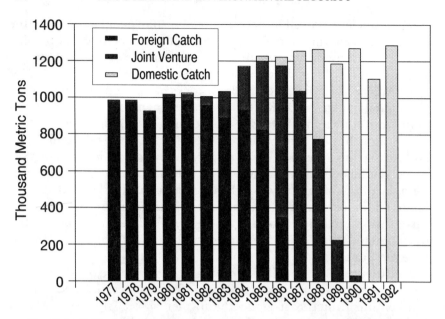

Fig. 3.3. Pollock catches in the eastern Bering Sea and Aleutian Islands areas by foreign fisheries, joint ventures, and US domestic fisheries, 1977–1992. (Data sources: Plan Team for Groundfish Fisheries 1989, 1992)

tions target the same stock of fish. Under the exclusive jurisdiction of private ownership, where property rights are well defined, all costs and benefits from using fishery resources accrue to the owner. Resource management will be undertaken with the long-term objective of maximizing the net benefit to the owner. Frequently, marine fisheries do not have clearly defined property rights. Since fish not caught by one fisher may be caught by another, there is no incentive to conserve the resource to achieve optimum harvest. The absence of private ownership or some other central regulating mechanism can preclude the achievement of long-term management strategies.

One alternative solution for this is to estimate the economic results arising from open access versus those from assigned property rights with formal management of the resource. The cost over time of catching the fish, in each case, should be compared to the long-term benefits from the harvest. The level of fishing effort to achieve an economically optimum yield, maximizing these net economic benefits from the resource over time, is typically *lower* than the level at which the maximum sustainable yield is harvested (Schaefer 1957; Clark 1976). In general, the economics of shared fishery resources is well understood theoretically, but the problem of straddling high seas stocks is more complicated (Kaitala and Munro 1993).

Efficient fishery management that exploits the long-term potential of shared resources can be achieved either by exclusive control (privatization) or by the establishment of some other management authority. In the case of the Central Bering Sea, effective ownership through exclusive jurisdiction could be assigned to Russia and/or the United States by extending their EEZs and closing the Doughnut Hole. Alternatively, with joint management of the Central Bering Sea, Russia, the United States, and other nations could assess the health of the existing fishery stock and set and enforce fishing quotas. The potential efficacy of both these alternatives, however, is called into question by the presence of straddling stocks in the high-seas area: the status of foreign fishers and the conditions for new entrants must be settled, and changing national preferences, technologies, market conditions, and stock structure will lead to changed economic incentives and management objectives for the participants (Kaitala and Munro 1993).

Although efficient management of the resource can be approached by either of these avenues, in short, all nations that have a stake in the fishery may not agree or may not continue to agree. Countries that are not members of a management compact, for example, may find their access to the fishery denied altogether. Although fishing nations may agree on the importance of conservation, they may not agree on ways that the resources should be distributed. Rather than lose their access to the resource, some nations may even prefer to let the fishery stock become depleted under the existing unmanaged system. The benefits of achieving fishery conservation objectives should, in short, also be measured against the costs of disputes over fishing rights and the potential damage to the international legal system implied by unilateral or bilateral solutions.

LEGAL OPTIONS FOR STRADDLING STOCKS

There thus seem to be three main options for long-term resource management of straddling stocks in the Bering Sea: (1) unilateral extension of the EEZs; (2) unilateral extension of coastal nations' exclusive management authority over straddling stocks; and (3) the establishment of a management regime by international agreement. Both option 1, originally proposed by fishery groups and the North Pacific Fishery Council, and option 2, originally proposed by the US fishing industry coalition, undermine the intent and stability of the international legal regime created by the 1982 United Nations Conference on the Law of the Sea. As unilateral or bilateral extensions of coastal state jurisdiction, the exercise of these options might weaken the Law of the Sea Convention and encourage similar creeping jurisdiction in other areas (Kwiatkowska 1991). The 1988 nonbinding Resolution 396, adopted by the US Senate, declared a joint moratorium with the Soviet Union on fishing within the Central Bering Sea, including enforcement measures against states that do not comply—an action that is also without clear basis in conventional or customary law.

The extent of interest in straddling stocks is the subject of ongoing debate among legal scholars. According to Miovski, fishing states have a duty to avoid compromising the rights of the coastal state over its stocks within its EEZ (Miovski 1989). Burke (1991) posits that, because of the coastal state's duty to conserve its straddling stocks in the EEZ, it has, by implication, the authority to impose conservation measures on states fishing the stocks on the high sea (Miles and Burke 1989). Miovski (1989) goes further, arguing that coastal states have the right to forbid fishing for a straddling stock on the high seas if it threatens the stock inside the EEZ.

Even before the Law of the Sea Convention's entry into force in 1994, some of its statements were accepted as customary international law and could be applied to the Bering Sea case (as described by a combination of Articles 63, 87, and 116–118). Article 63 provides that,

> where the same stock or stocks of associated species occur both within the EEZ and in an area beyond and adjacent to the zone, the coastal state and the states fishing for such stocks in the adjacent area shall seek, either directly or through appropriate subregional or regional organizations, to agree upon the measures necessary for the conservation of those stocks in the adjacent area.

Under Articles 117–118, fishing states are required to conserve high-seas fisheries exploited by their nationals and to cooperate with other states in these efforts. Thus, in the case of straddling stocks, the fishing states must cooperate with measures adopted by the coastal states if these measures are necessary to conserve the resource. This would include negotiations to establish agreed-upon measures. The high-seas fishing state is obliged to exchange information and release data with the coastal state as part of this process. If these statements are to be effective, they must be interpreted as providing for special responsibility and rights of coastal states.

Fig. 3.4. Soviet and American ships out in the Bering Sea off Alaska in 1988, hauling in catches of mintai, flounder, cod, and rasp. (Sovfoto)

The first step in realizing the approach that claims special responsibility and rights of coastal states for Bering Sea management was the 1988 Soviet-American Agreement on mutual fisheries relations, where, in accordance with international law, the parties took responsibility to cooperate in realizing their rights and obligations "to coordinate conservation, exploitation, and management of Bering Sea living resources" (fig. 3.4). In particular, the parties to the agreement will consult on measures connected with uncontrolled fisheries in the Bering Sea.

Among the other states, Japan seems to be the most interested in regional cooperation on the issue of straddling stocks. Japan is particularly dependent upon the domestic fishery policies of coastal states. It has a considerable investment in the pollock fishery of the Central Bering Sea (which became uneconomical to operate before the voluntary two-year suspension went into effect in January 1993), and decisions reached by the coastal states could have a bearing on its own domestic economy.

Although a long-term legal regime for the Central Bering Sea fishery problem, consistent with the goal of environmental security, has not yet been established, the choice of option 3—a management regime by international agreement—seems to be emerging from the course of events. The multilateral suspension of pollock fishing in the Doughnut Hole is entirely consistent with the spirit of a management regime by international agreement, as are major international fora that have emerged in the North Pacific region to foster multilateral cooperation on living resources.

The same series of multilateral conferences that in 1992 adopted the voluntary suspension of Central Bering Sea pollock fishing concluded a Convention for the Conservation and Management of Pollock Resources in the Central Bering Sea in February 1994 (before the expiration of the temporary suspension of fishing). The basic features of the agreement proposed jointly by the United States and Russia, clearly portend a multilateral system, open for accession by the two coastal states and those negotiating partners who had fished in the area before 1989: Japan, Korea, China, and Poland (Kwiatkowska 1993). The convention expresses consensus on such customary legal tenets as high-seas freedoms, the duty to cooperate in the conservation of living resources, and the value of scientific cooperation (Canfield 1993b). It is to be implemented through annual conferences with the help of a scientific and technical committee, without a formal secretariat. The conferences will annually establish the allowable harvest level (AHL) for pollock and individual national quotas (INQs). Decisions will be by consensus, and if this cannot be achieved the AHL will be set by a formula based on total biomass estimates from US and Russian institutions. Failing consensus on INQs, the conference will establish a season for the Doughnut Hole, which will end when the AHL is reached (*Ocean Policy News* 1993c). In the event of noncompliance by a nonparty, the parties are empowered to take "whatever measures, individually and collectively, which are consistent with international law and which they deem necessary and appropriate" to ensure compliance (Kwiatkowska 1993).

The 1992 United Nations Conference on Environment and Development set the stage for resolving the problem of the interests of coastal states versus high-seas fishing interests through multilateral negotiation (Burke 1993). The final draft of UNCED's Agenda 21 omits a number of proposals that would have added weight to coastal state interests (Burke 1993)[5] and instead calls for an intergovernmental conference on straddling and highly migratory stocks, whose work and results are to be fully consistent with the balance struck in the Law of the Sea Convention concerning "the rights and obligations of coastal states and states fishing on the high seas" (para. 17.50). The first meeting of the United Nations Conference on Straddling Fish Stocks and Highly Migratory Fish Stocks was held in New York in July 1993, with 105 countries participating. The conference planned subsequent meetings to complete a report with recommendations for the 49th UN General Assembly in late 1994. Participants at the conference discussed a draft "negotiating text" that attempted to express points of general agreement: the serious and increasing evidence of effects of overfishing, the need for cooperation in high seas fishery management, the interdependence of marine fisheries and the need for an ecosystem approach, the importance of the Law of the Sea Convention as a legal framework, and the need for improved scientific knowledge (*Oceans Policy News* 1993b). Progress in the conference will also contribute to resolution of the related controversy surrounding high seas driftnet fishing.

DRIFTNET FISHERIES AND INCIDENTAL TAKE

Serious environmental threats are created by passive fishing gear, especially driftnets, although some evidence (as cited in Burke 1993) indicates that driftnets can be used with very low side effects, depending on the place and time of harvest. Driftnets are "gillnet(s) suspended vertically from floats at required depths and left free to drift" with currents and tides (Tver 1979). Net snaring is accomplished through entanglement by the movements of fish attempting to escape. Modern driftnets are made of nylon mono- or multifilament with a diameter of about 0.5 mm. Because the material is both invisible and acoustically undetectable, catches by driftnets are indiscriminate. And because the nets are virtually unbreakable, almost nothing over mesh size can escape, making driftnetting a highly efficient mode of capture (Johnston 1990). Soak times for nets range from 4 to 15 hours, with retrieval times ranging from as little as 3 to 4 hours for salmon nets up to 24 to 36 hours for squid nets.

Driftnetting illustrates a common paradox: as fisheries gear becomes more technically efficient over the short term, it is more threatening in a long-term managerial sense, and thus regulatory measures are often designed to restrict the efficiency of the industry (Johnston 1990). Driftnets are tempting to use over very large scales, they are difficult to supervise, and they are easy to lose. The UN Law of the Sea Convention created an unintended incentive to expand this type of fishery in international waters by excluding distant-water fishing fleets from the EEZs of many states. To some extent, increases in large-scale driftnet fisheries for squid and albacore in the North Pacific Ocean were a direct result of expansion of the US and former Soviet EEZs in the Bering Sea. Indeed, many of the fishing vessels used for driftnetting were merely converted from earlier designs that had been formerly used in the Bering Seas fisheries (Yatsu 1990).

Before implementation of an international moratorium in 1992 (see "International Legal Responses to Living Resource Problems," below), seven driftnet fisheries operated in the Bering Sea and North Pacific Ocean: Japanese for squid, albacore, and traditional and nontraditional land-based salmon; Taiwanese for squid and albacore; and Korean for squid.

Driftnets are thought to pose at least five kinds of threats to international environmental security and protection:

- threats to existing systems for managing and conserving straddling stocks of tuna, salmon, and squid;
- threats to commercial fish species from by-catches of non–target size/age classes;
- threats to migratory seabirds, sea turtles, and marine mammals and to existing arrangements for their protection;
- threats to ecosystems due to the indiscriminate nature of the catch, particularly from lost nets continuing to drift, catch, and kill for indefinite periods; and
- threats to vessels whose rudders, engines, or other gear may become entangled and seriously damaged by lost or abandoned driftnets, thereby contributing to the general accumulation of debris at sea. (See FAO 1990; Garcia and Majkowski 1990; Johnston 1990)

Most of the concern over driftnet fisheries has focused on the first two or three threats, but incidental take has been arguably the most controversial. Incidental take has not been restricted to the driftnet fisheries operating in the Bering Sea and North Pacific Ocean. The first reports of incidental marine mammal mortality were made by Japan and the Soviet Union for the pollock trawl fishery in the mid-1970s (Lowry, Frost, and Loughlin 1989). Although the Alaskan trawl fishery for walleye pollock does not use passive fishing gear, this huge fishery generates massive quantities of lost and discarded net, which act as passive fishing gear. Such debris is one of the principal pollutants in the North Pacific–Bering Sea region.

Dall's porpoise (*Phocoenoides dalli*), harbor porpoise (*Phocoena*), and northern fur seals were the principal species of marine mammal formerly taken incidentally by the Japanese mothership salmon fishery inside the current US exclusive economic zone (Jones, Dahlberg, and Fitzgerald 1990). Annual catches for seventeen species of seabirds taken by the Japanese high-seas and land-based salmon gillnet fisheries have been estimated as high as 750,000 (Ainley et al. 1981; Jones and DeGange 1989). Many, if not most, of these seabird species originate from island colonies entirely within either US or Russian jurisdiction. Squid driftnet fisheries have taken northern right whale dolphin (*Lissodelphis borealis*), Pacific white-sided dolphin (*Lagenorhynchus obliquidens*), Dall's porpoise, common dolphin (*Delphinus delphis*), striped dolphin (*Stenella coeruleoalba*), and northern fur seals, as well as more than ten seabird species and several species of sea turtles (Jones, Dahlberg, and Fitzgerald 1990). Most entanglements occur in the middle and upper thirds of the nets (Amano 1990), and as few as 3 percent of animals are alive when taken (Gjernes et al. 1990).

The history of driftnet fishing in the North Pacific Ocean spans several decades, starting with high-seas salmon gillnet fisheries initiated by Japan after World War II. By the late 1970s, Japan had greatly expanded its salmon fishery. At its peak, this fishery involved more than four hundred catcherboats in the mothership salmon fleet and more than seventeen hundred in the land-based salmon fleet. The salmon mothership and the large-vessel land-based fisheries together conducted an average of 28,796 gillnet operations annually for more than twenty years over an extensive area of the North Pacific Ocean and Bering Sea. After passage of the MFCMA in 1976, this fishery declined substantially, and by 1989 it was conducting only 3,327 operations (Jones, Dahlberg, and Fitzgerald 1990).

Driftnet fisheries for squid had been established by Japan, the Republic of Korea, and Taiwan as of 1978, 1979, and 1990, respectively, and these soon became the largest of the driftnet fisheries in the North Pacific Ocean, with more than fifty thousand operations annually (Jones, Dahlberg, and Fitzgerald 1990). In 1987, the total catch of squid was 230,652 metric tons, with an ex-vessel value of $400–$600 million. Most or all of the catch has been marketed domestically. Vessels engaged in this fishery use up to 65 km of net per operation. Net units are called *tans* or *poks*, each of which varies from 30 to 60 m in length. Typical net depths range from 7 to 11 m. Traditionally, the nets extend from the surface, although subsurface nets are being tested for their ability to reduce by-catch (Hayase, Watanabe, and Hatanaka 1990). The target species for this fishery has been the neon flying squid (*Ommastrephes bartrami*), a large species (50 cm, 4 kg) widely distributed in the North Pacific in waters warmer than about

7°C (Gong, Kim, and Kim 1985). Some species of fish may also be retained for commercial use.

In 1981, Japan implemented time/area restrictions as well as a limited entry licensing system for the fishery (Yatsu 1990). These regulations were aimed primarily at minimizing by-catch of salmonids. Subsequently, Taiwan and the Republic of Korea established similar restrictions for their squid fisheries. There is some evidence that catch per unit of effort for squid may have decreased in the following years (Yatsu 1990). In any case, in 1993, with a global driftnetting moratorium in effect on a voluntary basis, the squid catch was negligible (Niemeier, pers. comm., 1993).

POLICY IMPLICATIONS OF INCIDENTAL TAKE

The unavoidable conflicts between the efficiency of fishing technologies and the unintended by-catch of nontarget species can make it very difficult to reconcile the economic and commercial interests of fishing states with the environmental protection interests of those coastal states concerned with protection of such nontarget species as marine mammals and seabirds. With strict enforcement and monitoring for by-catch (e.g., Wohl and Gould 1991), certain nations bear the cost in loss of fishing efficiency due to selective fishing.

Not only the allocation of fishery resources but also the constraints placed on the ability to achieve fish landings efficiently has an effect on each nation's ability to market its fish products in the international market. Conservation measures may be seen as advocated by some nations simply to handicap other nations whose fish products represent direct competition in the international market. Potential solutions to this problem are fundamentally different from those of conservation and of the optimal exploitation of fishery resources. Understanding both the gains and the losses to fishing nations due to proposed allocation of fishery resources is essential for successful negotiation to limit political friction and the disruption of normal relations. Compensation for economic losses may be necessary to reach an agreement.

In the case of salmon, nations seem to be harmed or to benefit economically in approximate proportion to their respective management practices. Because of high consumer demand in all countries, salmon are major targets of regional fisheries, and during the 1980s salmon catches exceeded 600,000 tons. Although Asian and North American stocks were considered to be equally abundant and exploited thirty years ago, their evolution to the current state of stocks has been quite different. As a result of intensive American and Canadian restocking programs and the closure of American stocks to high-seas fishing, the catch rates for these coastal states increased by a factor of 2.5 (Fluharty and Danson 1979). At the same time and under the expansion of high-seas fishing, Asian salmon stocks fell: both marine and coastal stocks decreased, shortages of male fishes occurred in populations of the most valuable salmon species, and the reproductive capacity of subpopulations diminished (Kolmogorov and Sinelnikov 1984).

Some fish species are valued differently among fishing nations, of course, because of variation in consumer taste. For example, halibut are protected and highly valued by the US consumer, and the United States is very sensitive to incidental take of halibut by other nations. Small flounder and sole are in high demand in the domestic markets of the former Soviet republics. Pollock as well as pollock roe are highly valued by Japan, which also consumes a large share of the sablefish caught in the North Pacific–Bering Sea region.

Because consumers attach different values to different fish species, the most highly valued species are, assuming the same fishing costs, always targeted first. As availability of the target species becomes lower, catch per unit of fishing effort falls, raising the market price. This causes two effects: (1) fishers' switching their efforts to relatively more abundant and less costly species and (2) consumers' shifting away from the high-priced fish to less expensive species. A similar effect may be expected with price increases that arise with restrictions on fishery practices, such as the avoidance of by-catch. To some extent, this tendency provides a self-regulating mechanism toward not depleting particular species because market prices as well as fishing costs are high for species close to depletion. However, given the nature of open-access, unregulated fisheries, this self-regulating mechanism cannot be expected to achieve the optimal level of fishing activities, or the most beneficial fishing practices, for the best utilization of either the target or the nontarget resources.

The issue of incidental take, however, extends beyond standard economic arguments of commercial losses and gains. By-catch of marine mammals and seabirds, for example, complicates ecosystem management of the North Pacific–Bering Sea region. Because these species are not marketed commodities, losses from their by-catch cannot be defined strictly in commercial terms (Thorne-Miller and Catena 1991). Although nonmarket values placed on the protection of these species may well vary across nations, the issue is not merely whether the citizens of some nations place greater value on the protection of certain marine mammals or seabirds than do the citizens of others. First of all, by altering the rates or kinds of predation, unregulated removal of apex predators from the food web can have profound and unforeseeable consequences for the entire marine ecosystem (see, e.g., Paine 1979, 1980; Estes, Duggins, and Rathbun 1989). Such alterations can then damage the ecosystem for any user. Second, as with salmon, the rookeries, haul-out sites, mating grounds, and colonies for many seabirds and marine mammals (especially pinnipeds) are located almost entirely within the national jurisdictions of the United States and Russia. Migratory populations of threatened marine vertebrates are not necessarily international populations; for many species, they are not the property, even in part, of the fishing states. Thus, international commercial enterprises are often at direct odds with domestic environmental protection. By preventing or complicating those protection measures mandated by domestic law, incidental take is liable to be viewed as a "hostile act."

NATIONAL LEGAL INITIATIVES ON DRIFTNETS

Commercial and sport fishers, fish processors, environmentalists, groups of indigenous and native peoples, and numerous states and provinces have all been part of a very effective lobbying effort at several levels of government to address the driftnetting controversy. Because of the efforts of these groups, several nations have implemented or at least considered bans on the sale of all tuna caught in driftnets unless the catch has been certified "dolphin safe," and US and Canadian states and provinces have urged their governments to take diplomatic actions against driftnetting through such formal statements as the North Pacific Driftnet Proclamation and the US Driftnet Impact Monitoring, Assessment and Control Act of 1987 (Jones, Dahlberg, and Fitzgerald 1990).[6] Access by such groups to government decisions has been longstanding and much better developed in the United States than in Russia, and consequently the United States has a much more elaborate set of domestic laws and regulations on driftnets. The United States and Canada have both banned driftnetting within their EEZs. These largely unilateral pressures then lead to problems of how victim states can deal effectively with culprit states within the limits of political, diplomatic, legal, and ethical acceptability (Johnston 1990).

Under the 1976 MFCMA, the United States regulates all foreign and most domestic fishing within its 200–nautical mile jurisdiction. The United States also regulates fishing for creatures dwelling on the continental shelf beyond the 200-mile limit and for anadromous species such as salmon and steelhead trout when they are beyond the EEZ. The US government allocates quotas, limited to that portion of the optimum yield not harvested by US fishers, to foreign fishers through bilateral agreements termed *governing international fisheries agreements*, or GIFAs. In 1993, the United States had GIFAs in place with Estonia, the European Community, Iceland, Korea, Lithuania, and Poland, as well as a special agreement with Russia.

For each of eight US regions, the MFCMA established a regional fisheries management council, which sets fishery management policy for federal waters within its jurisdiction. The North Pacific Fishery Management Council (NPFMC) consists of representatives from the states of Washington, Oregon, and Alaska. Membership on the council includes public officials from federal and state agencies and private citizens representing commercial, recreational, technical, and scientific interests who are knowledgable about the conservation, management, and harvesting of fishery resources in the region. Pursuant to its framework management plans for individual fisheries (chiefly salmon and groundfish), the NPFMC continuously proposes measures to achieve or maintain the optimum yield from each fishery. These measures are submitted to the Secretary of Commerce and, if approved, result in regulatory amendments adopted each season for implementation by the National Marine Fisheries Service, with enforcement assistance from the Coast Guard. Recent decisions submitted by the NPFMC for approval by the Secretary of Commerce have included gear allocations of Pacific cod in the Bering Sea/Aleutian Islands; time/area closures, salmon caps, and other salmon by-catch control measures; the establishment of total allowable catch levels for Atka mackerel in three areas within the Aleutian Islands management area; and a proposed framework to allow flexibility in setting each year's opening date for Bering Sea/Aleutian Islands Pollock 'A' fishing to coincide with peak roe maturity and value (*NPFMC Newsletter* 1993).

In late 1990, the US Congress passed the Fishery Conservation Amendments of 1990 (PL No. 101-627, 104 Stat. 4436 [1990]), which contained three measures aimed at ending the use of driftnets on the high seas. These amendments (1) ban the use of large-scale driftnets of 1.5 miles or longer in the US exclusive economic zone and ban the use of these nets by US fishers on the high seas; (2) direct the US Secretary of Commerce, through the US Secretary of State, to seek to negotiate international agreements to ban or restrict the use of large-scale driftnets on the high seas; and (3) amend the Marine Mammal

Protection Act of 1972 (16 USC 1371 (a) (2)) to provide enforcement authority in support of the 1989 UN resolution that initiated a voluntary moratorium on large-scale driftnet fishing in the Pacific.

As for agreements with foreign nations concerning US marine resources, the amendments direct the Secretary of State to include several important monitoring and enforcement provisions that apply to foreign vessels even when operating outside US waters and on the high seas:

- Each large-scale driftnet vessel of a country that is party to a US agreement is to be equipped with satellite transmitters to provide real-time position information when on the high seas;
- each driftnet fishing vessel is to employ on-board observers or other statistically reliable methods of monitoring all driftnet take or entanglements of target and nontarget fish species, marine mammals, sea turtles, and seabirds;
- in designated areas of the high seas, US officials have the right to board and inspect driftnet fishing vessels of nations that are party to the agreement;
- each vessel is to monitor and document all catch landed or transhipped at sea; and driftnets are to be tagged to identify vessel and flag nation.

The Amendments also encourage the use of biodegradable materials for driftnets and provide for time and area restrictions to prevent interception of anadromous species. The above provisions, however, apply only to those nations with whom the United States negotiates GIFAs that provide for allocations of excess US fishery or other marine resources to foreign nations. Nevertheless, all nations that permit their nationals to use large-scale driftnets on the high seas can still be made subject to sanctions under the Fishermen's Protective Act of 1967 (Pelly Amendment 22 USC 1978 (a)). Under the Pelly Amendment, nations whose actions diminish the effectiveness of international fishery or endangered species conservation agreements, programs, and resolutions to which the United States is a party or otherwise subscribes are subject to certification by the Secretary of Commerce. Such certification triggers the President's authority to embargo fish products from the violating country.

In the United States, marine mammals are protected under the 1972 Marine Mammal Protection Act (MMPA), which provides a general moratorium on catches ("takes") of marine mammals, subject to exceptions for aboriginal peoples and some fisheries with by-catch.

This law is based in a broad ecological perspective, requiring management action to ensure that populations of marine mammals continue to be functioning elements of their ecosystems. In practice, this perspective has been more narrowly implemented through regulation of fishing involving by-catch of marine mammals when such populations are thought to be "depleted," that is, below levels of abundance where maximum net productivity would occur.

The MMPA was amended in 1988 to provide an interim exemption to the general moratorium, allowing for unregulated by-catch in commercial fisheries for five years, while the Secretary of Commerce collected additional data on the problem. Based on these data, in 1994 the MMPA was amended again, to allow for observer coverage of fisheries involving significant levels of by-catch, and for regulation of fisheries having by-catch of populations or stocks that are classified as "strategic." Strategic stocks are those that are endangered or threatened under the Endangered Species Act or are judged likely to become depleted under the MMPA. The MMPA applies to all marine mammal populations that occur wholly or in part in US waters and requires accounting for takes outside of US waters in determining management actions. The law also allows for consultation with other governments on such by-catches.

The 1973 US Endangered Species Act (ESA) applies to species that are currently in danger of extinction throughout some or all of their range and to species that are threatened with the risk of extinction. The ESA can also protect habitat areas that are recognized as being essential to the conservation of the species. The ESA strictly regulates the taking of species listed as threatened or endangered and limits the ability of federal agencies to issue permits or provide funds that will jeopardize a listed species. The Steller's sea lion is one species in the Bering Sea–North Pacific Ocean region that falls within this classification (and is also designated as depleted under the MMPA). The globally endangered short-tailed albatross (*Diomedea albatrus*), a migrant to this region, has been incidentally caught during fishing operations (Gould and Mendenhall, pers. comm., n.d.) and is listed as protected under the ESA.

In contrast to this extensive body of US laws and regulations, the former Soviet Union had no domestic laws targeted specifically to individual environmental resources or practices. Instead, authority to issue and enforce regulations was delegated to various ministries under a very broad environmental protection law dating back to 1960. The Ministry of Fisheries tracked and conducted research on fish stocks; developed national strategies targeting particular stocks, individual fisheries, and even the types of fishing vessels to be built; and, in conjunction with the Foreign Ministry, established quotas and issued permits to foreign fishing concerns. The fact that most Soviet fisheries were government enterprises made enforcement of the ministry's regulations comparatively simple, but it also greatly exaggerated commercial considerations, to the point of nearly obliterating the aboriginal traditions of small coastal fishing communities and their emphasis on resource conservation.

Russia's new law "On Environmental Protection," which was enacted less than one week before the official collapse of the Soviet Union, is nearly as general as its predecessor, and the formulation of more specific regulations has been greatly complicated by the move toward privatization and reorganization of the government bureaucracy. (Fisheries is now a committee of the Ministry of Agriculture, for example). As in other sectors, most of Russia's fisheries will soon be privatized, and new regulations must take into account the unfamiliar problem of the government's authority and accountability with respect to the activities of private enterprise. The management of marine biological resources in the North Pacific is one of the environmental and socioeconomic concerns that will become the subject of future regulations.

Although organizational disarray associated with the former Soviet Union's economic transition has impeded and delayed more effective conservation measures, the importance of its political transformation for enhanced environmental security must not be underestimated. As Canfield (1993b) recognized, these changes opened the way for expanded Soviet (later Russian) cooperation on North Pacific living resources conservation in several ways: through heightened public awareness of and involvement in environmental affairs, through the elevation of environmental security to priority foreign policy status, in calls for comprehensive ecosystem management approaches, and in the greater attention being paid to Russia's role in the Asian Pacific region. The radical changes in the former Soviet Union thus contributed to an improved atmosphere for and accelerated movement toward a multilateral approach to the management of North Pacific living resources.

INTERNATIONAL LEGAL RESPONSES TO LIVING RESOURCE PROBLEMS

The historic 1911 Fur Seal Treaty and the 1957 interim agreement among the United States, Japan, Canada, and the Soviet Union were successful landmarks in multinational conservation (Busch 1985), but since the 1920s bilateral treaties have been the basic response to problems with international fisheries and other resources in the North Pacific Ocean and the Bering Sea (see Wallace 1994). A Canadian-US bilateral fisheries commission was established in 1923 for halibut. Driftnets used in the salmon fishery were first subject to regulation through a bilateral agreement between Japan and the Soviet Union restricting mesh size as early as 1953. The International North Pacific Fisheries Commission (INPFC) was established in 1953 by Canada, Japan, and the United States for the primary purpose of regulating Japanese high-seas salmon fishing in the North Pacific (Johnston 1990). Despite restructuring in 1978, the INPFC failed to prevent several serious disputes from arising in the region, especially between the United States and Japan. The subjects of these disputes included alleged illegal fishing in the US zone by the Japanese, alleged Japanese interception of salmon of US and Canadian origin on the high seas, and the scale of pollock fishing within the Central Bering Sea. Canada and the United States settled their differences largely by establishing a framework for joint management of five salmon species. The 1985 US and Canada Pacific Salmon Treaty addresses problems of foreign interceptions, as well as joint conservation measures (Twitchell 1989).

The US-USSR Agreement on Mutual Fisheries Relations was signed in May 1988. The level of fishing or processing allowed to Soviet (now Russian) vessels in US domestic waters would equal the level of fishing or processing allowed to US vessels in Russian waters (Miovski 1989). This agreement also encourages cooperation on a wide variety of other matters, including protection of marine mammals, management of anadromous species, boarding and inspection, and scientific research.

The Convention for the Conservation of Anadromous Stocks in the North Pacific Ocean was signed in Moscow in February 1992 by Canada, Japan, Russia, and the United States. The 1992 convention supercedes earlier fishery arrangements in the region, including the USSR-Japan agreement on salmon fishing in the Central Bering Sea and the parallel US-Canada-Japan agreement. Together with ongoing US-Soviet dialogue, these earlier agreements had had the effect of keeping "Japan trapped in the middle and on the way out of a high seas salmon fishery" (Johnston 1990). Japan was forbidden to fish in certain areas under one agreement but was allowed to fish those areas under the other (see fig. 3.5). The more comprehensive 1992 convention removes such inconsistencies. It establishes the North Pacific

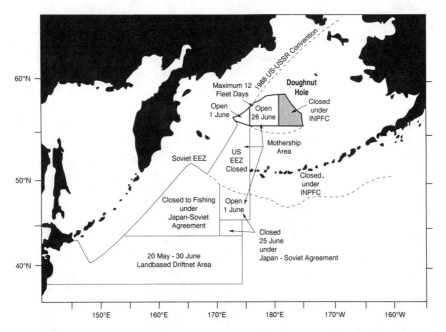

Fig. 3.5. North Pacific–Bering Sea locations of Japanese high-seas salmon fisheries in 1990.

Anadromous Fish Commission (NPAFC) as the successor of the INPFC to serve as "the venue for coordinating the collection, exchange, and analysis of scientific data" on anadromous stocks and ecologically related species within convention waters and as "a forum for promoting the[ir] conservation" (NMFS 1993b). The NPAFC will also coordinate high-seas fishery enforcement activities by member countries and thus serve as a forum to forestall the emergence of new tensions.

Concerning driftnetting, Johnston (1990) concluded that no hard law obligations directly resolve the problem and suggested a three-step approach based on the soft law principles enunciated within the Law of the Sea Convention. These are (1) the duty to conserve the living resource of the high seas (e.g., Articles 117, 119, and 194 [5]); (2) the duty to cooperate with other states (Articles 117, 118, 119, and 197); and (3) the duty to negotiate conservation arrangements.

In the fall of 1989, the UN General Assembly passed a carefully worded resolution (U.N. Doc. A/RES/44/225) urging all nations to cooperate in the sharing of scientific data on the effects of driftnetting. The resolution further called for an immediate cessation to the expansion of large-scale driftnet fishing on the high seas of the Pacific Ocean and a complete moratorium by June 30, 1992, unless acceptable conservation and management measures were taken by all concerned parties (para. 4). Subsequent resolutions noted progress and reaffirmed basic objectives (Res. 45/197, 1990) and then redefined June 30, 1992, as the target date for achievement of a 50 percent reduction in fishing effort, with the start of the global moratorium delayed until December 31, 1992 (Res. 46/215, 1992), and continuing until there is a statistically sound analysis upon which to base an effective conservation and management regime (UN Secretary General 1990).

Accordingly, the United States extended the enforcement provisions of its 1991 bilateral driftnet agreements with Canada, Japan, Korea, and Taiwan. In August 1991, the US Secretary of Commerce, acting on reports from the US Coast Guard and NMFS enforcement agents, certified Taiwan and Korea for violating their respective agreements "by allowing their nationals to conduct driftnet fishing operations . . . north of agreed boundaries in the North Pacific" (US Secretary of Commerce 1992). By April 1992, however, both countries had taken punitive actions against vessel owners, and the United States had decided to defer sanctions "pending their adherence to UNGA Resolution 46/215—particularly the global driftnet moratorium by December 31, 1992" (US Secretary of Commerce 1992).

Seabirds were already afforded some tentative protection from driftnetting under the 1976 US-USSR Convention Concerning the Conservation of Migratory Birds and Their Environment (Lyster 1985). More generally, the 1976 convention obliges both states to identify and list especially important habitat areas within their jurisdiction and to protect the ecosystem of their areas to the fullest extent possible. Several other bilateral treaties pertaining to one or more of the states involved in management issues within the Bering Sea and North Pacific Ocean also protect seabirds:

- the 1916 Convention for the Protection of Migratory Birds, between Canada and the United States;
- the 1936 Convention for the Protection of Migratory Birds and Game Mammals, between the United States and Mexico;
- the 1972 Convention for the Protection of Migratory Birds and Birds in Danger of Extinction and Their Environment, between the United States and Japan;
- the 1973 Convention for the Protection of Migratory Birds and Birds under Threat of Extinction and Their Environment, between Japan and the former Soviet Union; and
- the 1974 Convention for the Protection of Migratory Birds and Birds in Danger of Extinction and Their Environment, between Japan and Australia.

All of these treaties restrict the circumstances under which migratory birds may be traded or taken, and some have exceptions for subsistence purposes. The four conventions signed during the 1970s urge all parties to establish protected areas for migratory birds.

Many of the laws designed to protect marine mammals, seabirds, and sea turtles unfortunately have proved ineffective in practice, particularly when viewed at the international level (Harrison et al. 1990). The Convention on International Trade in Endangered Species (CITES), for example, precludes commercial trade of many marine vertebrates and associated products, but the issue of trade is irrelevant to incidental take.

In general, most international agreements dealing with resource conservation are limited by membership and by function. Agreements are not based on functional interactions within the food web and are thus not able to mitigate perturbations to the North Pacific–Bering Sea ecosystem. Since the early 1970s, there have been proposals for a new international organization to promote scientific investigation and information exchange concerning the oceanography and fisheries of the North Pacific and the Bering Sea. Given the proposed organization's functional similarity to the International Council for the Exploration of the Sea (ICES), which operates in the North Atlantic, it has long been referred to as Pacific ICES, or PICES.

The convention establishing the North Pacific Marine Science Organization, which refers to itself as PICES, came into force on March 24, 1992. The contracting parties were Canada, China, Japan, and the United States, with Russia, a party to the negotiations, expected to join. The convention is open to accession by other countries. The organization's scope and purposes are broad:

> (a) to promote and coordinate marine scientific research in order to advance scientific knowledge of the area concerned and of its living resources, including but not necessarily limited to research with respect to the ocean environment and its interactions with land and atmosphere, its role in and response to global weather and climate change, its flora, fauna, and ecosystems, its uses and resources, and impacts upon it from human activities; and
> (b) to promote the collection and exchange of information and data related to marine scientific research in the area concerned (PICES Fact Sheet, n.d.).

The new organization has four scientific committees, on biological oceanography, fishery science, marine environmental quality, and physical oceanography and climate. Among the six working groups established at the first annual meeting in Vancouver, British Columbia, in October 1992 are one on the dynamics of small pelagics in coastal ecosystems and another on the Bering Sea (PICES Fact Sheet, n.d.).

GAPS IN SCIENTIFIC INFORMATION AND FUTURE RESEARCH PRIORITIES

Current scientific knowledge is inadequate in several areas to support the selection and implementation of a comprehensive, long-term management approach for the Bering Sea and adjacent North Pacific fisheries (Canada Department of Fisheries and Oceans 1991).[7] First, the origins and movements of pollock stocks throughout the North Pacific are incompletely known. Such information is crucial for delineating any economic and legal solutions for disposition of transboundary species within an ocean commons. Although it is generally recognized that such stocks are both migratory and transboundary, assessing the magnitude and rate of depletion in any given stock is still a speculative exercise. Linkages between off-shelf (Aleutian Basin) and on-shelf components of the eastern and western Bering Sea pollock resource are not understood. A start at addressing these issues has been made by the National Marine Fisheries Service (NMFS) and other agencies through continued surveys and modeling of the physical processes that drive recruitment and dispersal (i.e., the cooperative research efforts of Russia, Japan, and the United States through the Bering Sea Fisheries-Oceanography Coordinated Investigations, or FOCI). A related problem is the reporting system and structure of the existing data set on world fisheries, which make it difficult to identify commercial species and the other species that constitute the ecological structures in which they operate. Although both biological and economic data are needed for the management of commercial stocks, only fragmentary information is available on the price of what is caught, and even less is known about the costs of the inputs to the productive process (Marine Policy Center 1991; Pontecorvo 1991). These gaps introduce a chain of uncertainty in which inadequate reporting of catch and effort amplifies uncertainty about stock size, which in turn increases uncertainty about actual depletion effects.

Second, from a conservation perspective, none of the legal and economic solutions pertaining to resource allocation and sustainable yield is biologically realistic unless it considers the issues of ecosystem function and structure, and these must be better understood.[8] One priority approach for the North Pacific–Bering Sea region would be to develop credible models of the consequences of ecosystem perturbations in the region, taking into consideration the results of removal of or reduction in the various trophic components or compartments. Estes, Duggins, and Rathbun (1989) showed, for instance, that local extinctions (or even population reductions) are not easily reversible through introductions and other conservation measures because of various behavioral and ecological barriers to repopulation. To be realistic, such models should be able to incorporate multiple anthropogenic disturbances (incidental take, oil spills), as well as natural ocean climate oscillations (see Takekawa, Carter, and Harvey 1990 for an illustration). Through empirical studies, special attention can be focused on identifying keystone species, ecosystem linkages (both biotic and abiotic), and structural stability, complexity, and resilience (*sensu* Pimm 1982).

Third, observer programs should be maintained to monitor by-catch and conservation compliance (Wohl and Gould 1991). A 1989 scientific observer program sponsored by NMFS and the US Fish and Wildlife Service in cooperation with Canada and Japan produced more than fifty documents about the squid fishery, including incidental take and net debris, for submission to the INPFC before it was succeeded by the NPAFC. Scientific observer programs should require at least (1) catch rates for incidentally caught animals by appropriate time, area, and stock; (2) fishing effort statistics by time and area; (3) biological characteristics of animals incidentally taken; and (4) stocks of major animals involved (Yatsu 1990).

Finally, more technical research is needed on opportunities for alternative technologies and practices. Hayase, Watanabe, and Hatanaka (1990), for example, reported successful reductions in the rates of incidental take of marine mammals and seabirds when nets were placed deeper in the water column rather than at the surface, with no loss in target catch per unit effort. Recruitment of ecological studies to help specify the closure of certain fisheries zones during limited periods may also significantly reduce by-catch or its effects.

A REGIONAL REGIME FOR ENVIRONMENTAL SECURITY

Depletion of walleye pollock and protection of salmon, marine mammals, and seabirds in the North Pacific–Bering Sea ecosystem are current and interconnected problems of international environmental security that are not being adequately dealt with through existing mechanisms. Current regulation of the most serious environmental threats in the North Pacific does not conform to the current and long-term requirements and interests of the coastal states. Nor does it guarantee the conservation of regional marine ecosystems or prevent depletion of common resources and associated stocks that are under the jurisdiction of the coastal states. Fisheries management is exercised not on a regional or ecosystem basis; rather, it is based on a few species and is hampered by the inadequacy of regional infrastructure. The policy of coastal states on the issues of conservation and rational utilization of living resources is marked by an absence of long-term strategy and a vagueness of national and regional priorities. National policy seems to be driven by discrete issues, plus the occasional interference of external and internal political considerations. Although there is ample evidence of the interconnection of resource conflicts, no serious attempt at complex regulation on a regional basis has yet been undertaken.

International fishery conflicts demand radical changes in the marine policy of coastal states by means of transferring their cooperation to the level and focus of regional environmental security. Within such a context, it seems reasonable to treat resource conflicts as particular manifestations of the general problem of conserving marine living resources and to attempt to solve them in conjunction with the creation of a new international regime. Such a regime could be instituted by means of consecutive steps:

1. elaboration, on a regional basis, by the joint efforts of scientists, lawyers, and diplomats, of guidelines for the conservation of the living resources of the North Pacific (along the lines of the Joint Resolution of the Fifth Bering Sea Conference, 1992);
2. creation of a scientific infrastructure for resource regulation in the region (combining efforts in monitoring resources and environment, elaboration of standards, creation of statistical database) and elaboration of an ecosystem approach to resource management (a task that could be accomplished within the framework of PICES activities); and
3. negotiation by all concerned states with a view to reaching an evolving agreement on the conservation of living resources of the North Pacific (building, perhaps, on the basis of an expanded and strengthened NPAFC).

The experience of conflict resolution in other marine regions attests to the fact that fishing states are more inclined to the idea of high-seas regulations if they are accompanied by some provision for access to the EEZ stocks. One can assume that culprit states who are parties to a number of regional resource conflicts would be less opposed to joint resolutions that have the potential of reciprocal compromises.

Although Japan will probably seek an end in 1995 to the voluntary moratorium on high-seas pollock fishing in the Doughnut Hole, it seems to be the nation most interested in a scheme of regional cooperation. Its position is determined by the scale of operations in the Doughnut Hole and Okhotsk Sea, its dependence on the domestic fisheries policies of the coastal states, and the potential effect on its economy of any decisions taken. Japan was a strong supporter of creating the regional organization NPAFC, whose membership and functions will supersede and extend beyond those of the INPFC. The expansion of fishing pressure for Japanese sardine and access to the stocks of other states' EEZs should contribute to Japan's self-interest in this new international management mechanism.

Korea has already agreed, under US pressure, to restrict its presence in the enclave and to other limitations of high-seas activities. It can be expected to react positively to international regulation in the region. Multilateral agreement covering a wide range of regional fishery issues creates opportunities for self-interested maneuvering by China and Taiwan as well.

The United States and Russia, as regional maritime powers and principal coastal states, should certainly lead in the process of creating the international regime of environmental security in the North Pacific–Bering Sea ecosystem. Their leadership has already shown good results in the voluntary suspension of Bering Sea pollock fishing and negotiations toward an agreed system for managing pollock fishing in the Doughnut Hole. Multilateral participation has been excellent. However, even if the system for Bering Sea regional fisheries management is achieved, it will neither encompass the full, relevant Bering Sea–North Pacific region nor constitute a comprehensive ecosystem management approach. To the extent that management measures for Central Bering Sea pollock are made consistent with the practices of the coastal states within their own jurisdictions, in keeping with emerging customary law (Kwiatkowska 1993), the way will be eased toward a broadened regional system with multilateral participation. For an ecosystem approach in the North Pacific to be consistent with coastal state practice, however, both the United States and Russia would have to move toward ecosystem management within their own EEZs.

Optimal management of fishery resources and protection of various marine species should be achieved by means of negotiations and treaties that are mutually agreed upon by the nations involved. Potential international conflict arising from nonaccepted treaties and unilateral or bilateral actions for closing the Central Bering Sea should be recognized as potential costs of moves to achieve improved fishery resource management. For some proposed solutions, gains from achieving a conservation objective of fishery resource management may fail to outweigh the associated losses in fairness or damages to the international legal system. A truly comprehensive regime for environmental security will require genuine international cooperation on fishery conservation and ecosystem protection in this subarctic region.

CHAPTER 4

Hazardous Materials Transport

SEABORNE TRANSPORT OF hazardous materials poses regular and substantial risks to coastal environments and communities. Well over 90 percent of all goods carried in foreign trade move by marine transportation, so inevitably a large portion of hazardous cargoes moves by sea. International institutions for dealing with pollution from other hazardous materials are much less comprehensive or effective than are those for petroleum; thus, this is a pressing issue of environmental security for most coastal states and a persistent source of uncertainty for most maritime states.

In contrast to the many large and small mishaps associated with the transport of petroleum by sea, as yet there have been few large-scale incidents arising from the shipment of hazardous materials. Nevertheless, accidents of smaller proportions are a stark and continuing reminder of the risks posed to human health and the marine environment by the many hazardous materials, other than oil, that are carried by ship (see table 4.1). In one deadly incident, four members of the five-man *Belgrave* crew were killed by fumes released from a load of scrap that had not been marked as dangerous. The March 1989 sinking of *Perintis* in the English Channel and the subsequent loss of a marine container carrying 5 metric tons of lindane raised a furor when the cargo of toxic pesticide could not be immediately located on the bottom. The chemical carrier *Anna Broere* sank in the North Sea after colliding with a containership, and there were fears of severe environmental damage before she was raised in pieces and sold for rebuilding. Other incidents, particularly involving the transboundary shipment of hazardous wastes, have been well publicized.[1]

Despite a measurable decrease in the number of ship casualties and injuries over the past two decades worldwide, the potential for such incidents will always remain, given the uncertainties of the marine environment, human lapses, and unforeseeable events. A large dis-

Table 4.1
Partial List of Accidents Involving Seaborne Transport of Hazardous Materials, 1979–88

Year	Vessel Name/Type/Flag	Accident Type	General Location	Chemicals Involved
1979	*Traugutt*, general cargo, Poland	Fire/grounding	Off Pakistani coast	Sulfuric acid, calcium carbide
1979	*Aeolian Sky*, general cargo, Greece	Collision/sinking	Off French coast	32 canisters arsenic trichloride; 10 drums liquid chlorine; 50 cylinders anhydrous ammonia
1979	*Sinbad*, general cargo, Iraq	Loss overboard	Off Dutch coast	51 cylinders chlorine gas
1982	*Craiganlet*, containership, Cyprus	Grounding/break-up	Off Irish coast	Dimethyl sulfate waste (one freight container)
1982	*European Gateway*, roll-on/roll-off ferry, UK	Collision/sinking	Off British coast	Various toxic chemicals in drums
1982	*Poinciana*, general cargo, Cayman Is.	Grounding/sinking	Miami River entrance (US)	900 kegs calcium hypochlorite; 239 drums hydrochloric acid
1984	*Forum Hope*, general cargo, Greece	Loss overboard	Bay of Biscay	200 drums flammable liquids
1984	*Dana Optima*, roll-on/roll-off containership, Denmark	Loss overboard	Off British coast	80 drums nitrophenol herbicide (Dinoseb)

(continues)

Table 4.1 continued

Year	Vessel Name/Type/Flag	Accident Type	General Location	Chemicals Involved
1984	*Mont Louis*, roll-on/roll-off containership, France	Collision/sinking	Off Belgian coast	450 tons uranium hexafluoride in 30 steel flasks; 32 empty flasks
1984	*Ariadne*, bulk vessel fitted for containers on deck and in hold, Panama	Grounding/break-up	Entrance to Mogadishu Harbor (Somalia)	2,000 tons packaged dangerous cargo, including tetraethyl lead, sodium pentachlorophenate, calcium carbide, organophosphorous compounds, and organochlorine pesticides
1985	*OK Menga*, barge, flag not applicable	Capsizing	Off Papua, New Guinea, coast	15 freight containers (2,700 barrels) sodium cyanide
1987	*Cason*, general cargo, Panama	Fire/grounding	Off Spanish coast	500 tons packaged dangerous cargo, including 500 drums o-cresol (marine pollutant), diphenylmethane diisocyanate, and aniline oil
1988	*Caribe*, containership, France	Loss overboard (jettisoning)	Between Azores and Puerto Rico	Various packaged dangerous goods and marine pollutants

SOURCE: Church and Jones 1988.

charge of hazardous materials such as persistent cancer-causing substances in an important coastal area could have dire physical and economic effects on a wide area of coastline. In response, it is quite likely that coastal governments would enact laws severely restricting the navigational rights of all ships carrying hazardous materials. Enactment of such laws by several countries, all desiring environmental security, would undermine a key aspect of international ocean law: the freedom of navigation.

The conflict between the navigational rights of vessels carrying hazardous materials and the interests of coastal states in environmental security is very real and growing. This conflict arose, for example, in the context of the 1989 Convention on the Control of Transboundary Movements of Hazardous Wastes and Their Disposal (Basel Convention) and of the 1990 Protocol to the 1983 Convention for the Protection and Development of the Marine Environment of the Wider Caribbean Region (Cartagena Convention). Another example arose in 1988 when Haiti announced the prohibition of entry to its ports, territorial seas, and exclusive economic zone of ships transporting wastes, refuses, residues, or any other materials likely to endanger human health or pollute the marine environment. The US government stated publicly that it viewed Haiti's announcement as violative of international law, and Haiti has not yet taken any action to enforce the prohibition.

Restrictions or regulatory regimes have been widely accepted if they have developed from an international consensus that reflects the balance of maritime and coastal state interests. Clearly, it is in the interests of the shipping industry to establish means to avoid accidents and minimize the damages arising from the accidents that do occur. It has also been necessary to institute minimum standards governing operational discharges from ships. Various international regimes, some of which are currently in negotiation under the auspices of the International Maritime Organization (IMO), address different aspects of prevention, response, liability, and compensation involving hazardous materials transportation. Nonetheless, these regimes leave many important issues unresolved.

PROBLEMS OF DEFINITION, CLASSIFICATION, AND INFORMATION

The terms *hazardous cargo, dangerous goods, noxious substances,* and *toxic cargo* have all been used to describe the commodities in question. Here we use the term *hazardous materials* to include a broad class of cargo whose release into the ocean would adversely affect human health or the marine environment. Cargoes of this type often have intrinsic properties that make them toxic, combustible, explosive, or corrosive. Because the maritime shipment of petroleum products—considered a hazardous cargo by many analysts—is managed largely through a separate set of national and international legal instruments, it is not included in this analysis.

Defining a Material as Hazardous

Thousands of hazardous materials are produced and move in seaborne trade, with hundreds of new materials entering the market each year, often before their properties are well known. The fact that many such materials are mislabeled (intentionally or otherwise) contributes to the difficulty of quantifying the total volume of trade in these cargoes and the number of accidents in which they are involved.

The difficulties of agreeing on a comprehensive international classification scheme, or even a standard nomenclature, for hazardous materials moving by sea are enormous. The materials' great number and variety challenge a unified treatment. Attempts to define or categorize such materials have led some analysts to extend the term *hazardous* to half of all seaborne cargo (Chircop 1988). Such difficulties in classification help explain why the development of effective international, regional, and even national regulatory regimes is an elusive task. Harmonizing existing regimes may be even more elusive.

Manufacturers, the shipping industry, national governments, intergovernmental organizations, and nongovernmental organizations use various classification schemes to identify cargo for purposes of labeling, safety, handling, and insurance. The most authoritative classification schemes are those adopted by the IMO and under the terms of the Basel Convention. Proponents of the precautionary approach suggest that all questionable substances should be labeled as hazardous until they are proven to be safe. Most lists or annexes of hazardous materials deal only with those known to be toxic, persistent, or bioaccumulative.

The IMO's International Maritime Dangerous Goods (IMDG) Code was initially developed as a uniform international code of guidance for the implementation of regulations under the 1974 International Convention for the Safety of Life at Sea (SOLAS Convention), which deals with safety standards for dangerous goods. The IMDG Code includes recommendations for the nine classes of dangerous goods addressed in Chapter VII of SOLAS: (1) explosives, (2) gases,

(3) flammable liquids, (4) flammable solids, (5) oxidizing substances and organic peroxides, (6) poisonous and infectious substances, (7) radioactive materials,[2] (8) corrosives, and (9) miscellaneous dangerous substances, including marine pollutants. The IMDG Code covers matters such as packing, stowage and segregation, labeling and placarding, classification, and documentation of dangerous goods. Ongoing changes in technology affecting safety measures for the carriage of dangerous goods, together with the introduction of new chemicals in the market, require continuous review and update of the IMDG Code. Although the IMDG Code is a set of technical recommendations, it enjoys de facto mandatory status in many nations. The Basel Convention does not parallel the IMDG Code in providing a list of hazardous materials per se, but it offers a set of forty broad classes of wastes to be controlled.

Data Inadequacies and Casualty Records

The lack of an adequate centralized incident reporting database has plagued efforts to quantify the scope and potential environmental security implications of hazardous materials transport by sea. The reporting systems described by the IMO under Annex III of the 1973 International Convention for the Prevention of Pollution from Ships, as modified by the 1978 Protocol relating thereto (MARPOL 73/78), and the Draft Convention on the Carriage of Hazardous and Noxious Substances by Sea (HNS) are voluntary and thus would capture only a portion of the incidents.

On the national level, trade-related databases compiled by private vendors and US Bureau of the Census foreign trade statistics data tapes offer general information on export movements by cargo type, volume, port pairs, and, in some cases, shipper, consignee, and carrier. From these it is possible to quantify and track physically the volume of trade in commodities that might meet hazardous criteria. The US Coast Guard routinely collects statistics on casualties and injuries occurring aboard US flag commercial ships, which are available on an annual basis and can be sorted according to cargo type and causal factor. The Marine Index Bureau and the Tanker Advisory Center, both private, nonprofit industry associations based in the United States, also maintain databases on accidents. Their records provide some detail on the nature of incidents and the cargo involved for selected types of ships and cargoes.

For data on shipping incidents involving hazardous materials, records administered by marine insurance companies and underwriters such as Lloyd's may be the most comprehensive. Unfortunately, the disaggregated nature of the data, the strict confidentiality associated with much of the information, and obvious incentives not to report[3] all point up the drawbacks of depending on private sources.

Reliable estimates of worldwide transboundary movements of hazardous wastes are not available at all, as most nations do not maintain a comprehensive monitoring system for waste transportation.[4] In the United States, exports and imports of wastes are tracked by the Environmental Protection Agency, as required by law. Virtually no data exist on illegal shipments of hazardous materials, including the illegal transboundary movement of wastes.

THE EXISTING LEGAL AND REGULATORY STRUCTURE

The current legal order articulated in the 1982 United Nations Law of the Sea Convention (see chapter 8) recognizes traditional navigational freedoms qualified by responsibilities to protect the marine environment. The balance between the world's interest in the free flow of commerce and the control of pollution is complex and delicate. In the context of the navigational rights of ships that pose an extraordinary danger to the marine environment, the balance becomes precarious.

The Law of the Sea convention, which enters into force in November 1994, goes further than any other international legal instrument in clarifying the balance between coastal state and maritime state interests and in codifying the general rights and obligations of those having an interest in the transport of hazardous materials (Panel on the Law of Ocean Uses 1993). Nearly all of the sixty parties that had ratified or acceded to the convention as of early 1994, however, were developing countries with little direct involvement in the transport of hazardous materials. How quickly the major maritime states will ratify the convention and thereby endow it with universal recognition remains in doubt. Moreover, the convention leaves open to interpretation several issues concerning the rights under international law of a coastal state vis-à-vis ships carrying hazardous materials: To what extent does a coastal state have the right to prohibit ships carrying hazardous materials from sailing through areas off its shores? To what extent does it have the right to impose regulations concerning design, construction, equipment, or manning standards? What other regulations may it enact to protect its environmental security?

Absent a universally accepted convention that establishes these basic legal principles, the answers to these questions will be found in the acts taken and the practice pursued by countries in the coming years. Their practice will serve either to solidify further the current order of the oceans as reflected in the Law of the Sea Convention or to signal a destabilizing departure toward a new international context characterized, at least initially, by a lack of consensus on the rights and responsibilities of coastal and maritime states. Lack of consensus on such fundamental international legal issues as navigational freedoms spells an increase in international tensions, diplomatic protests, and conceivably resort to gunboat diplomacy.

The Prevention of Safety Hazards and Pollution at Sea: SOLAS and MARPOL 73/78

Marine safety and pollution prevention measures are similar for bulk and packaged substances. Technological innovations in the chemical manufacturing and transportation industries have altered the risks associated with the carriage of hazardous materials, decreasing some risks while at the same time magnifying the consequences of others. Many preventive measures adopted under the IMO framework have evolved in close connection to technological innovations brought on by industry and have been widely adopted by member nations as a result of the need for standardization in highly technical areas and the perceived degree of protection afforded by standards.

The 1974 International Convention for the Safety of Life at Sea (SOLAS Convention) evolved from the first SOLAS Convention, adopted in 1914 and prompted by the sinking of the *Titanic* in 1912 (a classic example of international collaboration catalyzed by a highly publicized disaster). The 1974 SOLAS Convention covers a wide range of aspects relating to safety at sea, including survey and documentation; fire safety; construction, machinery, and electrical installations; life-saving appliances; communications; and rules of ship operation. The safety aspects of the transportation of dangerous goods are addressed in SOLAS Chapter VII. Contracting parties are required to issue detailed instructions on safe packaging and stowage of dangerous goods and to prohibit the carriage of dangerous goods by vessels flying their flags except in accordance with the provisions of the convention. The SOLAS Convention extends its application to bulk chemical tankers and liquified gas carriers by mandating compliance with the IBC or the BCH Code (both concerning dangerous chemicals in bulk) and with the IGC Code (concerning liquefied gases in bulk).[5]

The treaty system comprising the 1973 Convention for the Prevention of Pollution from Ships and its 1978 protocol and associated annexes (MARPOL 73/78) is the cornerstone of the vessel-source pollution control regime established under the auspices of the IMO. Inspired largely by concerns about petroleum carriage and focused originally on oil pollution, it covers all forms of pollution generated by ships. Five different annexes to the main text address the following categories of pollutants:

- Annex I, Regulations for the Prevention of Pollution by Oil;
- Annex II, Regulations for the Control of Pollution by Noxious Liquid Substances in Bulk;
- Annex III, Regulations for the Prevention of Pollution by Harmful Substances Carried by Sea in Packaged Forms, or in Freight Containers, Portable Tanks or Road and Rail Wagons;
- Annex IV, Regulations for the Prevention of Pollution by Sewage from Ships; and
- Annex V, Regulations for the Prevention of Pollution by Garbage from Ships.

Unlike Annexes I and II, Annexes III through V are optional. Consideration of a new annex to MARPOL 73/78 that would cover noxious solid substances is under way at the Marine Environment Protection Committee (MEPC) of the IMO.

Detailed standards in Annex II (1) restrict the routine discharge of chemical residues from cargo tanks, line washings, ballast water, and other residues; (2) set requirements for allowed discharges, such as maximum quantities to be discharged, vessel speed, distance to shore, water depth, and dilution of substances before discharge; and (3) mandate reception facilities in ports for certain categories of substances. Annex II entered into force on April 6, 1987. As of October 1993, discharges were prohibited in three special areas: the Baltic and Black Seas and the Antarctic Area.

Criteria for the categorization of hazardous substances under Annex II are based on a method devised by GESAMP to assess the "hazard profile" of substances carried in bulk.[6] These criteria function as a taxonomy of pollutants that range from most severe, Category A, to least severe, Category D. A list of substances appended to Annex II establishes pollutant categories for the operational discharge of substances that are routinely carried in bulk (e.g., sulfuric acid, hydrochloric acid, arsenic trioxide). Although Annex II itself is directly concerned only with operational discharge procedures, it addresses accidental pollution by mandating the IBC Code or the BCH Code.

As a way to implement Annex II, the IBC Code has been extended to cover marine pollution aspects; ship-type requirements under the IBC and BCH Codes are now allocated according to maritime safety as well as pollution prevention.

Annex III of MARPOL 73/78 regulates the marking, labeling, packaging, stowage, quantity limitations, exceptions and notifications, and documentation of marine pollutants. Some suggestions have been made for its amendment to improve the ability of port states to respond to potential accidents involving vessels carrying noxious or dangerous substances. Spain, for example, has drawn attention to the need for ports visited en route to be informed of any dangerous goods carried aboard a ship, even though such ports may not have to handle those goods (MEPC 1990).

Annex III entered into force in 1993, as ratified by the United States and the former Soviet Union (and confirmed by Russia).

The Response to Safety Hazards and Marine Pollution: Intervention, Salvage, and Preparedness

Two other international treaties help moderate the threats of hazardous materials to environmental security by codifying practices and cooperative actions that may be taken by states or their agents to intervene in accidents and salvage.

The Intervention Convention. The 1969 International Convention Relating to Intervention on the High Seas in Cases of Oil Pollution Casualties (Intervention Convention) provides coastal states with the authority to take measures on the high seas to prevent, mitigate, or eliminate the threat of oil pollution from ships that pose a serious threat to their coastline and the resources of the coastal environment. The origin of the Intervention Convention can be traced to the *Torrey Canyon* accident in 1967, which prompted reconsideration of the rights of a coastal state to intervene when maritime casualties threaten its interests.

The 1973 Protocol relating to Intervention on the High Seas in Cases of Marine Pollution by Substances other than Oil extends the authority provided by the 1969 Intervention Convention for coastal states to take action on the high seas in casualties involving substances other than oil. Substances covered by the 1973 Protocol to the Intervention Convention are listed in an annex and extend to chemicals that have characteristics similar to those listed. Consultations with the flag state, the owners of the vessel, and other interested parties must precede any action taken by the coastal state. Measures taken by the coastal state must be proportionate to the damage or the threat, and measures taken beyond those permitted by the convention result in making the coastal state liable to compensate for damages resulting from its actions.

The Salvage Convention. The 1989 International Convention on Salvage (Salvage Convention) governs assistance to vessels in danger; its aims are to preserve human life and property at sea and to protect the marine environment. It was motivated by widespread recognition that salvage practices provided inadequate financial incentives for salvors to take actions to secure the protection of the marine environment, presumably because environmental protection holds no particular value for owners of endangered vessels or cargoes. There was also widespread concern that market forces, compounded by existing salving practices, were not conducive to the maintenance of a stable salving industry.

The Salvage Convention recognizes the right to a reward for all salvage operations having a useful result; no payment is due for operations having no useful result. Criteria for fixing the reward due to a salvor include the following: (1) the value of the property; (2) the skill and efforts of the salvor in preventing or minimizing damage to the environment; (3) the measure of success; (4) the time, expenses, and losses incurred; (5) the risk of liability run by the salvor; (6) the promptness of the salvor; (7) the availability and use of vessels or equipment; (8) the readiness and efficiency of the salvor's equipment; and (9) the nature and degree of danger.

The Salvage Convention also establishes a mechanism for special compensation of a salvor who fails to earn a reward for a useful result but has prevented or minimized damage to the marine environment. This special compensation creates an incentive for salvors to assist endangered vessels posing a threat to the marine environment by entitling the salvor to a special compensation equal, at least, to the expenses incurred in the salvage operations. Compensation can exceed 100 percent of the expenses incurred, but in no case shall it exceed 200 percent. A salvor who is negligent by failing to safeguard the marine environment may be deprived of the special compensation in whole or in part.

The Salvage Convention recognizes the right of a coastal state to take measures to protect its coastline or related interests against the threat of pollution from maritime casualties according to recognized principles of international law, including the right to give directions in relation to salvage operations. However, when deciding upon admittance to its ports of vessels in distress (commonly referred to as

Fig. 4.1. Millions of starfishes, thousands of crabs, and thirty seals were found dead in the Arkhangel'sk Region of the White Sea in 1990. Thorough investigation has shown that a poisonous chemical agent containing sulfur compounds, which probably had been thrown overboard a ship, was responsible for the deaths. (Tass from Sovfoto)

leper vessels) or other salvage matters such as the provision of facilities to salvors, the coastal state is required by the convention to consider the interests of other parties and the need for cooperation among salvors to ensure the salvage of life and property and the prevention of damage to the marine environment (fig. 4.1).

The Transboundary Movement of Hazardous Wastes

Transboundary movements of hazardous wastes have garnered the attention of the Organization for Economic Cooperation and Development (OECD), UNEP, and the Commission of the European Communities (CEC). The OECD Council has long recognized that disposal trends in developed nations can result in an increase in the transboundary movement of wastes to developing nations and has adopted several Decision-Recommendations that are binding on its members. The OECD directives seek, among other things, to ensure that transportation of hazardous materials takes place only with prior consent of the receiving nation and with advance notification to any transit nations through which the wastes may be shipped. The broad definition of wastes used by the OECD includes substances defined as wastes by any of the countries implicated in the movement.

The Basel Convention. The 1989 Convention on the Control of Transboundary Movements of Hazardous Wastes and Their Disposal (Basel Convention), which entered into force in May 1992, regulates the transboundary trade and disposal of toxic wastes. Transboundary movements are allowed with prior informed consent of the importing and transit countries. Exports to and imports from nonparties to the convention are banned in the absence of bilateral agreements. Exports are to take place only if the exporting state does not have the capability to dispose of the wastes in an environmentally sound manner or if the wastes are to be recycled or recovered by the importing state. The convention also requires the adoption of a protocol on liability and compensation for damages resulting from the transboundary movement and disposal of wastes.

The convention sets forth forty classes of wastes that are to be controlled. A fundamental premise of the convention is the overriding need for waste minimization, recycling, and source-country responsibility. Curtailing the access of waste producers to inexpensive or unregulated disposal sites in developing nations is expected to increase waste minimization by producers in developed nations, thereby reducing transboundary movements and disposal of wastes.

The Basel Convention is the only global regime that provides a basis for the regulation of transboundary movements and disposal of wastes, and some critics doubt its effectiveness and point to controversial aspects of the convention that remain unresolved. For example, the Basel Convention does not specify standards for the environmentally sound disposal of wastes in the receiving country. Although the treaty requires exporters to ensure the adequacy of disposal rights, no specific mechanism is prescribed. Critics also point to the lack of a clear definition of wastes, which can result in mislabeling wastes as nonhazardous, placing the burden on the importing nation to discover the fraud, and possibly using the recycling provisions to facilitate toxic waste dumping.

Provisions for enforcing the Basel Convention have yet to be developed. Implementation will require the harmonization of requirements under existing conventions that have evolved under IMO auspices (such as MARPOL 73/78, Annexes II and III, which regulate the transport of hazardous and noxious materials) and other conventions (e.g., the 1972 London Convention, also known as the London Dumping Convention).

The fact that the movement of hazardous materials poses a threat to the international balance of state sovereignty represented in the 1982 Law of the Sea Convention was openly discussed in the negotiations of the Basel Convention. Specifically, the conference considered the possible infringement on navigation rights and freedoms at sea of the consent provisions for transit states. Some nations sought to codify in the Basel Convention the right of the transit state—the state off whose shores the vessel is passing—to deny passage through waters under its jurisdiction to vessels carrying hazardous materials. The United States vigorously resisted inclusion of such a right, which it saw as disruptive of fundamental concepts of navigation rights established in international law. The final text of the Basel Convention includes a general provision stipulating that anything in the convention shall not affect the sovereignty of states over their territorial sea, EEZ, or continental shelf, nor the rights and freedoms of ships and aircraft provided for in international law.

Regional instruments that provide more stringent regulations have already been developed. In December 1989, as part of the Lomé IV Convention on Trade and Aid, the European Community (EC) and sixty-eight African, Caribbean, and Pacific (ACP) nations banned the transboundary movement of wastes among EC and ACP signatories.

REGULATORY INADEQUACIES AND THEIR IMPLICATIONS FOR MARINE ENVIRONMENTAL PROTECTION

There is an unmistakable link between catastrophic oil spills and the evolution of treaties and national laws governing marine safety and oil pollution prevention. By contrast, the seaborne transport of hazardous materials has traditionally been considered a technical matter best left to those who ship, load, carry, and discharge such materials (Gold 1990). For the most part, the international regulatory regime for the carriage of hazardous materials by sea has failed to keep pace with more environmentally oriented concerns and lags behind the framework of rules governing the carriage of oil.

A related problem has arisen with the increasing amount of hazardous waste being produced and transported by sea for disposal elsewhere. Shipping and disposal of wastes has become a lucrative, specialized business that includes a gray market element. At the regional level, attempts have been made to prohibit such illegal trade altogether, and the problem has prompted an international response to regulate waste trade more strictly through the 1989 Basel Convention. The question of an appropriate liability and compensation framework was not addressed at Basel, and it can be argued that any framework established for hazardous materials may be insufficient to cover incidents related to the carriage of hazardous wastes—a far less valuable, but potentially more deadly, cargo.

Unratified and Uncompleted Conventions

Unratified, unsigned, and uncompleted conventions forestall effective implementation and acknowledgment of the need for improvements in the current international regime for seaborne transportation of hazardous materials.

The Draft HNS Convention. In 1984, a Draft Convention on Liability and Compensation in Connection with the Carriage of Hazardous and Noxious Substances by Sea (HNS Convention) was considered by a diplomatic conference but failed to gain acceptance, in substantial measure because of controversy over the definition of subject substances. The Draft HNS Convention attempted to establish the following elements in relation to liability and compensation for incidents involving hazardous materials: (1) a new, two-tiered strict and joint liability system for the shipowner and the shipper; (2) limitations of liability; (3) requirements for compulsory insurance; and (4) definition of substances covered by the convention that was to exclude packaged substances. Since 1984, work on an HNS Convention has proceeded within the framework of the Legal Committee of the IMO.

An important unresolved issue for the HNS Convention is the appropriate linkage between the 1976 Convention on Limitation of Liability for Maritime Claims (LLMC) and the Draft HNS Convention. The 1976 LLMC, which entered into force in 1986 without US participation, creates a virtually unbreakable system of liability limits for claims against human injury, loss of life, and property loss, such as damage to ships or harbors. The LLMC, however, does not provide a basis for addressing any other aspect of liability and compensation related to seaborne hazardous materials. Parties to the 1976 LLMC support a strong link and would require ratification of the 1976 LLMC prior to accession to the Draft HNS Convention. Maritime nations that are not party to the 1976 LLMC oppose its low limits for liability and compensation and therefore support an independent HNS Convention. At its March 1993 session, the IMO Legal Committee indefinitely postponed a new diplomatic conference on the Draft HNS Convention that had been scheduled for 1994–95 and accorded elevated priority to revision of the 1976 LLMC. Additional time was granted to a working group of technical experts to propose solutions to technical problems with the Draft HNS Convention (*Oceans Policy News* 1993a; see also *Oceans Policy News* 1992).

The Extended Global Convention on Oil Spill Response. The International Convention on Oil Preparedness and Response was adopted at an IMO Conference in November 1990, and the United States was the first to ratify it. Russia has not ratified the convention, which requires fifteen ratifications to enter into force. Achievement of that number is virtually assured. The convention calls for international assistance and cooperation in the preparation for and response to spills from oil, but efforts have been made through a resolution to expand its coverage to include hazardous and noxious substances. The convention covers (1) information exchange with regard to the capabilities of nations to respond to incidents; (2) preparation of oil spill response plans; (3) reporting of significant incidents or incidents affecting coastal states; and (4) research and development into ways of combating pollution in the marine environment. It further requires vessels and offshore installations to have oil spill response plans and to report any incidents involving discharges to the marine environment. Parties to the convention will be required to prepare a National Contingency Plan, with oil spill equipment and personnel to be ready on a national, bilateral, or multilateral basis for emergency purposes. Under the proposed resolution of extension, these measures would be required for hazardous materials as well.

The Lack of Coastal State Authority to Enforce Preventive Environmental Regulations

The Law of the Sea Convention recognizes a coastal state's authority to exercise jurisdiction to protect and preserve the marine environment within its territorial sea and EEZ, yet it limits the measures a coastal state can adopt to do so (see chapter 8). Within the EEZ, for example, a coastal state can adopt only generally accepted international rules or standards. Within the territorial sea, coastal state authority is limited by Article 24, which states that the laws and regulations of coastal states may not have the practical effect of denying or impairing the right of innocent passage. Thus, a coastal state may not regulate stopping and anchoring in fragile coral reefs. The Law of the Sea Convention also fails to define what kinds of laws and regulations are permissible, and major maritime nations have thus far been reluctant to give the issue serious consideration (Bodansky 1990).

To the extent that a proposed measure may merely impair innocent passage, it has been suggested that, rather than reject such a measure outright, there should be a requirement that it balance the severity of the environmental risk against the degree of navigational inconvenience. The alternative—unilateral action by coastal states—may be even more unpalatable to maritime interests (Bodansky 1990).

Within their EEZs, coastal states do have the authority to establish international rules and standards for the prevention, reduction, and control of pollution. Except in ice-covered areas, however, where a special Law of the Sea provision carved out an exception to the general rule, coastal states are given no complementary authority to enforce such standards. A coastal state may ask a ship for information when it suspects a violation in its EEZ waters, and it has the right to inspect when the information is unconvincing; but only when a violation causes or threatens major damage may a coastal state go so far as to detain a vessel (Article 220, paras. 5 and 6). Nonintrusive surveillance measures such as transponders, vessel registries, and reporting requirements have been resisted as vigorously as have more intrusive boarding and inspection methods, as an infringement on traditional navigational freedoms, although such measures do not seem to have impinged on the ease and efficiency of civil aviation operations (Bodansky 1990).

Port states and coastal states are further limited in their ability to protect their marine environments by the preemptive jurisdiction of flag states over vessels that have violated international laws. Contrary to developments in other areas of international law, under the Law of the Sea Convention, all ships, their activities, and their crew and cargo are subject to the exclusive jurisdiction of the flag state in all but a few exceptional circumstances.[7] By contrast, under the Convention for the Prevention of Torture, jurisdiction is shared by the state where the violator is located and the home state, and each is obliged to exercise its jurisdiction by either extraditing or prosecuting the torturer (Bodansky 1990).

The question of what measures a coastal state may take to ensure its environmental security is likely to become increasingly contentious. For example, a region that is a self-declared nuclear-free zone prohibiting the disposal of nuclear waste on land and sea may be prompted to take the obvious next step of declaring itself off limits to the transportation of hazardous wastes. Decommissioned chemical weapons and their constituent agents, which at present are not subject to the safety precautions clearly prescribed for nuclear materials, might fall into this category.

The Case of Chemical Weapons: Tradeoffs in Environmental Security

Efforts to enhance environmental security generally can entail, at least in the short term, heightened specific threats to some states. This problem is illustrated by a recent program to ship decommissioned US chemical weapons from Germany to Johnston Atoll in the Pacific Ocean, about 750 miles southwest of Hawaii.

In 1990, several South Pacific countries expressed unease about the US Army's Johnston Atoll Chemical Agent Disposal System (JACADS) for demilitarizing chemical weapons on the Johnston Island Group, including concerns about the security of shipments through the region. Should these sorts of concerns regarding vessel safety and navigational standards be left unattended until after an accident has occurred, unilateral action is likely to follow.

The 1990 chemical weapons shipments resulted from a 1986 agreement between US President Reagan and German Chancellor Kohl to remove from Germany the US chemical munitions stockpile of over 100,000 155-mm and 8-inch unitary projectiles containing the nerve agents GB (volatile) and VX (persistent).[8] The transport of the chemical weapons stockpile to the JACADS facility was managed by military officials as a classified operation. Unitary chemical weapons were removed from storage near Clausen, West Germany, beginning in July 1990. Using truck and train transport, American and West German military personnel and civilian contractors moved the weapons from their storage facilities to an intermediate site at Miesau and then on to the port of Nordenham. The weapons were loaded onto two Ready Reserve vessels operated by the US Military Sealift Command, SS *Gopher State* and SS *Flickertail State,* which sailed nonstop for forty-six days from Nordenham to JACADS (United States Army Program Manager 1988).

The transport ships were escorted by a US Navy guided missile cruiser, and each was refueled at sea three times during the voyage to obviate the need for intermediate port calls. Although the precise sea route taken by the ships remains classified, it is known that they used the Cape Horn route around the tip of South America. Military officials reported no chemical agent leaks or security incidents during the transport phase.

The total cost of the transport operation is estimated at $53 million, which represents a cost overrun of 26 percent. This total, according to the US General Accounting Office, does not reflect another $7.2 million paid by the German government and $1.4 million paid by the US Department of Defense for "retrograde container production and repair" costs not charged to the transport operation. The cost overrun was due largely to problems encountered in container and ship modifications and repair. The problems reported by federal inspectors are relevant to any future attempt to transport chemical weapons by sea. Among these were structural defects in many military cargo containers, failure of the containers to meet international standards for carrying hazardous cargo, and improper shipboard equipment installation.

The planning, construction, and operation of the JACADS testing and incineration facility (fig. 4.2) beginning in June 1990 generated widespread criticism among environmental advocacy groups, which have challenged the Army's chemical weapons disposal program on numerous grounds. One of these involves the classification of information related to the shipping route used during the transport phase and the lack of information provided to coastal and transit states that stand to be adversely affected should an accident occur either during shipment or as a result of testing.

The Department of Defense was required by US law to prepare an environmental impact statement (EIS) evaluating the risks associated with the JACADS operation, as well as a Global Commons Environmental Assessment (GCEA) of the potential impact of the chemical munitions shipment on sea routes, transit and coastal states, and territories in the immediate vicinity of the operation. Challenges to both reports by environmental groups attempting to halt the project were unsuccessful in US district court. According to a representative of the US Arms Control and Disarmament Agency, during the operational phase JACADS has met or exceeded all applicable environmental standards (Staples, pers. comm., 1994).

The obsolete chemical weapons stockpile transported from Germany to the JACADS facility in 1990 represents only about 1 percent of US chemical agents. (The stockpile of outdated chemical weapons of the former Soviet Union is thought to be of a comparable magni-

Fig. 4.2. US Army chemical weapons disposal operations on Johnston Atoll in the mid-Pacific Ocean. (Courtesy of the US Army)

tude.) Concerning the future destruction of unitary chemical weapons, the US Army's plans include destruction of existing stockpiles already at JACADS with no further shipments to the site, plus an $8 billion program to construct and operate incinerators of JACADS design at eight different US sites, which is slated for completion by 2004. One of these sites, the Tooele depot in western Utah, houses more than 42 percent of the 25,000 tons of chemical agent remaining in the US stockpile (National Research Council 1993a).

These plans are a result of a US Army testing and development program launched in 1978, well before the thaw in East-West relations began to ease the longstanding stalemate in efforts by the Conference on Disarmament to complete a convention banning not just the use of chemical weapons (as under the 1925 Geneva Convention) but also their possession. Along with a separate US Army program for the destruction of binary chemical weapons within the same time frame, the experience gained at JACADS places the US on track to meet international deadlines for the destruction of chemical weapons established by the 1993 Convention on the Prohibition of Development, Production, Stockpiling and Use of Chemical Weapons and on Their Destruction (the Chemical Weapons Convention, or CWC). The CWC calls for the destruction of all chemical weapons beginning not later than two years after its entry into force (which is expected in early 1995, based on a required waiting period of 180 days after deposit of the sixty-fifth instrument of ratification) and for completion of such efforts within ten years of that date.

Few would argue that the destruction of chemical weapons is not a step forward for environmental security—or, indeed, for global security in its very broadest sense. With 154 signatories and speedy ratification virtually assured, the CWC clearly shares the special status of its precursor, the Geneva Convention, as one of the few international norms that are shared by industrial and developing nations alike, in all regions of the globe (see Roberts 1992).[9] Nonetheless, its implementation must be expected to create other occasions for the shipment of chemical agents that will entail problems of environmental security similar to those of the JACADS experience (see National Research Council 1993a).

In fact, the CWC, even before its entry into force, has already figured prominently in one international dispute between two of its signatories, the United States and China, in which another signatory, Iran, was indirectly involved. In July 1993, the United States reportedly received intelligence that the Chinese cargo ship *Yinhe* had left the Chinese port of Dalian bound for Abbas, Iran, with a load of two chemical substances banned under the CWC. In the view of Chinese Assistant Foreign Minister Qin Huasun, US surveillance by aircraft and warship and serious disruption of the *Yinhe*'s planned navigational route, which caused the ship to remain adrift on the high seas for several days, were in violation of international law by virtue of being based on "fabricated information" ("China Bucks at US" 1993; Yue 1993). In support of this contention, the Chinese government stated that the *Yinhe* had never planned to harbor in the port of Abbas—which, the Chinese noted, cannot accommodate containerships—and that the ship's actual destination was a port in Kuwait (Yegang 1993). Repeated statements of US intention to board the *Yinhe* and inspect its cargo were finally satisfied by a joint team of US, Chinese, and Saudi Arabian inspectors, who ultimately issued a report stating that none of the chemicals mentioned in the US allegations was found aboard the *Yinhe* (Yegang 1993). Declaring itself vindicated, the Chinese government sought a formal apology by the United States as well as compensation for economic loss associated with the disruption of the *Yinhe*'s operations.

The JACADS experience offers a small-scale profile of the transport management issues and broader policy questions that will come to the fore as states begin implementing large-scale destruction programs in compliance with the CWC. Ensuring the environmental security of all involved states will require military authorities to balance the prudent protection of plans for routing and scheduling with the moral and practical advantages of fully disclosing their technical plans for the safe handling of chemical agents and for action in the event of emergencies. Building formal collaborative mechanisms into any future multilateral programs can help, too, by ensuring the transfer of technology and information among countries engaged in the destruction of their chemical weapons stockpiles.

Disparate Standards and the Issue of Open Registries

The regulatory deficiencies associated with the unfinished international agenda for hazardous materials shipping are compounded by the existence of disparate standards among national fleets. This disparity poses a threat to environmental security, especially to the extent that it encourages owners to register their ships in the so-called open registries of countries least able or least willing to impose and enforce strict environmental, safety, and manning standards. Compliance with such standards entails costs that increase overall fleet operating costs. Although obvious economic benefits accrue to a shipowner from operating a fleet with a well-trained crew, an adequate complement of safety equipment, and an ongoing program of inspections and shipboard maintenance, the business of shipping is such that bottom-line operating costs—particularly variable capital and crew costs—determine profit and loss. For better or worse, the decisions of most shippers are guided by tariff rates and charter prices rather than by the safety records or precautionary practices of the shipowners.

The Problem of Open Registry. Registration is how a state confers its nationality to a ship, which is then governed by the national laws of the registry state. International negotiations surrounding the UN Conference on Trade and Development (UNCTAD) Code for Liner Conduct and the UN Convention on Conditions for Registration of Ships (UNCCORS) focused attention on the practices and standards of open registries, often termed *flags of convenience, black flags,* or *tax haven flags* (Kasoulides 1989). A general definition has been developed for this broad, heterogeneous group:

> flags of those states with whom shipowners register their vessels in order to avoid, firstly, the fiscal obligations, and secondly, the conditions and terms of employment or factors of production

that would have been applicable if their tonnage was [registered] ... in their own countries (Metaxas 1981).

Open registries, such as those of Liberia, Cyprus, Panama, and the Bahamas, are distinguished in practice from the national registries of the traditional maritime nations and from the newer international registries of Germany and Norway.

Open registry fleets are poor performers in safety, training, and compliance (Titz 1989; Chadwick 1984; Rother 1980; Giziakis 1982; and IMO, various studies). For example, Titz (1989) analyzed the loss ratios of 32 flags between 1968 and 1986. Among those with the worst loss ratios were four open registry fleets accounting for over a quarter of the world's total gross registered tonnage: Cyprus (ranked worst at 32nd), Panama (28th), Singapore (27th), and Liberia (24th). A pollution likelihood analysis by flag identified as "potential polluters" several open registry fleets: Greece, Panama, Cyprus, Liberia, Singapore, and the Philippines (Titz 1989). Across several independent analyses conducted since 1980, the flag most often found to have a very high casualty rate, relative to the world average, is Greece, followed by Panama and Cyprus. Liberia can be considered a high-loss flag as well, based on past performance (Titz 1989).

Clearly, the ships of open registry fleets that fail to adopt or enforce international safety standards are more likely to have accidents. When hazardous materials are carried as cargo, the increased likelihood of a casualty constitutes a threat to environmental security that is perpetuated by the absence of adequate regimes of regional and international standards.

Institutional Responses to the Problem of Substandard Ships. The phasing out or outright elimination of open registries, with their problems of safety, manning, and regulatory enforcement, is often cited as a policy objective that would enhance the environmental security of coastal states and ease the controversy surrounding the rights of coastal versus flag states. It has also been argued that elimination or nationalization of open registries would give developing countries a better foothold in international commerce by encouraging the development of viable national fleets, the carriage of a proportionate amount of their own cargoes, and improved access to the international maritime infrastructure.

The UN Convention on the Conditions for Registration of Ships (UNCCORS), the result of an UNCTAD-sponsored diplomatic conference, represents a multilateral attempt to establish strict rules for ship registration and manage the gradual nationalization of existing open registries. Although finalized in February 1986, it has not yet come into force, having failed to gain the support of most traditional maritime countries or developing countries, with the notable exception of the former Soviet Union. Neither the United States nor Russia has signed or ratified UNCCORS. One critic has termed UNCCORS a simple de facto institutionalization of the open registry system (Kasoulides 1989). Widely criticized by diplomatic coalitions, shippers, labor interests, and shipowners, this convention is unlikely to gain many signatories in the near term or to strengthen compliance with international shipping standards.

Regional ship inspection and monitoring programs, such as the European port state control (PSC) system, are, in part, a response to fears about the effects of shipping market forces, the deficiencies associated with open registry ships, and the inability of coastal states to counter those deficiencies with measures that would adequately protect port state marine environments. The foundation of PSC is the Memorandum of Understanding on Port State Control, signed in 1982 by fourteen European states. It calls for inspection of 25 percent of all ships calling in each member nation's ports on an annual basis and the arrest of those found to be in violation of established safety standards. Vessels are inspected for twenty individual criteria, and member states' efforts to harmonize inspection procedures and to share inspection data have intensified over the past five years, resulting in an overall improvement in program effectiveness.

Port state controls were criticized during UNCCORS deliberations as being unable to address the "basic economic and social problems created by the open registry system" (UNCCORS Preparatory Committee Report 1984). Since 1982, however, the European PSC program has evolved into a functioning system that is more than a simple substitute for responsible flag state control. The European PSC program expanded with the adherence of Poland in 1992, and Japan, Croatia, Canada, and the United States have taken transitional status as cooperative authorities, with Russia expected to adhere in the near future (Nollkaemper 1993). Meanwhile, similar regional programs are being developed in South America and the Asian Pacific. Although the data are inconclusive on the question of whether PSC has led to a reduction in casualties involving ships of the PSC signatories or incidents in the territorial waters and ports of PSC members, the program's defenders are confident that time and program experience will prove their case.

With UNCCORS unratified and regional approaches of only recent vintage, it may make sense to examine other alternatives, such as the adoption of a compliance regime by trade lane (via the liner conference structure). Under such a program, with a uniform set of standards applicable to both conference and nonconference carriers in a particular trade lane (and, by association, to their ships regardless of registry), shipowners operating consistently substandard ships would be faced with two options: (1) meeting the universal standards/compliance regime and bearing the short-run cost of upgrades, periodic inspections, and so forth, or (2) leaving the universal standard/mandatory compliance trade for another and forfeiting the potential benefits of carrying high-value goods (which draw higher freight revenues). The long-run benefits of compliance might include increased shipper confidence, enhanced market share in more profitable trades, and lower insurance premiums.

Regardless of form or scope, any institutional regime for addressing these problems should consider the fragile position of developing nations, whose fleet assets make up a substantial share of total world tonnage. Strict international regulations concerning marine safety and, more specifically, the shipment of hazardous materials impose severe operating and financial constraints on young fleets. Moreover, not all developing nations operate open registry fleets, nor are their ships necessarily safety deficient. Several developing states have voiced concerns that stringent rules governing vessel-source pollution might discourage the development of their maritime industries (Chircop 1988). Any regime proposed to manage the problem of substandard ships (or to regulate seaborne transport of hazardous materials) that fails to address issues of equity and the disparate distribution of economic power in international shipping markets has little hope of widespread ratification.

From the perspective of many developing countries, the best outcome might involve a marriage of international ship safety standards (with compliance managed in a graduated fashion) with the phased elimination of the open registry system. Such a regime would allow for the mandatory transfer of open registry ships—many of whose owners can be traced to industrialized countries—back into national fleets, some of which would be those of developing countries. As part of the transfer process, the owners would be required to verify compliance with universally applied standards for the safe handling of hazardous material. Subsequent to the transfer, these ships would be subject to the highest of several possible levels of enforcement. Tonnage presently outside the realm of open registry (not subject to transfer) would also be subject to universally applied standards, but conversion and compliance might be managed according to a more liberal schedule. Under such a scenario, young fleets could be shielded for a time from excessive regulatory compliance costs. At the same time, a proportionately higher level of compliance costs would tend to be borne by shipowners and the national fleets of industrialized countries.

ECONOMIC PERSPECTIVES ON SEABORNE HAZARDOUS MATERIALS

A fundamental inadequacy in the management of international seaborne hazardous materials is the absence of consistent economic incentives for shipowners to observe rigorous standards for safety and environmental protection. A key is to create mechanisms through which the owners will bear the full cost of (i.e., internalize) potential environmental damages. Enhanced efficiencies can be achieved, too, by spreading the risk of such damages through well-devised insurance arrangements.

A crucial aspect of international marine environmental protection has, in fact, been the formulation of frameworks for liability and compensation. Although a comprehensive legal framework for the compensation of damages resulting from hazardous materials is not addressed under any international convention, the 1982 Law of the Sea Convention calls for the development and acceptance of an international legal regime for liability and compensation in the case of damages resulting from hazardous or noxious materials. Already there is an international regime that establishes limits to liability, the 1976 LLMC. The 1984 Draft HNS Convention established a two-tiered liability system for shipowners and shippers, but difficulties would arise in identifying the shipper of hazardous materials. Moreover, as discussed above, there are numerous other obstacles to the completion and ratification of that convention. Any framework for liability and compensation that is eventually developed, however, must recognize the basic economic principles underlying its purpose if it is to be effective.

Liability Regimes and Economically Optimal Levels of Precaution

Current regulatory regimes do not impose full liability, since the financial liability for clean-up and damage from marine accidents is limited by international convention. Figure 4.3 depicts the economic incentives for a shipowner to take precautionary measures in pollution control. The shipowner would minimize total costs (the sum of expected damages and the costs of precaution) by choosing the level of precaution that equates the marginal cost of additional precaution with the resulting reduction in marginal expected penalty. This mar-

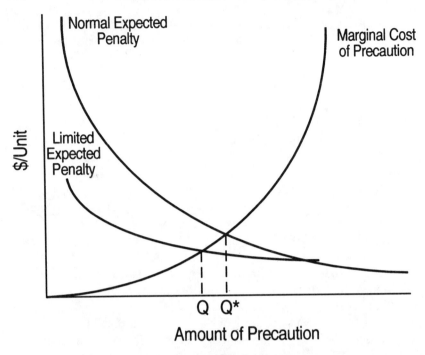

Fig. 4.3. Simple incentive effect of limited liability.

ginal reduction in expected penalty is a function of two factors: the likelihood of an accident and the magnitude of financial obligation triggered as a result. This function slopes downward because larger amounts of precaution are presumed to yield smaller marginal reductions in both the likelihood and the magnitude of accidents.

The shipowner's cost-minimizing choice with unlimited liability is depicted as Q^*. As long as the imposed penalty is equal to the actual damage and the probability of having to pay these damages once an accident has occurred is certain, this outcome will normally be efficient. The external costs thus would be internalized, and the shipowner would minimize its private costs by taking all possible cost-justified precautionary measures to reduce both the likelihood and the seriousness of any accident. Taking these precautions would simply be cheaper than paying for the clean-up.

Limited liability at current levels produces a different outcome. With limited liability the expected penalty function rotates inward for levels of precaution lower than the level that would otherwise produce an accident resulting in damages exactly equal to the limit. Lower levels of precaution imply damages that exceed the limit, but the shipowner would not have to pay anything above the liability level limited by international convention. In this range, increasing precaution does not reduce the magnitude of the expected financial payment.

As long as the liability limit is binding, the effect of limited liability on the shipowner's decision to comply voluntarily with guidelines is such that the owner will choose too little precaution (as indicated by Q in fig. 4.3). Both the number and the magnitude of marine accidents resulting from hazardous material discharges would thus be inefficiently large, and environmental security would receive an inefficient level of protection.

The message here is that insurance carriers cannot be expected to generate the cost-justified minimization of environmental risk until they bear the costs of marine accidents. To the extent that liability limits are too low, this mechanism will not be an effective instrument for protecting the marine environment. Of course, there may be offsetting social benefits provided through liability limitations, such as a reduction in costly and protracted litigation and enhanced expectation of timely compensation for victims. Strict unlimited liability may even result in perverse effects, such as exit by responsible shipowners, reduced capitalization, less investment in precautionary measures, and escape into bankruptcy, that would diminish environmental security.

The answer is realistic limits on liability. Current practice, which is often based solely on the value of ship and cargo, does not offer a viable option from an economic or environmental security standpoint. Unlimited liability causes other problems of its own, related to the ability of insurers to determine proper levels for their premiums. In the absence of legal limits on what they would expect to pay out, insurers have little basis for estimating the expected costs that form their basis for setting premiums. Liability limits must be established at reasonable levels, but most important is that they reflect total expected environmental damage (both immediate and long-term).

The Potential Role of Insurance Markets

As long as shipowners can shop around for registries that choose not to enforce effective standards, just as shippers can book freight on substandard ships, many standards can be effectively circumvented. Marine insurance meant to provide financial guarantees against accident or mishap, moreover, may be extremely expensive or difficult to obtain in the case of many hazardous materials.

Would it be possible to change the economic incentives of shippers so that voluntary compliance would be the norm rather than the exception? One opportunity to achieve this result may be found in the policies and practices of marine insurance companies. Shippers and shipowners depend on insurance companies to help them deal adequately with the risk of accidents and cargo loss. When cargoes and ships are lost, the resulting financial burden can be staggering. By spreading risk over time and among those involved in the transport chain through insurance arrangements, this financial burden is made more manageable.

But insurance companies do more than spread risk; they can also play an active role in promoting cost-justified measures to reduce that risk. By tradition, insurance carriers in many markets have actively promoted safety measures that hold premiums down. Inspections of premises and practices are common when a substantial amount of insurance is involved. After the inspections, specific recommendations are forwarded to the insurance purchaser on how premiums could be reduced by adopting certain practices. Companies that adopt these safe practices enjoy lower premiums, whereas companies taking fewer precautions do not.

The current approach to marine pollution depends heavily on the ability of the legal system to internalize the costs of spills through liability law. In principle, the approach is straightforward: by forcing the owner to pay for the costs of a spill—including, in some cases, compensating for natural resource damages—a powerful incentive to exercise due care is created. Questions arise, however, as to whether this outcome is likely to be efficient in practice. A necessary condition for adequate precaution to be taken is for the insurance carrier (initially) and the owner (ultimately) to bear the full costs of the environmental damage. Only then will adopting adequate preventive measures seem economically justified to shipowners. The problem remains, however, of how properly to assess the environmental damage itself.

Damage Assessment

Once damages from a hazardous materials incident are recognized, three questions must be answered for a damage assessment: (1) What resources are damaged? (2) Who suffered damages? (3) How long will it take for the natural system to recover? Monetary assessment of damages is significantly influenced by answers to these important questions. A spill of hazardous materials may affect not only current users of damaged resources (e.g., fishers or tourists), but also non-users who value an unspoiled marine environment. Frequently, uncertainty related to natural systems does not allow scientists to predict the duration of damages with a high degree of confidence. Nevertheless, damage duration and recovery path must be estimated to assess damages adequately.

In general, damages can be classified into two categories: economic impacts and lost economic benefits. The former includes market values of lost economic activities due to a shipping incident, for example, values of lost cargo, ship, and revenues to local tourism and commercial fishing. Measurement of these values is relatively straightforward, since they can be observed from existing market transactions.

It is far more difficult to measure the loss of economic benefits from a shipping incident. Lost economic benefits include, for example, the values people attach to the enjoyment of clean water and the beauty of the natural environment. The values people place on such amenities are not recorded in market transactions. They are called *nonmarketed environmental amenities*. Measurement of lost values in nonmarketed amenities is difficult because there are few market data from which economists can infer values.

Lack of market-based information has led economists to rely on either direct or indirect methods to estimate economic values of such environmental amenities. With indirect methods economists infer the values people place on environmental amenities from observed behavior, such as travel costs incurred or changed property values, under different environmental conditions. By observing people's visitation patterns to different sites with different environmental quality levels, the values they place on changes in environmental quality can be estimated (Clawson and Knetsch 1966; Smith and Kaoru 1990). By analyzing market values of different properties and conditions of environmental resources near these properties, economists performing hedonic property value studies estimate the economic values of changes in environmental quality (Brookshire et al. 1982; Palmquist 1990). With direct methods, on the other hand, investigators use a survey approach to elicit values that people place on changes in environmental quality. Typically, a hypothetical market for environmental amenities is described within a questionnaire, and respondents are asked how much they are willing to pay for certain improvements in quality. This approach is often called the *contingent valuation method* (Cummings, Brookshire, and Schulze 1986; Mitchell and Carson 1989). The contingent valuation (CV) method is subject to numerous criticisms and is controversial even among specialists. Nonetheless, a special panel of distinguished economists, asked by the US

National Oceanic and Atmospheric Administration to assess the method, encouraged its further development, having concluded that "CV studies can produce estimates reliable enough to be the starting point of a judicial process of damage assessment, including lost passive-use values" (Arrow et al. 1993). Attempts by economists to value marine resources provide important information for assessing damage from shipping incidents involving hazardous materials (NOAA 1983; Grigalunas et al. 1987, 1988; McConnell 1986; Mendelsohn 1986; Hanemann and Carson 1987; Cameron and James 1987; Bockstael, Hanemann, and Kling 1987). The quality of these studies varies, however, and they do not measure benefits or damages on comparable bases. As discussed by Kopp and Smith (1989) and Carson and Navarro (1988), a wide variety of assumptions used for conducting damage assessments will influence the final damage estimates.

Marine resources and environmental amenities generate two distinct types of value: use and non-use values. Users of resources through recreational activities and other means derive positive use values. Many people, however, value the preservation of such things as marine biological diversity or arctic and antarctic environments, even though they may never actually see such things or visit those places. These non-use (or passive-use) values are an important part of the total value that people derive from preservation of the marine environment. Damage estimates accounting for both use and non-use values typically will be larger than those based solely on use values. To be truly comprehensive, damage assessments must include all of these values. The treatment of the duration of damages, use and non-use values, the availability of substitutes, the recovery rate of the natural system, and the geographic range over which damages are felt greatly influences damage estimates. Although case-by-case analysis is the surest way to achieve accurate estimates, it can be time consuming, contentious, and quite expensive. A potential improvement in the compensation process might be made simply by standardizing some damage assessment methods using agreed-upon assumptions across cases.

Under current international practice, though, the liability and compensation framework for hazardous materials is simply not developed, and it is surely not developed enough to accommodate such innovation. Indeed, under current practice, neither non-use values nor nonmarket environmental values are recognized as valid, compensable damages. Until they are, economic inefficiencies and inadequate environmental security are virtually assured.

Technology for Hazardous Materials Transport

The application of new technology to ship operations has been continuing for hundreds of years. Much of the technological change since World War II has been to reduce crew sizes, increase shipboard efficiencies, and lower operating costs. For hazardous materials shipping, advanced technology offers solutions to problems in stowing and handling cargo, preventing illegal shipments, reporting incidents, and mitigating environmental damage in the event of a casualty or incident.

Enabling Technologies

Technologies that can enhance environmental security in hazardous materials transport run the gamut from individual instruments to global or regional systems of hardware, information flow, and practice. Several such technologies are already available and rapidly advancing. These can be harnessed to enable states to provide improved environmental security.

Information Exchange and Incident Reporting. Titz (1989) noted that the major shortcomings of the European PSC inspection system are operational rather than structural. Technology for improved ship inspections (consistency in procedures, in program application among participants, or in the attainment of inspection targets), information exchange (casualty reporting, inspection record dissemination, etc.), and contingency response could invigorate the European PSC system and similar systems that might be developed elsewhere. Existing classification societies could serve as a logical vehicle for information exchange, especially as they seek new roles in a world of expanded economic competition.

Mandatory Position Reporting. Comprehensive shore-based monitoring of all specified hazardous cargo carriers could be accomplished with mandatory reporting of position. Reporting can be automated by a device connected to LORAN-C and/or Global Positioning System (GPS) receivers, with transmission to shore-based stations via INMARSAT's standard-C system.[10] Position reports could be monitored by a computer, and ships carrying hazardous materials out of or approaching the limits of defined permissible navigation areas could be identified for interrogation via an INMARSAT standard-A voice channel (Pease 1989).

Vessel Traffic Systems. Vessel Traffic Systems (VTS) offer a related tool for position reporting. At short range, radar determines position relative to shorelines, buoys, and structures more accurately than can LORAN-C or GPS (Pease 1989). Moreover, radar images accurately superimposed on electronic charts show the position of traffic relative to depth contours, underwater hazards, and the boundaries of navigation channels. The optimal settings for VTS, which are expensive and administratively challenging, may be found in situations where complex navigational patterns (New Orleans), a high volume of hazardous materials traffic (Antwerp), or fragility of the marine environment (coral reefs, zones of coastal upwelling) warrant a system of shore-based monitoring and communication. In addition to the shore-based technologies, numerous shipboard systems, including full-fledged integrated navigation systems, make possible safety improvements that lessen the risk of incident. Some of the ships carrying nuclear materials today demonstrate the sophistication that is technologically and economically feasible (*Fairplay* 1990). No other ships carrying hazardous materials come close to attaining that standard.

Electronic Chart Technology. As shipowners move to reduce the size of their crews, fully automated bridge equipment, electronic watchkeeping, and artificial intelligence systems will become the norm. Electronic charts and shoreside route planning will replace today's navigational practices. Economic analysis suggests that the adoption of electronic chart technology is much more cost effective in preventing environmental damages from spills than is the construction of double-hull vessels (Jin, Kite-Powell, and Broadus 1993). Introduction of this technology into widespread use for hazardous materials transport is stalled pending the resolution at the IMO of international performance standards for electronic charts (Kite-Powell 1992).

Cargo Location Devices. At present, there are few guidelines or regulations for the location and retrieval of hazardous cargo containers that are washed overboard or otherwise released into the ocean. Several existing technologies could be applied to this problem, most notably acoustic devices of the type used in oceanographic research. Several ships engaged in the carriage of nuclear materials already utilize sophisticated transponder beacons on containers for tracking purposes. The attachment of acoustic pingers, which continuously transmit a bearing, could be required for cargoes that pose a severe threat. Transponders, a more advanced type of location device, are also a low-cost option. They operate only when interrogated from a surface vessel, thereby saving power, and they release both a range and a bearing.

Double Hulls. US domestic law, at the federal level and in some states, now mandates the introduction of a double-hull standard for tankers. The Oil Pollution Act of 1990 requires all tankers operating in US waters (territorial sea and EEZ) to meet double-hull standards by 2015, with few exceptions. Recognizing that many specialty ships of the type that carry hazardous materials are already equipped with such structural characteristics, the obvious question remains: Can the use of double-hull technology provide significantly enhanced environmental protection benefits for ships carrying hazardous materials? Further, should technological options related to ship design criteria be *required* of shipowners to lower risks posed by seaborne hazardous materials transport? The answer to both questions is no. Technical problems related to corrosion and tank cleaning have yet to be adequately solved for double-hulled ships (National Research Council 1991). Although estimates vary widely, it is generally acknowledged that the introduction of double hulls is not likely to affect significantly the number of oil spill incidents occurring in US waters, although the severity of individual incidents might be reduced. Only a fraction of the world's shipyards are equipped to construct or refit double-hulled ships, and it is doubtful that the extra costs of any universal requirement could be justified in economic terms (National Research Council 1991). Further research is needed to weigh refitting or new construction against the benefits of reduced environmental damages, fewer cargo losses, lower insurance premiums, and so forth. It is impossible to legislate the risk of pollution out of existence with unilateral standards such as those the United States has set in place for tanker owners. Improved regulatory compliance and efforts at universally raising performance-oriented standards offer better management options for environmental security than does the imposition of an unpopular, unproven, and costly design technology like double hulls.

Technology Gaps among Developing Countries

The international conventions applicable to hazardous materials transport by sea introduce complicated rules and technical requirements that leave some states not adequately prepared technically for compliance, even if they are sufficiently motivated by national priorities (Nollkaemper 1993). Many developing states are simply unable to participate at the many technical levels required by the IMO, and as a result many are underrepresented at IMO meetings or not represented at all (Gold 1988).

Efforts to improve access by developing countries to advanced technology at reasonable cost must be strengthened. Among the appropriate vehicles for such technology transfer—which includes expertise—are joint ventures between national maritime administrations or national marine research centers, bilateral maritime trade agreements, private commercial transactions, and intergovernmental organizations such as UNEP and IMO. Through its administration of the World Maritime University, as well as provisions included in MARPOL 73/78, IMO has established a model for technical assistance. Funding for such ventures remains scarce, however, and the gap between aspirations and reality is huge (Nollkaemper 1993).

CONCLUSIONS

Although there have been few large-scale incidents, the seaborne transport of hazardous materials other than oil poses a significant and ongoing threat to environmental security. Thousands of hazardous materials are shipped by sea, and the great differences in their characteristics and hazard impede the development of a coherent environmental security regime. A fundamental problem is the familiar tension between coastal state and navigational interests and the likelihood that unilateral coastal state preventive measures will undermine navigational freedoms, destabilize the balance achieved in the 1982 Law of the Sea Convention, and lead to escalating countermeasures.

Apart from the need to bring the Law of the Sea Convention into force with universal or widespread participation and thereby to stabilize its authority and bring forth its full powers of dispute resolution, there is other unfinished business on the hazardous materials agenda. This includes completion of the 1984 draft HNS Convention and extension of the 1990 Convention on Oil Preparedness and Response to include hazardous substances. A broader and more complex problem is the difficulty of devising and applying effective environmental, safety, and manning standards for international shipping, since shipowners can seek refuge under nondiligent flags of convenience in open registries. This broader problem, with its intrinsic implications for international distributional issues, is not likely to be solved within the context of hazardous materials transport, but measures to improve environmental security specifically with respect to hazardous materials transport will help.

Several measures show promise, and these are likely to be more effective to the extent they marshal market forces and shape economic incentives within the industry. One approach is to seek reforms through the insurance industry, which can profitably bring pressure on shipowners with enhanced inspection programs, improved casualty recordkeeping, standards-related negligence provisions, and finer resolution in premium rate structures related to safety records, standards compliance, and cargo hazard. Existing classification societies could also contribute to these measures and the collection, dissemination, and exchange of improved information on factors correlated with casualties and on the best available designs, technologies, and practices. Other institutional improvements would include expanded port state control systems and the promotion of accelerated adoption and diffusion of enabling technologies such as vessel traffic systems, electronic charts, position recording and reporting systems, and cargo location devices.

Probably the most important unfinished business for hazardous materials transport, however, is the formulation and adoption of an internationally accepted legal framework for the assignment of responsibility and liability and the determination of compensation for damages, as called for in the Law of the Sea Convention. To ensure environmental security, such a framework must include attention to all economic damages imposed by hazardous materials accidents, including lost benefits and nonmarket values; this will put a premium on improved methods of quantifying such damages.

CHAPTER 5

Radioactivity in the Oceans

IN THE POST-COLD WAR era, fears of a nuclear exchange that would devastate the planet have given way to concerns about the less immediate, more localized consequences of routine nuclear activities around the globe. To date, radioactive waste management practices have not posed catastrophic health or marine environmental consequences, although poor operational decisions and accidental releases have indeed led to increased levels of radioactivity in the oceans. Reports of such incidents feed the public's fears about nuclear contamination—fears that arise not from alarming scientific assessments but rather from a lack of certainty as to the nature, extent, and long-term environmental consequences of radioactivity in the oceans. Indeed, agreement on these matters remains elusive even among scientists, in part because of the secrecy that has traditionally surrounded nuclear activities and the longstanding official emphasis on the military, industrial, and medical benefits of nuclear technology to the detriment of research into its environmental effects.

The consequences of this tradition are apparent in current international efforts to assess the risks to human health and the environment stemming from the Soviets' extensive nuclear dumping in the arctic seas between 1959 and 1992 and continued Russian dumping of low-level wastes in the Pacific. Although the Russian government has been forthcoming with locations and basic information on source terms, it has been unable to provide detailed information on the radionuclide composition of most of the dumped wastes. Thus, it remains for the international scientific community to generate the data needed for reliable source-term calculations, by investigation of archive sources and, to the extent necessary, by reconstruction from currently available information (Sjoeblom and Linsley 1993).

These recent efforts signal a new climate of cooperation in understanding and controlling nuclear contamination of the oceans, but that is only part of the picture. Ironically, the end of the Cold War also will bring an increase in activities that can generate marine radioactivity, in the form of defense drawdown programs to reduce the overall number of nuclear weapons and nuclear-powered vessels. Although such programs lessen the threat of a nuclear showdown, they also introduce new threats to environmental security. The task of decommissioning, disassembling, and disposing of naval nuclear ships and weapons is of particular concern from the standpoint of marine environmental protection because it entails various proposals to store nuclear waste material on or in the ocean floor and because seaborne transport of these obsolete assets and their nuclear material is often necessary.

The post–Cold War era has also ushered in a renewed concern over the proliferation of nuclear weapons. The seabed has become a resting place for an estimated fifty warheads, twenty-three reactors, and an unknown number of unidentified nuclear devices (Arkin and Handler 1989b; Handler 1993a; Sjoeblom and Linsley 1993), some of which constitute opportunities for the unauthorized recovery of fissionable material that may remain in the debris. All recovery efforts, whether authorized or not, must be viewed as potential environmental security flashpoints, since salvage or transport operations could lead to the unintended release of hazardous radioactivity.

POTENTIAL SOURCES OF RADIOACTIVITY IN THE OCEANS

The level of natural and artificial radioactivity in the oceans is variable over time, dependent on changes in the number and quantity of inputs, the types of radionuclides involved, and the ability of the marine environment to dilute, disperse, and incorporate radioactive substances. The environmental risks and health threats associated with these emissions are dependent on several factors, including the half-lives of radionuclides, their characteristic form of radioactivity, and interaction with other elements in seawater and with oceanic processes.

In addition to fallout from nuclear weapons testing, artificial radioactivity has entered the oceans through the disposal of wastes associated with the use of nuclear products for power generation, defense, and medical and industrial applications. Accidents and unintended releases are another source of radioactivity in the oceans, minor in terms of total quantity and net effects, but significant as a source of public fear and as the motivation for unilateral policy actions.

Naturally Occurring Radionuclides

Radionuclides that originate from natural processes occurring on land and in the atmosphere enter the marine environment. These radionuclides are either cosmogenic in origin—that is, they result from the interaction between cosmic rays and the atoms of specific elements found in the atmosphere (such as argon, oxygen, and nitrogen)—or they are terrigenous, present in the earth's crust and traceable to the origin of the planet. Cosmogenic sources enter the oceans primarily as a result of precipitation. Terrigenous radionuclides enter the marine environment as a result of weathering and river runoff. Although all known radioactive substances are present in the oceans, the most significant sources are natural: carbon-14 and potassium-40. These two radionuclides account for well over 90 percent of the naturally occurring radioactivity in the oceans.

Uranium Mining and Milling

Low concentrations of radioactive material are released during uranium extraction and processing activities on land, and these can reach the marine environment via runoff. In general, radioactivity associated with the disruption of this naturally occurring source can be detected at the source of the activity (in mine tailings and milling waste piles) but becomes negligible and lost in the natural background at a short distance.

Nuclear Weapons Testing

Nuclear weapons tests began within one year of the atomic bombings of Hiroshima and Nagasaki in 1945 and have continued since then.[1] Artificial radioactivity resulting from these tests is present in the oceans in the form of particle fallout and contaminated sediments. Weapons tests are thus far the largest contributor of artificial radioactivity to the marine environment, at 150,000 to 200,000 peta-Becquerels, or PBq (GESAMP 1990). This quantity is well under 1 percent of the combined total of natural and man-made radioactivity in the oceans (see table 5.1).

Radioactive particles that enter the marine environment as a result of weapons testing are transported at various distances over different time scales. Their pattern of deposition and dispersal is dependent upon weight and prevailing natural processes, and, for the most part, their effects are not uniformly distributed. Nevertheless, low-level

Table 5.1
ESTIMATED RADIONUCLIDE INPUTS TO THE MARINE ENVIRONMENT

Source	Quantity (PBq)[a]
From cosmogenic or terrestrial sources	
carbon-14	8.0×10^3
tritium	8.5×10^2
potassium-40	1.6×10^7
uranium-238	5.6×10^4
COSMOGENIC/TERRESTRIAL SUBTOTAL	$1.6–2.0 \times 10^7$
From atmospheric testing	
carbon-14	$>3.9 \times 10^1$
cesium-137	6.1×10^2
tritium	1.5×10^5
strontium-90	8.2×10^2
plutonium-239/240	3.7×10^2
ATMOSPHERIC TESTING SUBTOTAL	$1.5–2.0 \times 10^5$
From fuel processing plants	4.8
From disposal and at-sea dumping[b]	$>1.5 \times 10^2$
From accidental releases and debris	unknown
CUMULATIVE MARINE INPUTS[c]	$>1.62 \times 10^7$

[a] A Becquerel is a measure of radioactive decay equal to 1 atom/second; 1 PBq (peta-Becquerel) = 1 Bq $\times 10^{15}$; and 1 GBq (giga-Becquerel, cf. table 5-2) = 1 Bq $\times 10^{12}$.
[b] Includes more than 90 PBq of low-, medium-, and high-level radioactivity from wastes, reactors, and reactor fuels dumped by the Soviets between 1959 and 1992 (Sjoeblom and Linsley 1993).
[c] A GESAMP survey of previous research concludes that upwards of 2.0×10^7 PBq of radioactivity is present in the oceans, with over 98 percent the result of naturally occurring sources and processes.
SOURCES: GESAMP 1990, except for disposal and at-sea dumping (GESAMP 1990; Sjoeblom and Linsley 1993).

concentrations of radioactive elements associated with testing are measurable in the oceans on a worldwide basis.

Radioactive Waste Disposal: Industrial, Defense, and Medical Inputs

The disposal of waste products from the use of nuclear products in the power-generation, defense, medical, research, and chemical industries has also introduced artificial radioactivity into the oceans. Although much of this disposal has been conducted according to legal, regulated standards, an unknown level has occurred illegally or without official sanction, making it impossible to determine the total marine component of disposed radioactive wastes.

According to the International Atomic Energy Agency, at least forty-seven ocean sites are known to have been used for the dumping of low-level radioactive waste before 1982 (IAEA 1991), the year before a nonbinding moratorium on dumping took effect under the terms of what was then commonly known as the London Dumping Convention (LDC) and today is referred to as the London Convention (LC). A 1991 inventory by the IAEA resulted in an estimate of 46 PBq of known radioactive material dumped in the marine environment between 1949 and 1982 (see table 5.2). A slightly earlier estimate developed by the UN Joint Group of Experts on the Scientific Aspects of Marine Pollution set the total amount of marine-disposed radioactive waste at 60 PBq over roughly the same period (GESAMP 1990). Since these estimates were developed, the Russian government has acknowledged more than thirty years of secret dumping by the Soviets of high-, medium-, and low-level radioactive wastes at dozens of locations in the Barents and Kara Seas, in forms ranging from low-level liquid waste to vessel reactors with and without fuel. The IAEA has determined that the total amount of radioactivity dumped at these recently disclosed sites exceeds 90 PBq (Sjoeblom and Linsley 1993), which is roughly twice the total amount known to have been dumped by all other nations combined through 1982. Russian dumping of low-level wastes in the Pacific also took place in 1993 (Lystsov, pers. comm., 1993; Sanger 1993).

Low-level and transuranic[2] radioactive waste products often comprise material generated as a result of nuclear fuel fabrication and reprocessing[3] activities, weapons production, packaged medical wastes, and industrial wastes in liquid form. High-level wastes differ in that, because of the persistent nature of the fissionable material present and the highly toxic effects posed by the ionizing radiation they emit, extensive shielding and protective long-term containment are required. Much of this high-level waste consists of spent fuel assemblies, the alloy casings that house reactor fuel rods, and radioactive solids yielded after fuel reprocessing and weapons production operations. With the entry into force of the LDC in 1972, the marine disposal of high-level radioactive wastes was prohibited under international law.[4]

Operational Effluent Discharges from Nuclear Power and Reprocessing Plants

Routine operational releases of effluent from nuclear power plants and fuel reprocessing facilities contribute measurable amounts of

Table 5.2
OCEAN DISPOSAL OF RADIOACTIVE MATERIAL

Area of Disposal/ Disposing Country	Quantity (GBq)[a]	Percentage Share (by Ocean Basin)
Dumping known in 1991 to have occurred in 1949–82		
Atlantic sites		
UK	3.5×10^7	77.5
Switzerland	4.4×10^6	9.8
US	2.9×10^6	6.5
Belgium	2.1×10^6	4.7
France	3.5×10^5	0.8
Netherlands	3.4×10^5	0.7
Other European	3.6×10^3	0.01
NORTH ATLANTIC SUBTOTAL	4.5×10^7	100.00
Pacific sites		
US	5.5×10^5	97.1
Japan	1.5×10^4	2.8
New Zealand	1.0×10^3	0.2
Korea	—	—
PACIFIC SUBTOTAL	5.7×10^5	100.00
WORLD TOTAL	4.6×10^7	
1993 assessment of Soviet dumping during 1959–92		
Arctic	$>9.0 \times 10^7$	

Low-level radioactive waste disposal in the oceans

Location	No. of Sites	Quantity (PBq)
Northeast Pacific	16	0.55
Western Pacific	5	0.02
Northwest Atlantic	11	2.94
Northeast Atlantic	15	42.31
Arctic	>20	>90

[a] A Becquerel is a measure of radioactive decay equal to 1 atom/second; 1 GBq (giga-Becquerel) = 1 Bq $\times 10^{12}$; and 1 PBq (peta-Becquerel, cf. table 5-1) = 1 Bq $\times 10^{15}$.
SOURCES: IAEA 1991; Sjoeblom and Linsley 1993.

radioactivity to the marine environment through direct, riverine, and atmospheric pathways. The first reported operational discharge of this type occurred in 1944 at the Hanford, Washington, atomic installation, where large volumes of radioactive cooling water were released directly into the Columbia River and subsequently entered the Pacific Ocean. Operational marine releases also are routinely conducted at two European nuclear fuel reprocessing complexes, at Sellafield in the United Kingdom and at Cap de la Hague, France, as well as at the Tokai nuclear reprocessing facility in Japan and one in Bombay, India.

By law, operational discharges are closely monitored and regulated. A variety of radionuclides, found in low-level concentrations, are present in these effluents, and it is thought that they pose little risk[5] to humans even when released in a shallow coastal environment. At Sellafield (formerly known as Windscale), which is located on the Irish Sea, the radioactive effluents are derived from storage pondwater, reprocessing waste streams, and plant-cleaning operations (Hain 1986).

Accidents and Unintended Releases

Few quantitative data are available on radioactive contamination in the oceans resulting from nuclear accidents. In the case of accidents at civilian power plants, which typically involve unintended discharges owing to human error, equipment failure, or repair mishaps, records are often incomplete. This was the case with data submitted by Soviet scientists to the IAEA soon after the April 1986 Chernobyl accident. Fortunately, levels of artificial radionuclides in the seas of northern Europe and the Arctic have been regularly observed since long before the accident occurred (primarily in connection with Sellafield monitoring), and a wealth of data was available for pre- and post-Chernobyl comparisons (e.g., Pentreath and Lovett 1976; Livingston and Bowen 1977, 1982a, 1982b; Livingston et al. 1984; Mackenzie and Scott 1982; Aarkrog et al. 1983, 1987; Livingston 1985, 1988; Dahlgaard et al. 1986; Aarkrog 1988, 1989; Nies 1988; Nies and Wedekind 1988; IAEA 1991; Baskaran et al. 1991; Dahlgaard 1993). Significant but very uneven depositions of cesium-137 and cesium-134 were recorded in the northern Baltic during the month after the accident, the highest concentrations being associated with contaminated precipitation. (According to the 1990 GESAMP report, marine deposition of cesium-137 accounted for only about 6.7 percent of the total Chernobyl release of this isotope into the environment.) Although overall cesium levels in the Baltic have diminished somewhat since May 1986 (fluctuating seasonally with the water flows), a return to pre-Chernobyl levels is expected to take longer than the twenty-five to thirty years needed for a complete water exchange with the North Sea (Nies, Herrmann, and Goroney 1993). By contrast, the current system prevailing in the North Sea has made for rapid dispersal of the very elevated cesium levels observed in the eastern English Channel and the Inner German Bight in June 1986 (Nies, Herrmann, and Goroney 1993). Within one year of the accident, measurements at these locations had dropped to below and just modestly above pre-Chernobyl values, respectively, and within two years the entire North Sea was judged to be essentially free of Chernobyl fallout (Nies, Herrmann, and Goroney 1993).

Similar approaches to data collection have been used to assess the nature and effects of marine accidents involving nuclear vessels or weapons, which also are frequently unreported or are treated as classified information. For both types of accidents, the scientific data indicate only a negligible contribution to oceanic radioactivity concentrations and few long-term health effects as a result of radioactive contamination (GESAMP 1990).

Until recently, the two most significant sources of accidental marine radioactive contamination were thought to have been those involving the atmospheric reentry and disintegration of a satellite power generator in 1964 and the loss in Greenland of two nuclear warheads after the crash of a B-52 bomber in 1968.[6] In both instances, distribution of artificial radionuclides was tracked, and measurable increases in background plutonium-239 levels were detected over short periods. No effects on marine organisms or humans were reported as a result of either accident.

Scientists are not ready to dismiss the possibility of ecological damage in an area of the mid-Atlantic, however, where a Soviet strategic submarine caught fire and sank in 1986. Although the accident was well publicized at the time, it was not until January 1994 that the extent of break-up and radioactive leakage from the two-reactor, *Yankee*-class submarine and its nuclear-tipped torpedoes and long-range missiles was made public. At an international symposium in St. Petersburg, Russian scientists revealed that it was "certain that the warheads are badly corroded and leaking plutonium and uranium" (Broad 1994). Upon returning from the symposium to the United States, Charles Hollister, a senior scientist at the Woods Hole Oceanographic Institution who has monitored radioactivity levels in the oceans since the 1970s, described the accident site, roughly 500 miles east of Bermuda, as containing "the highest concentration of radio-

Fig. 5.1. The attack submarine USS *Thresher* imploded and sank in 8,500-foot waters roughly 200 miles east of Boston, Massachusetts, on April 10, 1963. One fueled nuclear reactor was lost. (Courtesy of the US Department of Defense)

activity of any of the many sources of high-level radioactivity dumped accidentally or purposefully onto the sea floor anywhere by any nation" (Broad 1994). Noting that the debris is in an area where the speed of deep currents is unusually high (up to a knot, or "9 on a scale of 10"), Hollister did not see "any way out of damage to the local environment" (Broad 1994).

Clearly, the public record of confirmed nuclear accidents at sea and other incidents involving marine releases of radioactive contamination is incomplete. This was particularly true throughout the Cold War period, when it was routine for the Soviet and NATO navies to keep secret most technical data related to the nature, location, and potential environmental effects of nuclear accidents (fig. 5.1). Exacerbating the lack of public disclosure about specific military accidents is the fact that civilian scientific organizations and agreements typically do not have jurisdiction to include military assets and incidents in their monitoring and analysis activities.

Radioactivity in the Oceans 131

A Greenpeace inventory of 1,200 specific naval accidents between 1945 and 1988 identified more than 600 as having involved nuclear material (Arkin and Handler 1989a). Since the 1989 inventory was compiled, information on another six incidents involving nuclear ships or weapons through 1988 has come to light (Handler and Arkin 1990; Handler 1993b), and at least eight more naval accidents involving nuclear-powered vessels are known to have occurred since 1989 (see table 5.3). The April 1989 accidental sinking in the central Norwegian Sea of a Soviet prototype nuclear attack submarine, the *Komsomolets*, and its two nuclear-tipped torpedoes occasioned the most anxiety, but it seems not to pose a threat to the environment (WHOI 1993).

PROJECTIONS OF FUTURE SOURCES OF CONTAMINATION

Although a comprehensive quantitative forecast of each of the primary inputs of radioactivity to the oceans is beyond the scope of this chapter, it is worthwhile to consider how changing economic, institutional, and political conditions are likely to affect the generation of radioactive inputs in the future. Simple increases in the worldwide quantity of radioactive material or in the scope of nuclear-related activities do not necessarily translate into heightened threats to humans, marine species, or the oceans. Equally important is the management of these inputs and the existence of critical pathways, which can return oceanic radioactivity to particularly sensitive populations. Critical marine-to-human pathways include direct human consumption of contaminated seafood or other marine products (e.g., fish, shellfish, seaweed, sea salt, desalinated seawater); indirect consumption, as of agricultural products fertilized with contaminated algae or fishmeal; and various modes of contact with contaminated airborne and waterborne sediments—for example, contact with contaminated marine aerosols, sea spray, or beach sediments while swimming, boating, bathing, or handling fishing gear (Bewers and Garrett 1987).

The Expected Growth of Civilian Nuclear Power Generation

Although the civilian nuclear power industry in the United States is stagnant and many older facilities are being shut down, in Europe and Asia nuclear power is expected to contribute a greater share of total electricity output over the next twenty years. Japan, for example, plans to increase the share of domestic electricity output attributable

Table 5.3
WORLDWIDE STATISTICAL SUMMARY OF CONFIRMED NAVAL NUCLEAR
INCIDENTS, 1955–92

Years	SSBNs[a]	SSNs[b]	Surface Vessels	Ship Totals	Weapons[c]
1955–56	0	5	0	5	?
57–58	0	5	0	5	?
59–60	0	12	0	12	?
61–62	3	6	0	9	?
63–64	3	5	2	10	?
65–66	1	11	0	12	68
67–68	7	12	0	19	48
69–70	3	10	1	14	62
71–72	6	8	0	14	80
73–74	4	12	1	17	52
75–76	1	12	2	15	46
77–78	2	10	2	14	27
79–80	5	9	3	17	?
81–82	2	13	1	16	?
83–84	3	9	3	15	?
85–86	4	8	2	14	?
87–88	5	8	2	15	?
89–90	0	4	2	6	?
91–92	0	2	0	2	?
TOTALS	49	161	21	231	385

[a]Nuclear ballistic missile submarine.
[b]Nuclear attack/cruise missile submarine.
[c]Includes air-, surface-, and submarine-launched nuclear warheads and nuclear weapons components.
SOURCES: Arkin and Handler 1989a; Arkin et al. 1990; Handler 1993b.

to nuclear power from 27 percent in 1989 to 40 percent in the year 2000 (Berkhout, Suzuki, and Walker 1990). Among the reasons for the continued growth in civilian nuclear capacity outside the United States are a desire to build national prestige in a high-technology sector; institutional and political commitments to the industry; a desire to diversify energy sources and reduce reliance on fossil fuels; and, in some cases perhaps, a strategic objective to acquire the capacity for weapons-grade plutonium production.

Most important, of course, is the fact that worldwide demand for electricity continues to grow steadily throughout most of the industrialized world. To the extent that industrial development extends to the Third World, demand for electricity will increase and a portion of this demand will almost certainly be supplied through the introduction of new nuclear capacity. If developing countries such as Algeria, Brazil, and Pakistan achieve technical progress with their existing civilian nuclear programs, other developing countries are likely to follow suit.

Because management strategies for handling radioactive waste for this expanded capacity entail differing effects on the marine environment, it is worthwhile examining the practices in place and the proposals that have been advanced, particularly since the overall magnitude and technical requirements of the waste management problem are changing rapidly.

Issues of Fuel Reprocessing, Transport, Storage, and Disposal

Plutonium reprocessing activities in the United States, concentrated at the installations in Hanford, Washington, and on the Savannah River in South Carolina, have been reduced in scale over the past ten years. As a result, fewer high- and low-level wastes are being generated, although temporary storage and containment problems persist at each facility. No long-term plan has as yet been adopted for the permanent storage or disposal of the wastes generated from reprocessing activities in the United States. In Europe, the world's largest operational commercial reprocessing plant, known as UP-3, went on line at La Hague in Normandy, France, in August 1990. The plant has a nominal capacity of 800 metric tons per year of spent oxide fuels from light-water reactors (Cogema, n.d.). As of July 1993, the long-awaited THORP facility at Sellafield, a thermal oxide reprocessing plant that will enable the recycling of fuels from the UK's advanced gas-cooled reactors and overseas light-water reactors, had begun low-power start-up testing and was expected to become fully operational within several months ("BNFL Gets Okay" 1993). Over the coming decade, these two plants will reprocess most of the spent nuclear fuel generated by civilian power stations in Europe and Japan, yielding an estimated 200 tons of plutonium (Berkhout, Suzuki, and Walker 1990).[7] According to one source, this will far exceed the total short-term demand for plutonium fuel in these countries, given existing reactor technologies and prevailing low prices for uranium (Berkhout, Suzuki, and Walker 1990).

One of the more urgent reprocessing issues of relevance to the oceans involves the shipment of spent nuclear fuel from Japan to Sellafield and Cap de la Hague and the return delivery of recovered plutonium. With air transport no longer an available option, present practice involves the use of specialized vessels under armed escort. It is not known how many vessel transits are planned by Japan, but contracts have been let for the shipment of 4,500 tons of spent fuel to Europe and the subsequent return of reprocessed plutonium. As of July 1993, Japan had received one shipment of recovered plutonium, which arrived from France on a vessel newly built for the purpose (Lockwood, pers. comm., 1993).

Military Nuclear Disposal Problems

For all the study and controversy civilian nuclear practices have generated, few environmental issues have so riveted the attention of the international community in recent years as the military nuclear disposal practices of the former Cold War superpowers. In late 1991, revelations of a thirty-year history of secret radioactive dumping by the Soviet Navy and Ministry of Marine Transportation caused a tremendous stir of public alarm. Investigations and reports by the Russian government, the IAEA, other scientific teams, and Greenpeace subsequently disclosed a litany of violations of international agreements to which the former Soviet Union was a signatory, notably the 1972 London Dumping Convention and the 1983 voluntary moratorium by its signatory parties on low-level radioactive dumping. Solid and liquid medium- and low-level radioactive wastes from nuclear icebreakers and submarines had been routinely discharged at dozens of sites in the Barents Sea through at least 1991 (Sullivan 1992; Perelet 1993). In the Kara Sea, the island archipelago of Novaya Zemlya, for years the site of underground nuclear tests, had also been a dumpsite for contaminated naval reactors and debris irradiated as a result of operational accidents (Roginko and LaMourie 1992). In all, eighteen vessel reactors are known to have been dumped, seven of them still containing fuel, in the shallow bays off Novaya Zemlya at depths ranging from 12 to 135 meters and in the trough of Novaya Zemlya at a depth of 380 meters (Sjoeblom and Linsley 1993; see map 5.1).

Beyond the direct consequences for the arctic marine environment, the Soviets' longstanding policy of illegal dumping meant the abandonment of plans to build solid and liquid waste treatment and storage facilities and a neglect of radionuclide monitoring in the arctic seas for more than twenty-five years (Perelet 1993). The Russian government has conceded that the inadequacy of its facilities leaves no satisfactory alternative, at least in the near term, to continued ocean dumping of low-level liquid wastes. In the Russian Far East, where

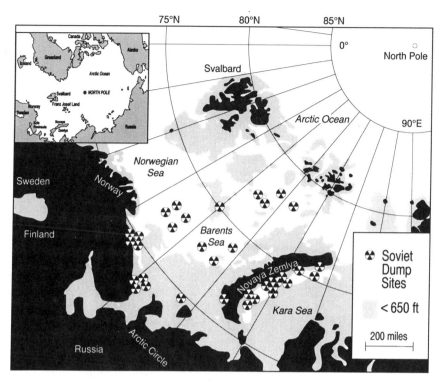

Map 5.1. Approximate locations of nuclear dump sites used by the Soviet and Russian governments between 1959 and 1992. Low-level liquid wastes were discharged primarily in the open Barents Sea. Materials dumped in the shallow coastal waters of Novaya Zemlya, at depths ranging from 12 to 135 m, and in the trough of Novaya Zemlya, at a depth of 380 m, include packaged solid wastes, nuclear vessel reactors with and without fuel, and a reactor shielding assembly. (Data source: Sjoeblom and Linsley 1993)

the shortage of facilities is most acute, tensions with Japan have flared over Russia's continued dumping of low-level liquid waste into the rich commercial fishing waters of the Sea of Japan (Sanger 1993). In October 1993, just weeks before the Sixteenth Consultative Meeting of the London Convention, at which the parties voted to replace the indefinite moratorium on low-level radioactive ocean dumping with a ban that is subject to review at least once every twenty-five years, Russia was forced to discharge the contents of a filled storage tanker in the Sea of Japan rather than risk an uncontrolled release of the low-level liquid wastes in its own shallow coastal waters. Citing this incident as confirmation that Russia's limited waste treatment and storage capacity constitutes a critical international problem, Vitaliy

Lystsov, Deputy Director of the Department of Ecological Security of the Russian Ministry of the Environment, also noted that the timing of the incident was unfortunate (although preferable, certainly, to an occurrence of this type soon *after* the Consultative Meeting), as were certain procedural aspects of the operation (Lystsov, pers. comm., 1993). The most important of these, Lystsov noted, were the failure to conform to the minimum depth standards recommended by the IAEA (the wastes having been discharged at a depth of approximately 3,300 meters rather than the recommended 4,000 meters or more) and the fact that the dumping occurred not in the open ocean but in a semi-enclosed sea (Lystsov, pers. comm., 1993).

The Soviets' failure to monitor radionuclide levels in the arctic seas, meanwhile, has meant a tremendous data collection burden and analytic challenge for the IAEA and others who are now working to assess the long-term effects of the dumping. Although a consensus emerged from a June 1993 international scientific conference that radioactive contamination in the Arctic and North Atlantic Oceans does not currently pose a regional-scale threat to human health or the environment, conference participants did recommend the continuation of international efforts "to assess and compare data and develop models for better understanding of the fate of radioactive material in the ocean" (WHOI Conference Statement 1993). They expressed particular concern over the potential for Arctic contamination by radioisotopes released from land-based weapons plants and storage facilities, as well as from the nuclear reactors of decommissioned submarines moored in shallow coastal waters (WHOI Conference Statement 1993; see table 5.4) Within Murmansk Harbor on the Kola Peninsula, home to the Russian Northern Fleet and the world's largest concentration of nuclear submarines, spent reactor fuel is held in floating storage barges for up to three years before being shipped by railcar to the reprocessing center at Chelyabinsk ("Soviets' Secret Nuclear Dumping" 1992). This center was the site of a 1957 explosion of stored radioactive waste (the Kishtim accident) that released some 2 million curies of radioactivity into the atmosphere (Protsenko, n.d.).

The concerns about Arctic contamination from land-based facilities also have disturbing implications for the United States, where military nuclear wastes have exceeded those generated by civilian industry since the late 1970s. Although some radioactive material has been permanently disposed of, a much larger quantity is under temporary storage pending the development and authorization of a long-term containment and disposal strategy. Recurring leakage and contamination problems with storage facilities at the Savannah River installation

Table 5.4
RISK SUMMARY FOR VARIOUS SOURCES OF RADIOACTIVE CONTAMINATION IN THE ARCTIC AND NORTH ATLANTIC

Impact	European Reprocessing Plants	Northeast Atlantic Dumpsites	Deep Point (1500 m)[a]	Shallow Arctic[b]	Shallow Other[c]	Wind-blown Dust[d]	Arctic Rivers	Potential Future Release[e]
Individual dose rate	0.1 mSv/yr	10^{-5} mSv/yr	0.001 mSv/yr	?	0.1 mSv/yr	Probably minor effects	?	?
Collective dose commitment	5×10^3 man Sv	4×10^4 man Sv (C-14)	10 to 100 man Sv (Cs-137)	100 man Sv + local	?		?	?
Local biota		Negligible	?	?	?		?	?
Confidence level	High	Adequate	Factor of 10	Factor of 10	Low			
Verification priority	Extensive studies complete and continuing	Low (in progress)	Depends on biota assessment	High	High?		High	High
Remediation	In progress (source reduction)	No	Unlikely, but depends on assessment	Evaluate options	Evaluate options		Evaluate options	Evaluate options

[a] Sites limited to sunken nuclear submarines and their nuclear weapons.
[b] Primarily waste disposal in Kara and Barents seas.
[c] Low-level wastes dumped in non-Arctic locations (e.g., off Massachusetts, Farallon Islands).
[d] Wind erosion and transport of soil materials.
[e] Primarily radioactive storage in land (civilian and military) or afloat in shallow waters (vessel reactors).

SOURCE: "Report of the Working Group on Risk Assessment and Remediation" (WHOI 1993).

and elsewhere are formidable obstacles to the development of politically acceptable, cost-effective management solutions. A 1977 estimate of the cost of preparing US military wastes for safe, permanent disposition was $20 billion. By 1990, the US Department of Energy (DOE) had a five-year plan for environmental restoration and waste management that called for expenditures of more than $30 billion from 1992 through 1996, but that was already widely viewed as "only the discovery phase of a program that could require hundreds of billions of dollars to complete" (OTA 1991).

The signing of weapons reduction treaties with the former Soviet Union and Russia has expanded DOE's task of managing nuclear materials to include a growing inventory of plutonium from dismantled weapons. A July 1993 notice of proposed changes in DOE's nuclear weapons program included mention of a new long-term storage site for plutonium, which was characterized in the notice as a "national asset." The proposal drew immediate fire from environmentalists and nuclear nonproliferation advocates, who argue that the material should instead be treated as nuclear waste (Associated Press 1993).

Decommissioning Military Assets

Although the new impetus to reduce force levels serves to lessen the threat of a nuclear showdown, at the same time it may introduce new threats to environmental security. The decommissioning, disassembly, and disposal of naval nuclear assets are particularly relevant to the marine environment, for several reasons.

First, both the Russian and US governments have proposed marine disposal of radioactive material generated in the decommissioning of nuclear weapons and submarines. (Nearly all of the submarines in the Russian nuclear fleet have two reactors, whereas US submarines have only one.) In 1984, the US Navy released a final environmental impact statement that outlined scenarios for scuttling disassembled submarines, including reactor components, irradiated bulkheads, and shielding material and piping. The analysis concluded that at-sea disposal options were more cost effective and posed fewer environmental and human risks than did land-based ones (US Navy 1984). Russian naval authorities have developed a program to dismantle nuclear submarines and retrieve spent fuel at Severodvinsk, the last remaining shipyard in the former Soviet Union designated for new nuclear construction (Kraus 1992). One proposal calls for radioactive waste materials to be shipped to underground chambers on Novaya Zemlya for permanent disposal, another to Chelyabinsk. Yet another plan advocates sending more irradiated reactor vessels and bulkheads to the bottom of shallow embayments on the archipelago.

Sheer numbers argue that there is an immediate need to manage the radioactive material generated in the decommissioning of nuclear ships and weapons. Since the commissioning of the first nuclear-powered vessel, the submarine USS *Nautilus*, in 1954, some 480 nuclear-powered ships (450 of them submarines) have been placed in service by the navies of the United States, the former Soviet Union, Britain, France, and China. Nearly 250 were built by the former Soviet Union and 190 by the United States during three decades of virtually constant expansion of both nations' nuclear fleets (Handler and Arkin 1990; Handler 1993a). In 1992, 450 naval nuclear reactors were in operation; by the end of the decade, however, the number of nuclear ships at sea is expected to decline by slightly more than half, to just over 200 (Handler 1993a).

Currently, the Russian Navy plans to decommission as many as one hundred submarines by the year 2000. Sixty of these are in the Pacific Fleet, where decommissionings began in 1987 and a total of thirty-five submarines were already inoperational as of 1992 (Handler 1993b). To minimize the danger of serious ecological harm if a decommissioned submarine were to sink at dockside, the Russians off-load fuel to service ships at the Chazma Bay and Bolshoi Kamen shipyards for transfer to a storage site at the tip of Shkovtovo peninsula, where it remains for up to two years, undergoing additional radioactive decay and cooling before being shipped to Chelyabinsk for final disposal. Since these procedures were introduced in 1987, however, lack of service ship capacity has been creating tremendous delays in the off-loading of reactor cores (Iakimets 1993). The scrapping of submarine hulls has proceeded even more slowly. Russian Navy officials have determined that current scrapping capacity and methods can process only 1.5 submarines per year, meaning that it would take thirty to forty years to complete the sixty decommissionings planned for the Pacific Fleet (Handler 1993b).

The United States has at least fifteen nuclear submarines in protective storage awaiting disassembly, and many others are approaching the end of their service lives. In the 1970s the US strategic fleet consisted of forty-one Polaris submarines, but by the end of the 1990s only eighteen ballistic missile submarines in the successor *Ohio* (Trident) class will remain operational (Handler 1993b). Based on data on the fleet sizes age profiles, and projected decommissioning plans of the US and other nuclear navies, it has been estimated that perhaps as many as fifteen nuclear submarines per year could be proposed for scrapping, with their radioactive waste materials subsequently disposed of in the marine environment (Davis and Van Dyke 1990).

The third critical environmental implication associated with military cutbacks and conversion arises from the need to ensure the security of seaborne shipment (or at-sea storage) of any nuclear material that might have economic or strategic value. Any increase in the volume of shipping or temporary storage of fissionable material contributes to an increase in the possibility of accident, theft, or sabotage, all three of which could have damaging consequences for the marine environment were radioactive contamination to result.

The Potential for Future Accidents and Unintended Releases

It is impossible to forecast such catastrophic events as a reactor mishap or the loss of a nuclear submarine that could release sizable levels of toxic radioactivity into the marine environment. Nor is it possible to glean convincing conclusions from published time-series data on civilian or military nuclear accidents. The main problem is the potentially large number of unreported or underreported events, evidenced in the way disclosures have been handled by US, NATO, and Soviet naval officials, private utilities, and even national regulatory agencies. Another critical obstacle involves the difficulty of assessing the long-term health and human safety consequences of nuclear accidents.

At best, future trends in contamination accidents and unintentional releases of radioactivity might be inferred by coupling information about the worldwide distribution of civilian and military nuclear capacity and existing modes of fuel production, testing, transport, and waste management with knowledge gained from policy analysis of the effectiveness of domestic and international regulatory regimes.[9] Analysis of these trends, in conjunction with research contributions about critical pathways, can yield insight into where the threats to environmental security are most serious.

NUCLEAR PROLIFERATION AND THE THREAT TO THE OCEANS

Perhaps the greatest environmental risk to the oceans from radioactivity is that posed by the transport of nuclear material at sea.[10] Twenty-two countries possess or control a worldwide estimated total of 1,000 metric tons of separated plutonium in various forms for use in both military and civilian applications (Perkovich 1993). The strategic value of plutonium could give rise to hijackings of ships transporting nuclear materials, as well as to efforts to recover fissionable material from sunken weapons and ships. In either case, the result could be an unintended release of radioactivity into the marine environment.

Two scenarios help illustrate the concern. The first involves the presence of an estimated fifty warheads, twenty-three vessel reactors, and an unknown number of unidentified nuclear devices on the bottom of the ocean (Arkin and Handler 1989b; Handler 1993a; Sjoeblom and Linsley 1993), some of which could contain intact fissionable material.[11] Many of these devices and vessels were military, and their fate and precise locations have been identified by authorities. In some instances, however, information related to the exact physical nature or final resting site of the nuclear material associated with these devices is either unknown or classified. Moreover, there is little public information about the status of long-term monitoring, if any, of these sites. These uncertainties contribute to the difficulty of assessing the strategic and environmental risks associated with the presence of nuclear devices on the seabed. The risks are considered minimal, but the presence of fissionable material could invite unauthorized recovery attempts, especially as commercially available marine technologies rapidly advance in capability.

Periodic shipments of spent nuclear fuel, plutonium, and enriched nuclear material by the Japanese provide a second illustration of how the transportation of nuclear material threatens both nonproliferation goals and international environmental security. US officials and environmental groups first took issue with the Japanese nuclear establishment in 1984, when an unescorted cargo vessel carrying 253 kilograms of reprocessed plutonium from France applied for a permit to transit the Panama Canal.[12,13] Passage was approved after provision was made for armed naval escort. The next transit of reprocessed plutonium from Japan to Europe was not until 1992. Japan's nuclear efforts have been heavily reliant on European technology and expertise, and the Japanese program to move toward future use of plutonium-fueled reactor systems has meant long-term arrangements with European reprocessors.[14] Japan's near-term domestic nuclear capabilities and its reprocessing contracts are such that far more plutonium than needed will be generated over the next decade unless some action is taken (Berkhout, Suzuki, and Walker 1990). Politically, meanwhile, Japan is legally bound under the terms of a 1988 bilateral agreement to get approval from the US government for any plan to transfer reprocessed plutonium from Europe.[15] Notwithstanding the concern of US policymakers over the threat of weapons proliferation and their recent proposal to eliminate all domestic reprocessing activity, Japan did receive a shipment of reprocessed plutonium from France in early 1993 (Lockwood, pers. comm., 1993).

PUBLIC PERCEPTIONS AND RESULTING COUNTERMEASURES

In the past decade, there have been numerous instances in which states or communities have perceived their environmental security to be threatened as a result of potential nuclear contamination of local marine areas. The resulting policy actions have run the gamut from barring a nuclear-powered cargo vessel from a Siberian harbor after a flurry of protest to a regional treaty declaring an area of high seas to be a nuclear-free zone on the ground that strict control measures were required for adequate protection of the marine resources and marine environment of the region.

Such risk-averting policy may be perfectly rational in the face of extreme uncertainty and potentially catastrophic consequences, especially when the risks are imposed by outside agents. Often, however, policy is driven by emotional responses largely uninformed by valid scientific research and uncoupled from relevant scientific knowledge. In those cases, the policy may do more harm than good by foreclosing opportunities for scientific learning and well-tuned adaptive responses, by undermining normal relations and norms of international law through unfounded unilateral action, and by provoking international countermeasures.

The first and perhaps longest-standing controversy over radioactive discharges into the marine environment involves the Sellafield nuclear power and reprocessing complex on the British coast at West Cumbria.[16] For many, Sellafield is a potent symbol of the environmental security threat to the oceans posed by the nuclear fuel cycle. Routine operational discharges as well as unintended releases of radioactivity at Sellafield date from the early 1950s. Levels of artificial radioactivity in the Irish Sea are among the highest in the world, although human dosage levels associated with the plant are within accepted limits as defined by the International Commission on Radiological Protection (ICRP). The historical record shows discharge levels that have steadily decreased from those of the 1950s, including marked decreases after 1986. At the same time, Sellafield has contributed measurable amounts of radionuclides into the marine environment, among them americium-241, cesium-134 and cesium-137, plutonium-239, ruthenium-106, and strontium-90 (Livingston and

Bowen 1977, 1982a, 1982b; Nies, Herrmann, and Goroney 1993). The primary sources of these radionuclides at Sellafield are nuclear fuel storage pondwater, plant washings, and reprocessing streams.

Public inquiries into the health and environmental effects associated with operations at Sellafield have become political fora for a variety of regional, environmental, and industrial groups. The transboundary nature of the controversy is illustrated by active roles in research and independent analysis by Irish, Scottish, and West German government agencies. The scientific data indicate that radioactivity attributable to the Sellafield discharges has indeed returned from the marine environment to human populations along various critical pathways. Although exposure levels have been well below recommended maximum limits,[17] recurring outcries about potential public health hazards have motivated a steady reduction in the level of routine operational discharges. Nevertheless, operations at Sellafield continue to expand, increasing the chances for unintended releases of radioactivity into the Irish Sea and its vicinity. Despite discharge reductions and measurable decreases in radioactivity levels in the water column, the public is not likely to forget the operational and public relations debacles that have beset Sellafield management. Public mistrust of British Nuclear Fuels Ltd. has become the norm where the company's use of nuclear technology intersects the marine environment.

Antinuclear sentiment also swept the island groups of the South Pacific in the 1980s, fueled by objections to French nuclear testing on Mururoa Atoll, a sense that the United States was undervaluing its historical ties with these fledgling states, and concern for the future viability of fishery stocks. In 1985, the Labour Government of New Zealand threatened the future of the Australia-New Zealand-US (ANZUS) defense pact when it banned US military overflights and restricted all naval vessels powered by or equipped with nuclear devices from making port calls in that country. Almost overnight, the United States found itself in conflict over nuclear passage issues with Vanuatu and the Solomon Islands. Also in 1985, the Greenpeace vessel *Rainbow Warrior* was sabotaged and sunk in Wellington Harbor, allegedly the target of a group backed by elements of the French intelligence service. The vessel was in the area to protest French nuclear testing on Mururoa Atoll and to collect data on radiation levels.

These and other incidents led to swift political action to restrict the presence of any nuclear device or radioactive substance throughout the South Pacific. Political entities from several of the island nations of the South Pacific quickly seized on the contamination issue to build a powerful regional forum that could extend its influence into other critical areas such as fishery regulation, resource licensing, and trade negotiation. In 1986, the South Pacific Forum adopted a treaty establishing a South Pacific Nuclear-free Zone from which nuclear-powered and nuclear-armed vessels are banned. The treaty was signed by Australia, New Zealand, and most of the Pacific island governments, and nations with nuclear navies were invited to sign as well. Only the Soviet Union and China obliged; the United States, France, and the United Kingdom all declined (Gardner 1988). The South Pacific nations justified strict control measures as necessary for adequate protection of marine resources and the marine environment, but the measures achieved little recognition among those states whose activities in the region would have the greatest potential impact. Failing such recognition, the regional agreement was an essentially symbolic gesture.

The United States did, however, sign the 1986 Convention for the Protection and Development of the Natural Resources and Environment of the South Pacific Region, an instrument of potentially far greater consequence for evolving issues in international law. Negotiated under the auspices of the South Pacific Regional Environmental Programme (SPREP), the convention extends jurisdiction over certain activities beyond the limits recognized in the United Nations Law of the Sea Convention. For example, it bans all nuclear waste dumping not only within the 200-mile EEZs of the Pacific island nations but also in areas of the high seas beyond that are "enclosed from all sides" by these countries' EEZs (Van Dyke 1988). The ban explicitly "includes emplacement 'into the seabed and subsoil of the Convention area of radioactive wastes or other radioactive matter'" (Article 10(1)). By the early 1990s, the South Pacific Forum had become an important force for strengthening coordinating mechanisms and expanding the scope of dialogue about security concepts within and beyond the region, as evidenced by the involvement of SPREP with the Forum's Regional Security Committee on issues of environmental security (UN Secretary General 1992).

Controversies surrounding the presence of nuclear-powered or nuclear-equipped vessels have arisen outside the South Pacific as well. In 1990, in the port city of Recife, a local judge, responding to a petition by environmentalists, ordered out of Brazilian waters a US naval submarine that had been engaged in maneuvers with Brazilian naval vessels ("Judge Orders U.S. Sub Out of Brazilian Waters" 1990). Questions about nuclear safety and access likewise complicated recent US naval base lease renewal negotiations with Spain and the Philippines.

In Russia, *Vodnyi Transport* reported in 1989 that public outcry led local officials to ban temporarily the *Sevmorput*, a nuclear-powered container vessel, from docking at several Arctic ports. Longshoremen in Magadan had refused to work the vessel for fear of contamination. Although the vessel had been built to operate in ice, its hull reportedly cracked during initial trials, a fact noted by groups opposed to its presence in the region.

Since the mid-1980s there has been halting movement, led by Iceland, toward the declaration of a Nordic nuclear-free zone. In May 1985 the Icelandic Parliament passed a resolution calling for a study to determine whether there was regional consensus on the declaration of such a zone. (This resolution was passed one month before Iceland ratified the Law of the Sea Convention.) Although the issue has been raised repeatedly since then in various international fora, especially within NATO, no such consensus has yet emerged. Despite considerable sympathy for the idea among the Nordic states, particularly Finland, agreement on the appropriate reach of such a zone has remained elusive. Iceland has maintained that it should include the Arctic Ocean (which many Icelanders know as the Northern Icelandic Ocean), while opponents of the idea have argued that it is unrealistic to propose a nuclear-free zone that extends all the way to the Kola Peninsula.

It is generally assumed that Iceland's chief concern is to protect its fishery, which accounts for 70 percent of export revenues ("Iceland and the Environment" 1991); counterbalancing unfavorable perceptions of Iceland as a whaling nation has been suggested as another important objective (Horgan, pers. comm., 1993). At the 1992 UN Conference on Environment and Development, Iceland garnered environmentally favorable press with its call for further restrictions on shipping by nuclear-powered vessels and an end to all ocean dumping of low-level radioactive material by 1993 (e.g., "Iceland Takes Initiative on Marine Conservation" 1992). By contrast, the government of Iceland has shunned publicity for a March 1993 Special Committee report prepared for the Ministry of Foreign Affairs, which includes the following passage concerning "Other Security Interests:"

> Pollution of the ocean and environmental accidents, not least those involving radioactivity but also toxic substances, wastes or spills from the mining of fossil fuels or other resources on the ocean floor, could cause immense damage to the foundations on which Icelandic industries are based. Protection of the ocean against any such danger, far beyond the 200 mile economic zone,

is an important aspect of safeguarding enduring security interests, which must be guaranteed by international agreements. ("Iceland's Security and Defense" 1993)

Elsewhere in the North Atlantic and its marginal seas, the ominous presence of a large concentration of nuclear submarines has prompted the governments of Norway and Sweden, at different times, to demand information from the Soviet and Russian governments about the status of disabled naval vessels that might be outfitted with nuclear weapons. Unauthorized intrusions into Swedish waters by nuclear submarines, for example, including the grounding in 1981 of a diesel-powered submarine allegedly carrying nuclear weapons, soured Swedish public opinion on regional arms control proposals and motivated bilateral negotiations aimed at establishing groundrules for naval operations in the Baltic and procedures for emergency response to nuclear accidents. Only in 1987, after the Chernobyl disaster, did those negotiations culminate in an agreement. Swedish dissatisfaction during continuing bilateral dialogue has been recorded as recently as 1992 (FBIS 1992b).

The widely publicized loss of the Soviet prototype nuclear submarine *Komsomolets* in the Norwegian Sea in April 1989 provoked tensions and prompted inquiries by the government of Norway about the potential for radioactive contamination. As information about the accident was made public, several alarming items emerged, including indications that the submarine's final resting place at a depth of about 1,700 meters had not been pinpointed for forty-five days, that its titanium hull had been breached shortly after impact, and that the reactor vessel contained 2 kilograms of plutonium-239 (Handler, Wickenheiser, and Arkin 1990). After initial silence on the matter, Soviet officials stated in January 1990 that they had found no sign of radioactive leakage at the site of the wreck. Less than a month later, however, a Canadian observer made public a videotape taken by the crew of the Soviet research submersible *Mir* which appeared to record positive readings of radiation from a pipe aboard *Komsomolets* (figs. 5.2 and 5.3) (Handler, Wickenheiser, and Arkin 1990; "Underwater Risks" 1990).

By 1993, a private Russian group seeking funds to raise the submarine was itself showcasing the videotape and spinning scenarios of plutonium leaking out of torpedo warheads and into the food chain ("Russians, U.S. Differ on Arctic Sub Threat" 1993), even as scientists gathered at an international conference on radioactive contamination in the oceans concluded that submerged nuclear submarine reactors

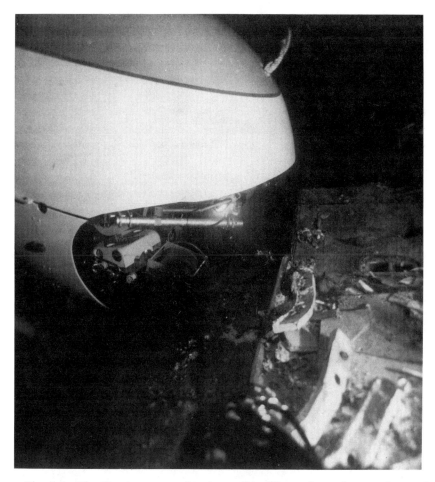

Fig. 5.2. The Russian research submersible *Mir* explores the wreckage of the attack submarine *Komsomolets*, which caught fire and sank in the central Norwegian Sea on April 7, 1989. In addition to the submarine's fueled nuclear reactor, two plutonium-tipped torpedoes were aboard and sank to a depth of roughly 1,700 m. (Courtesy of Anatoly Sagalevitch, Laboratory for Deep Manned Submersibles, P.P. Shirshov Institute of Oceanology, Russian Academy of Sciences)

and weapons did not pose any "significant present or future threat to human health or the environment" (WHOI Conference Statement 1993). According to one participating Russian scientist, in only one of three *Mir* expeditions to the site in 1991 did sensitive spectrographic equipment detect the presence of cesium-137, and in that case readings were taken in very close proximity to a ventilation window in the

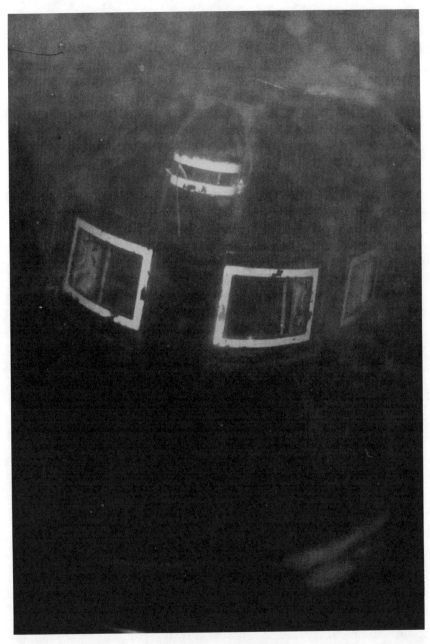

Fig. 5.3. The fore and sail of the sunken *Komsomolets*. (Courtesy of the P.P. Shirshov Institute of Oceanology, Russian Academy of Sciences)

submarine's reactor compartment (Nezhdanov 1993). A Norwegian scientist at the conference stated that "the enormous focus from the media" posed a far greater danger to the health of the fisheries in the region than does radiation from the *Komsomolets*, so long as the submarine remains at a depth where the water masses are effectively closed off from the upper layers by hydrographic conditions (Føyn 1993). Given "the enormous water masses available for dilution," Føyn explained, the waters immediately surrounding the *Komsomolets* can be expected to surface to biologically active layers only after about one hundred years and then in the southern part of the Atlantic Ocean and sufficiently diluted as to render their radioactive content "insignificant." The official Norwegian response to the revelations of Soviet nuclear dumping in the Barents and Kara seas has been somewhat more cautious, however. Prime Minister Gro Harlem Brundtland characterized the alleged dumping as a "security risk to people and to the natural biology of northern waters," a clear reference to the environmental security implications of the situation ("Soviets' Secret Nuclear Dumping" 1992). Yet official sanctions have been limited to the imposition of a ban on Russian nuclear icebreakers from picking up tourist passengers at Norwegian ports ("Soviets' Secret Nuclear Dumping" 1992).[18]

Clean-up planning is under way, but scant funds are available for this purpose to Russian environmental and naval ministries. Liability for the near-shore Soviet waste dumping environmental or health damages potential has scarcely begun to be discussed in a public forum. A beginning was made at the 1993 WHOI Conference, where discussion centered on the conceptual and practical difficulties of assigning liability in light of the acceptance of low-level dumping under the terms of the London Convention (Curtis 1993) and of regional agreements that recognize "most advanced technologies" as determining acceptable levels of low-level effluent discharges (Barsegov 1993).[19] Since then, however, the Annexes to the London Convention have been amended to include low-level radioactive wastes among the Annex I "black list" substances, with the proviso that every twenty-five years the contracting parties shall complete a scientific study on radioactive wastes and other radioactive matter and shall review the positioning of such materials on the Annex I list. At the same time, the LC amendments adopted at the Sixteenth Consultative Meeting in November 1993 call upon the contracting parties to "endeavor to co-operate in assisting countries with special problems relating to the safe disposal of radioactive wastes to meet effectively their international obligations" under the London Convention.[19a]

For the scientific community, the first step in the long process of compiling a complete inventory of the Arctic dumpsites has been to assess the overall profile of the dumping. In addition to sampling, data collection has entailed interviews with responsible officials and review of vessel logs and manifests, military records, and the accounts of witnesses. Building on these efforts, scientists are developing models that characterize the persistence of the radioactivity, the transport of radionuclides in the surrounding marine environment, and the movement of contamination into the marine food chain and, ultimately, to human populations.

Identifying and locating the responsible Soviet officials has been an especially difficult aspect of data collection because such sensitive decisions were typically entrusted to a very small circle of individuals whose involvement could not necessarily be deduced from their official titles and responsibilities (see "Ecocide in Russia" 1992). This circumstance is evident in the charter and work of Russia's Governmental Commission for Nuclear Waste Disposal in the Sea (the Yablokov Commission), which was established by President Boris Yeltsin in October 1992 to conduct both scientific research and an investigation of nuclear dumping activities by the former government. Although Yeltsin's decree specifically directed that "Ministries and Departments should provide the Commission with all available documents, materials and information, including materials of secret character," (Yablokov 1993) over a year later it remained difficult to determine the extent and status of any records the Soviet government may have kept.

The Uncoupling of Policy Responses from Scientific Data

Despite the lack of a complete scientific record on the input and effects of radioactivity in the oceans, the available data are adequate to form the basis for national, regional, and global technical control responses. Nonetheless, environmental policy formation in this sphere continues to be rooted in uncertainty and extreme public apprehension. The specter of widespread nuclear contamination of the oceans has become embedded in the public consciousness throughout many parts of the world. Anxiety about radioactivity in all of its forms far exceeds that associated with pollution threats that are far more serious, thanks to misinformation by both official and nonofficial sources,[20] inadequate understanding of scientific findings concerning quantities and risks, and heightened fear of threats that cannot be directly perceived by the senses. Anecdotal evidence strongly suggests that this apprehension transcends economic status, culture, and geography.

A good example can be found in the debate surrounding proposals to dispose of high-level radioactive waste in the ocean floor at geologically stable locations (Hollister, Anderson, and Heath 1981). For nearly twenty years, such proposals have been offered as alternatives to long-term land-based containment strategies. Available oceanographic, geological, and economic data support further investigation of this option. In the face of substantial public aversion to the idea of seabed burial of high-level radioactive wastes, however, researchers have found it difficult to fund the sophisticated theoretical work and optimization analyses necessary for an improved understanding of this disposal alternative relative to others that might place high-level wastes closer to home.[21] In contrast to other aspects of the marine radioactivity problem, this is one in which the scientific record will have to be augmented before it is sufficient to support informed policy making.[22]

The emotional impulse to uncouple policy decisions from scientific knowledge has recently found an outlet in the development and application of the *precautionary principle*. In its various expressions, this principle is intended to answer the question: What rule should we follow to make prudent regulatory choices in the face of scientific uncertainty? In prescribing that protective measures need not await full scientific certainty of harm, the principle does no more than state prevailing, sensible practice and gives no useful guidance on what measures, under what conditions (Bodansky 1991). In its more extreme version, requiring proof of safety before a regulated activity is permitted, the principle altogether voids any reasonable comparison of the expected benefits and costs or of the relative risks of alternatives.

On the question of marine radioactivity, the precautionary principle was invoked in Agenda 21 for the 1992 UN Conference on Environment and Development (Chapter 22, para. 22.5(b)), which called on states to

> encourage the London Dumping Convention to expedite work to complete studies on replacing the then-current voluntary moratorium on disposal of low-level radioactive wastes at sea by a ban, taking into account the precautionary approach, with a view to taking a well informed and timely decision on the issue.

Agenda 21 made no mention of the possibility of lifting the voluntary ban to allow for a resumption of low-level dumping or of simply extending the current moratorium (Curtis 1993).

Existing Legal Rights, Obligations, and Controls

To date, so-called general principles of international law that pertain to the environment are few,[23] but there are two that could be usefully applied to the question of radioactivity and the regulation of its discharge into the marine environment: (1) some environmental standard of duty or care related to use of the oceans and their resources imposed on all nations and (2) some measure of responsibility and liability imposed for damage resulting from use of the oceans and their resources. Regardless of specific treaty obligations, the claim could be made that all nations should be held to these principles, at a minimum, as they relate to the intentional or accidental release of radioactivity into the oceans.

Building upon and stemming from these basic principles are a number of international and regional regulatory frameworks that specifically address rights and obligations of signatory parties. Those pertaining specifically to marine radioactivity or nuclear contamination in the oceans are listed in table 5.5.

Table 5.5
INTERNATIONAL TREATIES AND PROTOCOLS WITH RELEVANCE TO MARINE RADIOACTIVITY

Convention on the Liability of Operators of Nuclear Vessels (1962)
Convention on Civil Liability for Nuclear Damage (1963)
Treaty on the Non-Proliferation of Nuclear Weapons (1968)
Convention on Civil Liability in the Field of Maritime Carriage of Nuclear Material (1972)
Convention for the Prevention of Marine Pollution by Dumping of Wastes and Other Matter (1972)
Convention for the Protection of the Mediterranean Sea against Pollution (1976)
Convention on Physical Protection of Nuclear Material (1980)
Convention on the Law of the Sea (1982)
South Pacific Nuclear Free Zone Treaty (1986)
Convention for the Protection and Development of the Natural Resources and Environment of the South Pacific Region (1986)
Convention for the Protection of the Marine Environment of the Baltic Sea Area (1992)
Convention for the Protection of the Marine Environment of the North-East Atlantic (1992)

The London Convention

The 1972 Convention on the Prevention of Marine Pollution by Dumping of Wastes and Other Matter was popularly known as the London Dumping Convention until November 1992, when the signatory parties changed its informal name to the London Convention. The LC sets forth a regulatory framework to control ocean dumping and disposal of a wide range of substances, including liquid and solid radioactive material. The genesis of the LDC can be traced to the UN Conference on Human Environment held in Stockholm in 1972. General principles of marine environmental protection as elucidated at Stockholm formed the basis for preparation of the LDC, which entered into force in 1975. From the beginning, Annex I of the LDC (the "black list") banned the dumping of high-level radioactive wastes, along with such substances as mercury and cadmium compounds, persistent plastics, and crude oil and other petroleum products.

Under the terms of the LDC, responsibility for management and regulation of issues associated with radioactive substances was assigned to the International Atomic Energy Agency. The IAEA was authorized (1) to define for purposes of setting regulations and restrictions on dumping those substances that should be classified as radioactive, (2) to provide guidance on the effects of and acceptable limits to ocean disposal of radioactive substances, (3) to monitor all dumping and disposal of all allowable radioactive substances at specified locations, and (4) to review periodically the global status of radioactive release and disposal activities and to make recommendations for future policy actions among the contracting parties of the LDC. In subsequent action by LDC signatory parties, the IAEA was authorized to conduct a baseline inventory of all sources of radioactive substances entering the marine environment. Other IAEA activities in this sphere have involved dosimetric risk assessment (in conjunction with the ICRP), the development of radioactive waste packaging requirements, and the support of oceanographic modeling and analysis related to radionuclides.

In 1983 and again in 1985, parties to the LDC agreed to a voluntary moratorium on all ocean dumping of low-level radioactive wastes. (Both times the United States voted against the moratorium and the Soviet Union abstained, with the latter asserting that it never had engaged in such dumping and had no plan to do so in the future [Yablokov 1993].) Before 1983, the United States, Japan, Korea, and several European countries conducted dumping operations at forty-seven sites in the Pacific and Atlantic Oceans; dumping operations by the

Soviets were also presumed to be occurring, although the location and true scale of those operations were not suspected.[24] The moratorium was originally proposed as a temporary measure pending better scientific data related to the effects of radioactive inputs on marine species. It represented a compromise between nonnuclear signatories who had sought a permanent cessation of all radioactive waste dumping and parties that had either engaged in dumping operations or wanted the option to initiate such operations in the future.

Since the moratorium was adopted, the LDC-chartered Intergovernmental Panel of Experts on Radioactive Waste Disposal (IGPRAD) has conducted several reviews of the scientific, technical, legal, and economic aspects of low-level radioactive waste dumping. Many of their conclusions have lent credence to the opinion, on scientific and technical grounds, that continuation of the moratorium is not necessary to protect human health or the marine environment adequately. Nonetheless, the moratorium was extended through the convening of the Sixteenth Consultative Meeting in November 1993, when the contracting parties voted to move low-level wastes from Annex II (which covers special dumping permits) to the Annex I list of banned substances. According to a Greenpeace analyst, "Ocean dumping of radioactive wastes has been the subject of decade-long critical review, yet the nuclear industry has failed to convince the vast majority of Parties to the LC that there are valid reasons to allow the resumption of this method of disposal" (Curtis 1993). Application of the London Convention to the *burial* of radioactive wastes in the seabed is not clear under the terms of the convention,[25] but the parties voted overwhelmingly in 1990 that, despite American and Soviet objections, such subseabed disposal was covered by the moratorium on low-level wastes (Curtis 1993). In 1993, when the moratorium was replaced by an initial twenty-five-year ban, subject to scientific study and review, the question of seabed disposal was once an open subject for consideration by the Contracting Parties.

The Law of the Sea Convention

The 1982 Law of the Sea Convention, which enters into force in November 1994, is increasingly guiding state practice on matters related to the marine environment. Many of its provisions have the effect of customary law, specifically those dealing with marine environmental protection, territorial principles, the conduct of marine scientific research, and freedom of navigation.

The Law of the Sea Convention is a so-called umbrella framework. In general, the convention addresses the issue of marine environmental protection by authorizing signatories to take measures to minimize and control pollution and dumping activities to the fullest extent possible. The tone and intent of the convention is that signatories establish rules, standards, and recommended practices on a regional basis to manage transboundary pollution threats, including threats posed by radioactive releases from vessels and land-based sources. Provisions of the convention that might pertain to radioactive releases are only broadly outlined, without specific reference to the control or regulation of marine radioactivity per se. Although dumping is not prohibited, the convention obliges signatory parties to conform to the regulatory requirements set forth in relevant international treaties such as the London Convention and the 1973 International Convention for the Prevention of Pollution from Ships and its 1978 Protocol (MARPOL 73/78, discussed at length in chapter 4). Specific protocols covering low-level radioactive waste dumping and other discharges of radioactive substances, found in some of the regional seas conventions developed under the auspices of the United Nations Environment Programme,[26] are an example of how regulatory practice related to this area has evolved directly from the Law of the Sea Convention.

Regionally Based Protocols and Regimes

On a supranational level, perhaps the most effective regulatory programs adopted to date for the management of radioactive contamination are a set of regional instruments that define specifically the types and sources of pollution to be controlled, as well as acceptable legal and technical measures for doing so. In three cases, the regional treaty contains a separate protocol addressing issues related to radioactive discharges into the marine environment.

In the mid-1970s, regional conventions aimed at protecting the Mediterranean and Baltic seas were adopted and entered into force in 1978 and 1980, respectively. The Protocol for the Prevention of Pollution of the Mediterranean Sea by Dumping from Ships and Aircraft was adopted in Barcelona in 1976. It expressly prohibits the dumping of "high- and medium- and low-level radioactive waste or other high- and medium- and low-level radioactive matter to be defined by the IAEA" into the waters of the Mediterranean. The 1974 Convention on the Protection of the Marine Environment of the Baltic Sea Area and its 1992 successor (both known as the Helsinki Convention) likewise prohibit the discharge of radioactive waste. So, too, does the 1992 Convention for the Protection of the Marine Environment of the North-East Atlantic (the Paris Convention), which combines and supercedes the 1972 Convention for the Prevention of Marine Pollution by Dumping from Ships and Aircraft (Oslo Convention) and the 1974 Convention for the Prevention of Marine Pollution from Land-based Sources in the North-East Atlantic (also known as the Paris Convention).

The South Pacific is likewise subject to stringent antidumping provisions. Article 10 of the Regional Convention for the Protection of the Natural Resources and Environment of the South Pacific Region obliges parties to prohibit dumping and seabed disposal of radioactive substances in the region. In 1986, signatories to the South Pacific Nuclear Free Zone Treaty agreed "not to dump radioactive waste and other radioactive matter at sea" within the geographic area defined in the terms of the treaty. In 1989, the South Pacific Permanent Commission adopted a supplementary protocol to the Convention for the Protection of the Marine Environment and Coastal Areas of the South-East Pacific (the Lima Convention), which specifically prohibited all disposal or release of radioactive matter on and in the seabed. The prohibition applied to national EEZs as well as to areas of the continental shelf outside the national jurisdictions of commission member states.

The Salvage Convention

The 1989 International Convention on Salvage (Salvage Convention) governs assistance to vessels in danger. Its aims are to preserve human life and property at sea and to protect the marine environment. Among the motivations for the convention was widespread recognition of the inadequacy of traditional salvage practices to provide financial incentives for actions taken by salvors to secure the protection of the oceans.[27] (These are discussed more fully in the context of hazardous materials transport in chapter 4.) Previously, because environmental protection held no special value for the owner of the vessel or the cargo, compensation was not available for salvors whose efforts served to minimize environmental damage.[28]

The relevance of the Salvage Convention to threats resulting from radioactivity in the oceans is narrow but significant. It involves the salvage, either authorized or unauthorized, of nuclear devices that rest on the seabed. Many types of salvage activity can pose a risk to the marine environment, but these risks are multiplied in the case where attempts are made to recover devices or wrecks containing nuclear material.

CONCLUSIONS

The physical and chemical characteristics of radionuclides in the marine environment, as determined through scientific observation, analysis, and interpretation, can have profound international legal and political significance.[29] Unfortunately, the scientific record on the input, pathways, and long-term effects of artificial radionuclides in the oceans is incomplete. Still, the oceanographic and radiological communities have made significant contributions to understanding the transport and persistence of certain radioactive elements in seawater, particularly those associated with the dumping of low-level waste products and effluent from power generation and reprocessing facilities.

In terms of the environmental security risks to the health of the oceans, the presence of radioactivity in the marine environment probably represents a markedly lesser threat than do crude-oil shipping, overfishing, and land-based sources of pollution. This does not obviate the need for satisfactory national, regional, and global control measures, but rather provides an opportunity to construct sensible and effective frameworks before widespread contamination could occur. As for threats to human health, public attitudes about the presence of radionuclides in the oceans seem to be out of synch with scientific data, which suggest that concentrations of marine radioactivity are of minor consequence compared to other sources of radioactivity that are land based or naturally occurring. For marine biota, the scientific record is inconclusive. Nonetheless, at present concentrations, including those attributable to unintended releases and marine accidents, existing scientific evidence does not indicate significant threats to living marine populations. ("Report of the Working Group on Risk Assessment and Remediation" 1993).

Setting Technical and Safety Standards

Perhaps the most important means of reducing public misinterpretation of the environmental threats posed by marine radioactivity is to improve the way in which national governments and the international community apply scientific knowledge in a thorough evaluation of various policy options related to discharge regulation, radiological protection, long-term waste management, and the use of nuclear material in general. A critical part of this process involves the development, adoption, and universal application of appropriate technical standards. In a marine setting, standards should be applied to radioactive substance classification, individual and collective dosage limits, shipment and packaging procedures, and environmental assessment methods.

Since the first Conference on the Law of the Sea recommendation in 1958 that IAEA assume responsibility for managing the problem of radioactive contamination of the oceans on a global scale, much of that agency's efforts have been aimed at determining and publishing acceptable levels of radioactivity in the marine environment, both from individual sources and cumulatively. These efforts represent an important contribution to the management of radioactive dumping and disposal activities, but discharge levels have not been compared with actual observed concentrations.[30] Rather, "best estimate values" for various radioactive inputs are based on data provided by numerous sources of varying reliability that are periodically revised by IAEA to reflect ongoing data evaluation efforts and improved calculation techniques. Greater emphasis on measuring marine radioactivity in critical locations (in harbors or near sensitive ecosystems, upwelling areas, regions of high productivity, etc.) and improvements in overall data reporting would add substantially to public awareness of how actual concentrations of radioactivity at specific sites compare with national and internationally acceptable standards. Obviously, it is important not only to refine continuously the store of quantitative knowledge about past radioactive sources and inputs, but also to track ongoing releases with greater precision.

There has been much progress as well in quantifying the radioactive properties of various substances for purposes of definition and in technical standard-setting for individual and collective dosage levels associated with exposure to radioactivity from oceanic sources. Maximum dosage limits for humans' exposure to radiation have been developed and refined by the ICRP, through mathematical modeling of radionuclide dispersal and exposure pathways, and have been offered as the basis for regulatory work to limit or otherwise manage future releases of radioactive substances into the oceans (see Bewers and Garrett 1987).[31] The ICRP's ongoing efforts at calculating collective dose levels,[32] which seek to estimate the net exposure effect from all sources of marine radioactivity, should be supported at the national level. Although estimates of collective limits are not intended to be used as absolute indicators of actual harm from a specific practice (e.g., waste dumping), they are of great utility in the optimization process (Calmet and Bewers 1991). Additional research in this area is of critical importance because, in general, hazardous effects tend to become concentrated as radioactive contamination moves up the marine food chain and because available data do show that radioactivity can have widely varying effects on the health of different species.

Radioactivity in the Oceans 159

The setting of technical standards also involves coordinating the terms of national control regimes for managing land-based release of radioactive wastes and the physical transport of nuclear substances with the corresponding terms of multilateral regimes. In some cases, disparities in standards constitute substantive differences between national regimes, as evidenced by the land-based effluent discharge levels associated with coastal reprocessing facilities in the United Kingdom and France, which show wide variations in the levels of effluent allowed, the criteria used to determine acceptable discharge levels to begin with, and the overall responsiveness of regulatory authorities in enforcing compliance.

A uniform Code of Practice covering the transborder movement of radioactive wastes was adopted at the 34th Session of the IAEA General Conference in 1990. The code defines a number of guiding principles for signatories in the areas of export notification, transportation, and disposal of wastes. The code can be viewed as only a first step toward the development of an internationally acceptable transport regime, especially since a much broader legal instrument covering transboundary movements of hazardous waste was adopted less than a year earlier. In particular, much work remains on the development of appropriate packaging standards for seaborne nuclear cargoes.[33]

Monitoring, Reporting, and Information Exchange Responsibilities

The terms of the London Convention do not oblige signatories to report data on radioactive sources and inputs (although the IAEA recommends that they do so). Stricter reporting requirements would allow more reliable assessments of the quality of numerical models and would help policymakers determine the key parameters around which disposal or discharge operations should be built.

The foundation for accurate information reporting and data exchange is the monitoring of radioactivity in the marine environment. Effective monitoring, whether from vessels or land-based sources, requires monitoring both at the source location and in the surrounding environment (GESAMP 1988). Moreover, measurements must be capable of capturing and sorting the distinguishing properties and effects of each radionuclide.

Unlike the relevant articles of the Law of the Sea Convention (Articles 204 to 206), the regional protocols developed as control regimes for marine pollution have not emphasized the importance of continuous monitoring. They do advocate monitoring programs, but generally only for purposes of determining baseline or current levels of contamination (Gjerde 1992). Compliance monitoring, as is called for in the 1992 Helsinki Convention in the context of land-based point-source emissions, should be considered as a means of strengthening national and supranational control regimes concerning marine discharges of radioactive effluents (Gjerde 1992; see chapter 8).

Despite the reputation of Sellafield as the nuclear industry's laundry,[34] the facility's operational history is significant for the fact that long-term near- and far-field monitoring has been as comprehensive as at any nuclear facility in the world.[35] The scientific knowledge gained through twenty-five years of research on the radionuclides released into the Irish Sea has been invaluable in determining the physical, chemical, and biological properties of marine radioactivity. Equally important have been the insights into the design of the monitoring programs and the analytical techniques developed by Sellafield researchers (Pentreath and Lovett 1976; Livingston and Bowen 1977), which could be adopted for use in other coastal settings where radioactive effluent discharges are to be monitored.[36]

IAEA efforts to develop comprehensive oceanic inventories of sources and input levels should be strengthened. Signatories to the London Convention could improve their responsiveness to IAEA requests for data, and IAEA, for its part, should provide guidance and incentives to signatories that would promote greater homogeneity in data formats and recordkeeping. The recent cooperation of the Russian government and the intensity of international cooperative efforts to understand the nature and scale of the Soviet dumping and its long-term consequences signal welcome advances in these areas. They should not be allowed to obscure a fundamental weakness that plagues many of the international and regional frameworks established to protect the marine environment, however, namely that the Law of the Sea Convention looks to responsible international authorities to set reporting requirements and to disseminate data and research findings, yet few such authorities have the necessary institutional structure, funding, or legal mandate to carry out these functions (see Gjerde 1992).

Another shortcoming is the fact that the IAEA inventory does not extend to stockpiles of naval nuclear weapons or nuclear-powered vessels. History suggests that it is unrealistic to expect the release of strategic information. Recently, however, the United States and Russia have signaled a greater willingness to discuss openly the future status of their nuclear arsenals, which may open the prospect of a database on naval nuclear sources of potential radioactivity. This might entail verification procedures, for which a framework would have to be developed. Provisional models are available in the verification provisions set forth in the SALT II and START bilateral arms control treaties (Ausubel and Victor 1992).

Cooperative Multilateral Action

Cooperative action to improve the existing patchwork of national, regional, and global control regimes aimed at marine radioactivity encompasses several areas: efforts to harmonize rules and laws between and among national jurisdictions and at the international level; the development of multilateral recovery, clean-up, and enforcement programs; and the establishment of liability and compensation frameworks covering environmental and health damage caused by radioactive contamination.

As the Soviet/Russian case makes clear, the question of liability for environmental damage caused as a result of a nuclear accident or radioactive discharges in violation of international standards is presently beyond the reach of the international legal framework. In part, this is attributable to the lack of universality of the Law of the Sea Convention. In addition, national legal systems have not proven, until recently, to be technically suited to the complex cause-and-effect relationships that are often associated with long-term health effects and that form the basis of damage claims in product liability and toxic tort law (Van Dyke 1988).

As in the context of hazardous wastes, many legal scholars have advocated the application of strict liability provisions to damage situations where adverse health effects can be traced directly to radioactive contamination. Given that commerce involving nuclear material continues to be limited and strictly regulated, the application of strict liability standards might be practical from the standpoint of both commercial indemnification and international law. To be meaningful, such a legal standard would require wider applicability of the Law of the Sea Convention, stricter enforcement mechanisms, and supplemental programs through which funding pools for long-term health monitoring or damage minimization efforts might be underwritten by the economic beneficiaries of nuclear material or operations (see Van Dyke 1988).

From a liability standpoint, the remaining hurdle would then be to establish standards for allowable releases and requirements for compliance in the course of military operations. Before military operations can be brought under the umbrella of national and supranational environmental protection rules, however, a balance must be secured in international law between freedom of navigation and the rights of coastal states to protect their marine environments. Until that is accomplished, some degree of environmental security risk will continue to threaten the marine environment, particularly in areas such as the Arctic, the Mediterranean, and naval homeports. Moreover, public anxiety and misunderstanding about the risks posed by radioactivity in the oceans will probably persist despite further gains in scientific understanding and improvements in the civilian management of radioactive wastes.

CHAPTER 6

Environmental Protection for the Arctic Ocean

LONG DISMISSED AS A frozen wasteland of interest only to a handful of explorers, traders, missionaries, scientists, and indigenous peoples, the Arctic has emerged in recent years as an international region whose military, economic, and environmental importance rivals that of the world's other major regions (Young 1989) (see map 6.1). In part this is a consequence of increased human activity within the region, especially the launching of projects to extract vast quantities of nonrenewable resources. It also reflects the growth of linkages between arctic phenomena and human activities centered in the mid-latitudes—in environmental terms, the effects of airborne and waterborne pollutants that originate elsewhere but find their way into the Arctic via long-range transport mechanisms (Griffiths and Young 1990). Because of the operation of prevailing sea currents, the Arctic Ocean acts as a sink for a wide range of industrial pollutants, including, among others, heavy metals, toxic substances, and PCBs (see Soroos 1992).

The notion of environmental security is particularly relevant to the Arctic, for several reasons. The first concerns the fragility of northern ecosystems and their extreme vulnerability to any human disturbance. Second, the area has a profound influence upon global (or at least hemispheric) environmental processes, such as atmospheric and oceanic circulation, global warming, and ozone layer depletion.[1] Finally, environmental factors are closely linked to longstanding, but now changing, strategic military objectives in the Arctic.

ENVIRONMENTAL VULNERABILITIES IN THE ARCTIC BASIN

Recent years have brought sharp increases in arctic population, in the number of population centers and settlements, and in the rate of nat-

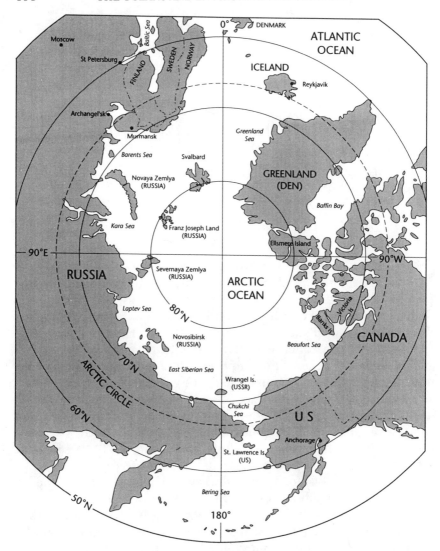

Map 6.1. Polar projection of the Arctic.

ural resource development. Expected discoveries of offshore oil and gas in the Arctic and the subsequent exploitation and shipment of these commodities will further threaten the environment. The continuing presence of military and naval forces in the region poses another set of risks.

For example, the testing, transport, storage, and decommissioning of weapons of mass destruction have had serious environmental consequences. Atmospheric and underground nuclear tests, conducted by the Soviet Union on Novaya Zemlya, have caused extensive radioactive contamination on the Chukotka Peninsula. The result is a cancer mortality rate that is 2 to 10 times higher among native Chukchi than the world average (MacKenzie 1989). Military maneuvers, treetop aircraft overflights, and other "normal" military activities disturb fish spawning grounds, reindeer pastures and herds, and other components of arctic ecosystems, including humans. The probability of accidents (collisions, groundings, etc.) involving nuclear-powered or nuclear-armed vessels is particularly high in the Arctic, as certain areas of the Arctic Ocean and its adjacent seas (the Barents Sea, the Greenland-Iceland-United Kingdom [GIUK] gap, etc.) are crowded with nuclear submarines (Heininen 1990). The dangers were dramatically illustrated by the sinking of a Soviet nuclear-powered submarine in the Norwegian Sea in April 1989 and by the collision of a US and a Russian nuclear submarine on patrol in the Barents in March 1992 (FBIS 1992b). Most recently, revelations of a thirty-year history of the secret dumping of radioactive waste by the Soviet Navy and Ministry of Marine Transportation have stirred tremendous concern over nuclear contamination of the Barents and Kara seas (see chapter 5).

Escalating human pressure on arctic marine ecosystems has caused a gradual deterioration of environmental quality in the Arctic Ocean. Data are scarce, but there is ample evidence of elevated levels of certain volatile chlorinated organic contaminants (i.e., hexachlorobenzene, toxaphene) and heavy metals (particularly cadmium) in arctic biota.[2]

Oil pollution levels, especially in the coastal and estuarine zones of Russian arctic seas, have been rising, with local concentrations (i.e., in the Ob estuary) exceeding national standards by as much as sixfold ("Sostoyanie prirodnoi sredy i prirodookhrannaia deyatelnost v SSR v 1989" 1990). Despite this scientific evidence, the open seas of the Arctic Ocean, including snow and ice cover, still remain the least polluted of all the world's oceans.[3] Scientists are therefore able to use some of its zones (along with others in the Antarctic) as baseline sites for global monitoring of environmental quality. Nevertheless, a single major oil tanker accident or a blowout of one oil platform could drastically change the environmental situation in the entire Arctic Basin.[4]

The high latitudinal position of the Arctic Ocean determines specific features of its abiotic environments (biotopes). These features include low water temperatures, the presence of ice cover throughout most of the year, and sharp contrasts in the amounts of solar radiation received during the polar day and night. The harsh climatic conditions that form "tough" biotopes also determine specific features of polar biocenoses: shorter food chains, poor species composition, dependence on ties with other oceans, and pronounced fluctuations in the production of organic matter. Life in the high Arctic often approaches the limit of possibility. Even slight pollution of arctic waters can be extremely serious, in part because of the slowness of natural processes that determine the assimilation of contaminants into the marine environment (MacKenzie 1989).

In addition to having direct toxic effects on marine biota, pollution of water or snow/ice cover—especially pollution by oil—causes significant changes in ice structure and its chemical and physical properties and can entail a number of serious environmental consequences. Surface oil films affect heat, moisture, and gas exchange between ocean and atmosphere and thus reduce the rate of evaporation of seawater by approximately 50 percent and the rate of ice growth by roughly 25–50 percent. Oil films also reduce the albedo of ice and snow by 10–35 percent, which results in more rapid melting. In spring and summer, ice and snow melt 5 to 10 times faster in polluted areas than in unpolluted ones (Izmaylov 1986). In addition, microbes capable of oil decomposition occur in the Arctic in such small quantities as to be insignificant in large-scale oil spills.

The ice and snow cover of the arctic seas has a high absorptive capacity with respect to many pollutants. This absorptive capability may contribute to the purification of ocean areas, but it may also be a factor promoting the long-distance transfer of pollutants. The general pattern of water and ice circulation in the Arctic Ocean is such that the path of pollutants from most of the likely pollution sources (e.g., the Barents Sea and the continental shelf and coast off Alaska and Canada) runs across almost the entire Arctic Basin. At the low temperatures characteristic of arctic waters, destruction of most pollutants is slow; they can continue to move from the ice or snow into water throughout the entire period during which the ice and snow drift across the ocean. Any pollution of the aquatic environment in the Arctic could, therefore, have a much more prolonged and widespread effect than would be the case with pollution in lower latitudes, possibly resulting in serious disruption of the environmental balance across vast areas. For these reasons, no arctic rim state can be secure against the effect of such pollution, even when the pollution source is a thousand miles away from its coasts.

CURRENT ECONOMIC DEVELOPMENT ACTIVITIES IN THE ARCTIC

Extraction of hydrocarbons, minerals, and living species in the Arctic Basin represents both the basis upon which the region's economic development will be built and the greatest threat to its environmental security. Management of the environmental risks associated with arctic development has been difficult historically, for several reasons. Many of these are related to what Osherenko and Young (1989) have described as *internal colonialism*, the inability of arctic peripheries to influence key decisions regarding their economic and political destinies. Hard evidence suggests that environmental protection efforts in the region by both the United States and the former Soviet Union have not been adequate to address the special requirements of this severe environment. Many commentators have begun to acknowledge publicly, for example, that years of Soviet indifference to reckless, aggressive exploitation of the northern environment by the industrial ministries have resulted in a deep conflict between the economic interests of industrial civilization and the functioning of arctic ecosystems at critical levels (Vartanov and Roginko 1990). Russian estimates in 1989–90 of the total costs of environmental damage as a result of hydrocarbon development in the Tyumen region alone were 450 billion rubles (at the time, in the range of $80–$120 billion).[5] In the U.S., an assessment of the Alyeska consortium's pollution response program in the wake of the *Exxon Valdez* disaster revealed a wide range of deficiencies in planning, equipment deployment, and program administration (State of Alaska Office of the Governor 1989).

Hydrocarbon Resource Development

Reasonable estimates of recoverable offshore oil reserves in the Arctic Basin are between 100 and 200 billion barrels (Young 1986). An estimated 2,000 trillion cubic feet of natural gas deposits also exist in the region. Approximately 25–50 percent of these total crude oil reserves may be in North American onshore and offshore areas, with most of the rest thought to be in Russian arctic regions.

For the North American Arctic as a whole, both off- and onshore, it has been estimated that perhaps 50 billion barrels or more of recoverable oil exist, located primarily in the Beaufort Basin, the North and South Chukchi Sea, the Norton Basin, the Mackenzie Delta, and the Bent Horn fields (see table 6.1). North American producers have taken a phased approach to hydrocarbon development in the Arctic, gaining the requisite technical capabilities in nearshore and coastal areas before moving further offshore, where design engineering and daily operations will be much more costly (particularly in the outer continental shelf (OCS).[6]

Although exploration activity throughout the Gulf of Alaska and the Beaufort and Bering Seas has increased during the past decade,

Table 6.1
MARGINAL PROBABILITIES OF COMMERCIAL HYDROCARBON RESOURCES BY OCS PLANNING AREA

OCS Planning Area	Marginal Probability
Beaufort Sea	0.70
Navarin Basin	0.27
St. George Basin	0.22
North Aleutian Basin	0.20
Chukchi Sea	0.20
Norton Basin	0.15
Gulf of Alaska	0.08
Kodiak	0.05
Cook Inlet	0.03
Shumagin Basin	0.03
Hope Basin	0.02

SOURCE: Minerals Management Service 1990.

the only actively producing offshore oil and gas fields in Alaska are found in lower Cook Inlet. Were large fields to be found in Alaska, they would probably be in the National Petroleum Reserve between Prudhoe Bay and the Chukchi Sea or in the coastal plain of the Arctic National Wildlife Refuge (ANWR) between Prudhoe Bay and the Canadian border. In Prudhoe Bay, perhaps as much as 9–10 billion barrels of recoverable oil reserves may exist, as well as 26 trillion cubic feet of natural gas reserves. In the Canadian Arctic, production activities are concentrated on Cameron Island,[7] with exploration primarily in the Beaufort Sea–MacKenzie Delta region. Some exploratory activity has also been undertaken to the east, in Baffin Bay and the Labrador Sea.

It is difficult to project whether fluctuations in world crude oil prices will make increased arctic offshore exploration (and eventual production) an economically feasible option for producers anytime soon. In any case, concern over national energy dependence in the wake of the Persian Gulf war led US policymakers to revisit proposals advocating increased domestic production, with much of the potential reserves thought to be located in offshore arctic regions. In February 1991, President Bush unveiled a new legislative plan for a national energy strategy (NES) that would rely on market-based incentives and deregulation to spur domestic production. Included in the NES was a provision that called for opening the Arctic National Wildlife Refuge to exploratory drilling. The 1.5-million acre coastal plain of the ANWR, bounded by the Canning River Delta and the Canadian border, is estimated to hold as much as 9.2 billion recoverable barrels of oil.[8]

ANWR-related provisions of the NES renewed a contentious debate among federal officials, the petroleum industry, environmental interests, and Alaskans over drilling in one of the largest tracts of pristine land in North America and an important summering site for the porcupine caribou herd. At present, development is prohibited by Congress under the terms of the Alaska National Interest Lands Conservation Act (ANILCA). Oil firms have reported that extensive operations in the coastal plain will require the full utilization of 5,000 to 7,000 acres of land. In its conditional scenario for exploration and development, the Minerals Management Service (MMS) has projected the modification of approximately 5,120 hectares of land, or 0.8 percent of the total coastal plain area (Minerals Management Service 1990).

Oil fields in central Asia and western Siberia have reached their production peaks, according to Russian experts. Many of these are older fields, which require more wells and deeper drilling and therefore pose greater risk of ecological damage (Kvint 1990). Partly in recognition of this fact, new fields are being sought in offshore arctic areas, eastern Siberia, and the Pacific coastal regions of the Russian Far East. Because petroleum exports represent Russia's main source of hard currency, accelerating exploration and development of arctic hydrocarbon reserves has become a priority for ministry officials.[9]

The geology is most promising for the development of three Russian oil and gas basins: the Barents Sea, where development activities are under way in several fields;[10] the Kara Sea, where one of the world's largest gas discoveries was made recently ("Major Hydrocarbon Structures" 1993); and the Northeast Siberian arctic shelf, where extensive geologic and seismic surveys have yet to be undertaken (Clark 1991). Also promising are the Sea of Okhotsk and formations near the Pechora River region (Kvint 1990), although a project to develop petroleum reserves on Sakhalin Island in the Sea of Okhotsk has run up against public concern over the recent acknowledgment of a 1987 incident in which a nuclear reactor containing 6 kilograms of strontium-90 was jettisoned in the area by a Soviet military helicopter in distress ("Search for nuclear wastes may hold up Sakhalin program" 1993). Geologists discovered tremendous potential between the eastern shore of the Yenisey River and the Lena, especially through an oil belt dividing the Taymyr peninsula from the Siberian mainland (Kvint 1990). Estimates of recoverable oil resources in the Chukchi and East Siberian Seas are 1.5 billion barrels and 2 billion barrels, respectively, and natural gas reserves for those basins have been estimated at 18 trillion cubic feet and 40 trillion cubic feet, respectively (Kvint 1990). Moderate levels of production, approximately 20 million barrels per year, are being achieved in east Siberia (Young 1986).

Arctic Commercial Shipping

For centuries, the Russian maritime Arctic has been used as a commercial link between Europe and the Far East (Brigham 1991). At present, however, roughly 95 percent of routine ship traffic across the Arctic Basin services Siberian river ports and resource extraction enterprises in Russia, moving supplies and equipment in and commodities out (Granberg 1992). Harsh natural conditions, technological constraints, strategic concerns, and market considerations have served to limit seaborne commercial transportation across what has become known as the Northern Sea Route (NSR) or Northeast Passage. Waterborne cargo volumes crossing the NSR in recent years have been estimated at 5–6 million tons annually.[11]

Many of these constraints are disappearing, however, as Russia acquires the full suite of scientific and technical resources necessary for maintaining year-round shipping service across the entire basin. Coordinated Russian efforts to attract dry bulk cargoes to a commercial NSR service connecting ports in Europe and the Far East included several test voyages during the 1989 and 1990 seasons. These were conducted under the auspices of the Soviet Ministry of Marine Transportation and the Murmansk Shipping Company, using the formidable icebreaking assets of the former Soviet fleet. Burgeoning oil consumption rates in Asia suggest that Japanese or Korean interests might be willing to underwrite a significant portion of future arctic offshore exploration activities and a regional shipping link for the transport of crude oil from offshore production fields to Asian refineries.[12] This would obviously boost the commercial viability of an NSR service. Japanese research and development of an ice-strengthened tanker capable of making such a voyage has been ongoing for over a decade (Østreng 1991). The location of potential oil fields in the Russian Far East—off Sakhalin Island and the Kamchatka Peninsula—might also make crude oil delivery to Alaska (Port Valdez) economically feasible. Some commentators have openly discussed the idea of Russian oil flowing through the Trans-Alaska Pipeline System (TAPS) (Kvint 1990).

Despite these developments, there is little evidence that a Russian-managed NSR service could attract large volumes of freight traffic, notwithstanding the favorable transit times. Even small increases in ship traffic along the NSR will heighten the risk of environmental degradation for large areas of the Arctic. At present, there are no waste oil reception facilities in place at Russian arctic ports, nor have adequate oil spill contingency plans been established in the event of an accidental release along the NSR (Latukhov 1990; Hansson 1992; *International Challenges* 1992).[13]

For the North American Arctic, predicting the transportation modes associated with future offshore oil development is equally speculative. The use of subsea pipelines, transshipment terminals, and icebreaking tankers has been proposed, both independently and in conjunction with inland shipment through TAPS (State of Alaska Office of the Governor 1989; OTA 1985). As for natural gas, no transport system is presently available. Proposals include extension of an existing gas pipeline from Alberta north to Port Valdez, as well as the possible use of liquified natural gas (LNG) tankers, a technology that requires state-of-the-art ship design and safety protection measures (State of Alaska Office of the Governor 1989). The volume of Cook Inlet–produced oil currently exported to markets in Asia is restricted by law to approximately two tanker trips per year (3,000 barrels per day). Were these restrictions removed by Congress, it has been estimated that an average of twelve additional tanker trips would be made to the Far East per year (State of Alaska Office of the Governor 1989).

Mineral Extraction

Most mineral extraction activities in the arctic regions of North America and Russia are land based. Immense coal, ore, and gold deposits are mined extensively in Siberia and other frontier areas of Russia, and metal-processing complexes are situated along the northflowing Lena and Yenisey rivers. Substantially less mining occurs in Alaska, the MacKenzie River Delta, and other parts of the Canadian Arctic. Direct environmental effects in offshore and coastal areas have been measurably small, but the level of threat associated with land-based sources needs to be further analyzed (Harders 1987). The logistics of staging equipment and material for inland mineral exploration poses environmental risks to offshore areas from ship traffic, cargo handling, and even inland transportation (from road construction and runoff, for example). Likewise, the environmental effects of large-scale development in offshore arctic regions—construction of roads and powerlines, ports and pipelines; increases in ship traffic and helicopter overflights; the disposal of muds, cuttings, and tailings; and

There are many incentives for comprehensive multilateral environmental cooperation in the Arctic. To begin with, there is a need for cooperation to avoid mutual losses stemming from the disruption of the shared ecosystems in the Arctic, which is liable to produce effects on the earth's biosphere as a whole (Young 1989). Second, duplication of costly research and development (R&D) efforts can be avoided through international exchange of information gained by each country in the protection of fragile polar ecosystems and the development and use of environmentally safe technologies in the Arctic. Third, because the international pool of engineering personnel with arctic expertise is limited and existing facilities to simulate arctic conditions and test designs are few, the expediency of resource pooling on an international basis is obvious. Fourth, the organization of such a system of international cooperation would demonstrate to the world community the sincerity of commitment on the part of arctic rim states to preserve those arctic ecosystems affecting climatic and other environmental conditions on a global scale.

The impetus for regional cooperation among Nordic and Arctic states was provided by Soviet President Mikhail Gorbachev in his October 1987 speech at Murmansk, where he called for the establishment of an "arctic zone of peace" (Gorbachev 1987). Gorbachev put forth a six-point program for cooperation, including a proposal for "a united, comprehensive plan for protecting the environment of the North." The implementation of the Murmansk program is well under way, at least in the sphere of scientific and environmental cooperation.

In December 1988, an International Conference of Arctic and Nordic countries on coordination of research in the Arctic took place in Leningrad with the aim of preparing for the establishment of an International Arctic Science Committee (IASC), a nongovernmental scientific organization that may, in time, become affiliated with the International Council of Scientific Unions (ICSU), emerging as a structural and functional counterpart to the Scientific Committee on Antarctic Research (SCAR). As this initiative was motivated, in part, by a desire to link arctic science with the global change movement emerging under the auspices of the International Geosphere-Biosphere Programme, the importance of IASC's formal establishment in the summer of 1990 for environmental cooperation in the Arctic cannot be overstated. Moreover, successful conclusion of the IASC negotiation seems to be a precondition for completion of an arctic environmental accord.

Quite in the spirit of the Murmansk proposals, the government of Finland initiated in January 1989 a multilateral process aimed at reaching international agreement on cooperative action to protect the arctic environment. Popularly known as the Finnish or Rovaniemi Initiative, the process began with consultative meetings in Rovaniemi, Finland; Kiruna, Sweden; and Yellowknife, Canada. These meetings led to the First Ministerial Conference on the Protection of the Arctic Environment, held in June 1991, also in Rovaniemi.[15] The conference signaled the end of the preparatory process and yielded an Arctic Environmental Protection Strategy, a nonbinding political declaration that commits the eight signatories to implementing a four-point action program:[16]

- an arctic monitoring and assessment program (AMAP) to monitor the effects of anthropogenic pollutants in the region through the establishment of an organizational task force;
- protection of the arctic marine environment through coordinated preventive and other measures; application of the principles of customary international law concerning protection and preservation of the marine environment; and establishment of special standards in arctic areas concerning vessel discharges, marine fuels, emission limits, and ship design and equipment features, for application both within and outside national jurisdiction;
- establishment of a regional emergency prevention, preparedness, and response capability, and;
- coordination of research and facilitation of data exchange on conservation of arctic flora and fauna.

At this stage, the declaration entails no legally binding obligations on the signatories, although the document does include language that commits each party to "full implementation and consideration of further measures to control pollutants and reduce their adverse effects to the arctic environment." The progress of the Finnish Initiative, then, should be viewed as preparatory, in effect laying the groundwork for an as-yet-unidentified environmental protection and management regime that might take any number of forms.[17]

A range of legal, technical, and management options is available to the Arctic Eight. The Arctic Environmental Protection Strategy states explicitly that implementation "will be carried out through national legislation and in accordance with international law, including customary international law as reflected in the 1982 United Nations Convention on the Law of the Sea" ("Arctic Environmental Protection Strategy" 1991). However, the strategy also acknowledges that many existing regulatory instruments pertaining to marine environmental

gating in ice-covered areas within the 200-mile economic zones, with due regard to the interests of nonlittoral countries (these requirements, however, may vary from region to region and from season to season)[21]; establishment of uniform (and the most stringent possible) norms and standards for the discharge of pollutants from vessels;
- coordination of pilotage services and weather and ice forecasts;
- determination of the extent of liability for pollution caused by vessels of one country within the exclusive economic zone of another;
- cooperation and mutual assistance in combating pollution emergencies and in ameliorating their consequences;
- cooperation in training and proficiency testing of arctic ship personnel (possibly at the World Maritime University in Malmö, Sweden, or in Russia, where there is considerable experience in this field);
- joint use in the Arctic of satellite systems for navigation, communications, and search and rescue;
- cooperation in design, construction, and testing of arctic-class vessels, and;
- conduct of joint research or exchange of data on the effects of pollution on arctic marine ecosystems.

The MARPOL "Special Areas" Regime for the Arctic Ocean

Protection of the arctic marine environment would be increased, simultaneously with the elaboration of the agreement (protocol) mentioned above, if the 1973 International Convention for Prevention of Pollution from Ships and its 1978 Protocol (MARPOL 73/78) were amended to include the Arctic Ocean and its adjacent seas as a designated special area where more stringent vessel discharge standards for most pollutants (up to a complete prohibition of discharge) have been established. Although the discharge standards for special areas designated in MARPOL 73/78 are less stringent than, say, those embodied in Canadian legislation governing shipping in arctic waters, the MARPOL designation would nonetheless offer some potential advantages. The MARPOL requirements are clearly defined, whereas those in Article 234 of the Law of the Sea Convention are not, and several nations that are refraining even from signing the convention (e.g., Germany and the United States) have already ratified MARPOL and comply with its standards.

The establishment of special areas under MARPOL, however, does not grant coastal states any special powers to control compliance by foreign ships. On the other hand, the Arctic clause in the Law of the Sea Convention permits any arctic rim country to exercise a substantial degree of control within its exclusive economic zone over any foreign navigational activities potentially threatening the marine environment.[22] Consequently, adequate protection of the arctic environment from shipborne pollution depends on having both international legal regimes in force.

In addition to measures to regulate pollution from vessels and to improve navigational safety, an action plan for the Arctic could also include specific national obligations in such areas as the conduct of joint physical, chemical, and biological oceanographic research; the development of common environmental impact assessment procedures for projects with potentially significant transboundary effects (e.g., assessing the influence of large-scale industrial and hydroprojects on arctic climate and global weather); measures to prevent pollution of northern seas through the dumping of wastes,[23] pollution as a result of exploration and exploitation of mineral resources in the continental shelf,[24] or pollution as runoff from industrial operations in coastal areas; cooperation in organizing natural reserves as specially protected areas[25]; protection of rare and endangered species; and assistance in developing environmentally safe land-use methods in the Arctic, including recultivation of derelict landscapes.

In any case, an array of additional cooperative supporting measures along the lines suggested by Griffiths and Young (1990) will undoubtedly be helpful in saving the region from environmental degradation. These may include, for example, the introduction of joint monitoring arrangements (e.g., of offshore oil and gas fields), the formation of unofficial teams of environmental experts,[26] and the creation of an arctic environmental consulting service to provide technical assistance to government and private developers on the most up-to-date methods and technologies available for minimizing environmental damage associated with the development of arctic resources (Griffiths and Young 1990). What is more important, however, is that the success of an arctic action plan depends on taking full account of the rights and concerns of arctic native peoples, including the protection of marine and other resources upon which they depend and provision for their active involvement (e.g., through the Inuit Circumpolar Conference) during at least the policy-making and implementation stages.[27] The region's aboriginal people, who have lived in harmony with their natural setting for ages, have much to contribute to achieving a sustained and equitable way of life in the Arctic (Osherenko and Young 1989).

link to a US destination port (in Alaska or the Pacific Northwest), but if so, all ships sailing that link would be subject to double-hull standards and other ship safety and design regulations imposed by the Oil Pollution Act of 1990, the Port and Tanker Safety Act of 1978, and other applicable US statutes.

Ensuring that arctic shipping services are viable and safe will entail the imposition of design standards and marine pollution regulations that are consistent across arctic coastal states and flag states where possible. Although such consistency will depend on action in appropriate fora (such as the International Maritime Organization) and through treaty agreements, technology transfer can serve to promote standardization and the adoption of consistent regulations. As a specific example, an opportunity for transfers from Russia and Finland to the West may exist in the form of icebreaker ship design expertise or capacity. Russia now operates the world's largest fleet of icebreaking ships and possesses over fifty years of experience operating in severe ice conditions. In the 1980s, efforts to coordinate the construction of new icebreakers in both the United States and Canada, either as research platforms or for military purposes, encountered difficulties. Although budgetary and institutional conflicts contributed greatly to this problem, deficiencies in domestic technical capacity were also a factor.

Oil Spill Research and Development. Coordinated national efforts in arctic oil spill research and development have measurably advanced the present state of the art in mitigation and prevention, which nonetheless remains limited.[30] In Alaska, an oil spill response R&D program has been funded since the early 1980s by North Slope operators under a state-imposed regulatory requirement. Under the terms of the program, operators are allowed to drill offshore under broken-ice conditions if they satisfy the state's oil spill planning requirements and participate in the R&D program (Smith 1986). Canadian efforts in arctic oil spill research began in the 1970s, with most R&D projects centering on fate and behavior analysis, surveillance and monitoring, mechanical containment and recovery techniques, in situ burning, subsea containment, shoreline clean-up, and/or chemical dispersants (Hume et al. 1983). Private industry conducts the work, which is funded in part or in whole by Environment Canada.[31]

Bilateral or multilateral efforts, unlike national programs, have generated little in terms of tangible outputs.[32] Greater pooling of technical and financial resources among the arctic rim states and their private firms, perhaps through the auspices and under the coordination of a single entity, might remedy this situation. A shared private pool of response equipment, underwritten by all the arctic rim states and deployed across the basin where exploration and development activities are most heavily concentrated, offers a partial solution to the problem of clean-up logistics. In any case, progress on spill prevention, containment/clean-up technologies, and deepwater response planning will become mandatory over the next ten to twenty years, as firms move from nearshore fields to those farther offshore.

Information sharing, contingency planning, and joint clean-up operations are provided for under the US-Soviet (now Russian) bilateral oil-spill contingency agreement covering the Bering and Chukchi Seas, which took effect in October 1989. Progress to date has included a readiness plan for emergency oil spill clean-up, developed jointly by the US Coast Guard and the Soviet Ministry of Marine Transportation. The necessity for and the practical significance of such a plan was proven during *Exxon Valdez* clean-up activities with the deployment of the Soviet-operated oil skimmer *Vaidagubski* in Alaskan coastal waters. By American request, a Soviet coordinator of on-site operations was dispatched to Anchorage to assist and direct cooperative efforts (EPA 1990d).

Environmental Impact Assessment, Information Sharing, and Scientific Research. Soft technology—scientific and environmental data, impact assessment methods, and program knowledge and expertise—may be more amenable to transfer. Precedent for such transfers is established in the US-USSR Agreement on Cooperation in the Field of Environmental Protection, originally signed in 1972 and renewed in 1985. Although this bilateral program is not Arctic specific, during its two decades of existence it has opened channels of communication between US agencies and their Soviet and Russian counterparts on several key arctic environmental issues. Detailed program experience on the *Exxon Valdez* clean-up, profiles of US pollution surveillance and monitoring systems, and technical information on response and control equipment available from Western vendors have all been exchanged under the auspices of this bilateral agreement (EPA 1990d). At present, five joint research initiatives are covered under its Area X, "Arctic and Subarctic Ecosystems." Activities among the eleven other areas frequently relate to research in marine environmental protection and marine ecosystems (EPA 1990d).

Bilateral marine pollution contingency plans with mechanisms for information sharing and joint clean-up operations are also in place between Canada and the United States and between Canada and Denmark, in both cases covering areas in the western Arctic (Harders

versal in the relative weight given to environmental versus military-strategic factors. As a Soviet republic, Russia had always remained much more attuned to its own pre-Soviet culture and history than to the ideology or political structure of the USSR. Now that Russia has a new, independent leadership, elected to power by a democratic movement that gives high priority to socioeconomic concerns, demilitarization, and the promotion of international cooperation, expectations for the Murmansk initiatives are running higher than ever.

It is still too early to assess contemporary Russian policy in the Arctic, which thus far has entailed no official departure from Gorbachev's course. There are those who criticize the sluggishness and inconsistency with which the Murmansk initiatives have been pursued, but one can equally argue that international cooperation in the Arctic has progressed more in the few years since Gorbachev's speech than in all of the previous seventy years of the Soviet era. Bilateral agreements have been concluded with Norway, the United States, Canada, and Finland; a conference held in St. Petersburg to discuss the coordination of scientific research in the Arctic laid the groundwork for the establishment in 1990 of an International Arctic Science Committee; the Soviet Union actively participated in the development of a comprehensive regime for the protection of the Arctic environment; essential barriers to cooperation among native peoples of the Russian Arctic and those of neighboring countries have been removed; and concern for the Arctic environment is an essential ingredient of Russian foreign policy (as it had become, belatedly, for the Soviet Union).

Despite these very significant developments, however, there continue to be powerful counterbalancing forces that argue for a cautious optimism at best. The economic and military elite who so energetically resisted *perestroika*, essentially by exploiting the inconsistency and lack of professionalism of the Soviet leadership, today are joined by a range of antidemocratic groups (nationalists, fascists, etc.) bent on unseating Russia's new leadership by exploiting the profound socioeconomic crisis that grips the country. Given such an unstable, highly polarized situation, it is impossible to know which of two sharply divergent scenarios is more likely to unfold: a strengthening of democratic principles and continuing development of democratic institutions or the seizure of power by a totalitarian regime of quasi-communist, nationalist, or other stripe.

Under either scenario, the commercial use of arctic resources is certain to be given priority over environmental concerns because of the country's dire economic condition. If democracy prevails, the environmental factor will no doubt steadily gain in importance as the decision-making process continues to include the indigenous population, environmental NGOs, the scientific community, and other groups with compatible concerns. Under the competing scenario, we would expect the new leadership vigorously to advance the so-called sectorial principle (or sector theory) with respect to an arctic legal regime. As propounded by the Soviet government in 1926 (after a comparable Canadian pronouncement in 1907), the sectorial principle asserted that all land areas within the Soviet portion of the arctic circumference were the legal territory of the USSR; over time this doctrine came to include arctic waters as well (Franckx 1993).[34]

Should democracy in Russia fail, the new leadership is likely to advance the sectorial principle even more aggressively than when the Soviet parliament declared its jurisdiction over all resources within its territory, the adjoining continental shelf, and the 200-mile EEZ. Since then, the collapse of the Soviet Union has meant for Russia the loss of more than half of its European ports to five other former republics, essentially making it a geographically disadvantaged country and forcing it to turn increasingly to the Arctic and the Pacific for its transportation and other marine needs.

Russia's recent admissions concerning nuclear dumping in the Arctic and North Pacific are an encouraging sign that, should democracy prevail, an emerging arctic regime will have a distinctly socioecological profile. Nonetheless, an integrated Russian policy on the Arctic remains to be formulated, reliable information on which to base environmental policies and practices is still lacking, and the ultimate fate of the emerging regime depends on sustained improvement in the international political climate and in the socioeconomic reforms under way in the former Soviet republics, especially in Russia.

PROSPECTS AND PREREQUISITES FOR SUCCESS

There is no doubt that regional environmental cooperation in the Arctic faces a number of obstacles. At present, environmental issues do not rank very high among the priorities for arctic development for most states in the region. Apart from military interests (which are outward oriented), the primary emphasis is on industrial development—mainly raw materials and energy resources extraction—which, as multiple examples suggest, is not always compatible with environmental protection. Sound arctic development policy at the national level, with due attention to the management of fragile arctic

CHAPTER 7

The Southern Ocean

THE SOUTHERN OCEAN differs from the rest of the world's oceans in important physical, political, and economic ways.[1] Physically, only its southern limits are defined by landfall; its northern limit lies at the Antarctic Convergence, where the cold polar waters submerge beneath the warmer edges of the Atlantic, Pacific, and Indian Oceans (see map 7.1). Politically, most of the Southern Ocean has been removed from military competition among states while national claims to maritime jurisdiction have been put into abeyance by the Antarctic Treaty of 1959. Economically, the Southern Ocean supports a different mix of activities; there is some fishing, but navigation is minimal, and there is no exploitation of offshore oil, gas, or mineral resources. Though the climate is harsh, humans have adapted to it sufficiently to navigate, fish, and undertake marine scientific research. The Southern Ocean, like other oceans, is sensitive to human influence; the belief that it is also particularly vulnerable is widespread.[2] Provisions for environmental security in the Southern Ocean are, for the most part, closely bound to international arrangements for management of the Antarctic continent.

THE EXISTING LEGAL RULES

The legal regimes that apply to various activities in the Southern Ocean differ in their geographic reach. South of 60 degrees, the provisions of the 1959 Antarctic Treaty relating to nonmilitarization, nondisposal of radioactive wastes, and prohibition of nuclear tests apply. An inspection system established in Article VII of the Antarctic Treaty applies to all ships landing or embarking persons and things. A provision in Article IV places territorial claims in abeyance so that there can be no exercise of coastal state jurisdiction over what otherwise

The Southern Ocean 191

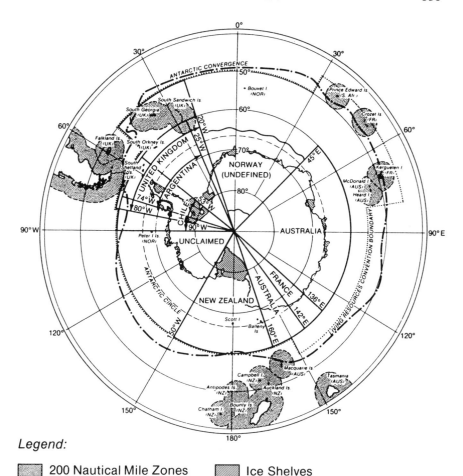

Legend:

▨ 200 Nautical Mile Zones ▨ Ice Shelves

Map 7.1. Claims and jurisdictions in the Southern Ocean. (From Joyner 1992, by permission)

would be territorial seas, contiguous zones, and exclusive economic zones. Navigation is governed by the general law of the sea both north and south of the 60-degree line.

Prevention and abatement of pollution is addressed not only by the relevant global treaty rules but also by provisions of recommendations adopted by the twenty-six Antarctic Treaty Consultative Parties[3] and by the Antarctic Treaty's 1991 Protocol on Environmental Protection, which consolidates, clarifies, and eventually will replace those earlier recommendations. The global treaties addressing such matters as operational oil discharges, pollution as a consequence of accident, disposal of ordinary wastes, and handling of hazardous wastes apply to

affairs, it failed to deal in any substantive manner with issues regarding resource exploitation, management, or ownership" (Joyner 1984). The Antarctic Treaty Consultative Parties negotiated a proposed regime between 1975 and 1988, but it has been set aside in favor of a moratorium lasting at least fifty years (see "Which Regime?" below). The 1988 Convention on Regulation of Antarctic Mineral Resource Activities (CRAMRA) would have covered mineral and fuel resources on the Antarctic continent, offshore islands south of the 60-degree line, and the surrounding continental shelves, even when they extend north of the line (*International Legal Materials* 1988; *Polar Record* 1989).[14] Neither it, nor the moratorium, covered possible exploitation of icebergs as a source of freshwater for desert parts of the world.[15] The Consultative Parties have omitted deep seabed resources from all their discussions and make no claim to regulate their exploitation.[16] Rather, those resources have been left to be managed under Part XI of the 1982 Law of the Sea Convention (*International Legal Materials* 1982; United Nations 1982) or some other set of rules accepted later. The Antarctic Treaty states do, however, regard protection of the whole Antarctic environment as within their competence.

This responsibility was expressed most forcefully and succinctly in the 1991 Protocol on Environmental Protection, which supplanted the proposed CRAMRA and explicitly banned mineral resources activities for at least fifty years beyond the protocol's entry into force. The protocol, which will enter into force only after ratification by all twenty-six Consultative Parties, was adopted in Madrid on October 14, 1991. It designates Antarctica as "a natural reserve, devoted to peace and science," and details, through five annexes, a comprehensive system for protection of the Antarctic environment.

The annexes to the Protocol on Environmental Protection specify obligations and requirements: (1) for environmental impact assessment, (2) for conservation of Antarctic flora and fauna, (3) for waste management, (4) for prevention of marine pollution, and (5) for special area designation and management. For the most part, these annexes are based on and are designed to replace earlier recommendations agreed to by the Consultative Parties (National Research Council 1993b). Annex I of the protocol establishes three categories of activity, two of which require prior environmental assessment: (1) activities having "less than a minor or transitory impact" do not require a formal prior assessment; (2) activities likely to have only a "minor or transitory impact" require an "initial environmental evaluation"; and (3) activities likely to have "more than a minor or transitory impact" require a "comprehensive environmental evaluation." The criteria for "minor or transitory" impacts are not clearly established. Annex II restates measures agreed to in 1964 by the Consultative Parties for conserving Antarctic fauna and flora, essentially prohibiting the taking of any species without permit, which shall be issued only for specific and limited reasons. Annex III (based on a 1989 recommendation) establishes strict guidelines for waste management, setting forth a broad category of wastes that must be removed from the treaty area altogether, but it does not allow for ocean disposal of sewage and domestic liquid wastes. Annex IV, on marine pollution, speaks to vessel-source pollution and closely follows the strictures of the 1973 Convention for the Prevention of Pollution from Ships and its 1978 protocol (MARPOL 73/78) and annexes (see chapter 4). Annex V provides for the establishment of two types of specially protected areas: (1) Antarctic Specially Protected Areas, possessing outstanding environmental, scientific, historic, aesthetic, or wilderness values and requiring a permit for entry, and (2) Antarctic Specially Managed Areas, helping to facilitate planning and coordination, avoid conflicts, and minimize environmental impact but without requiring an entry permit (National Research Council 1993b). A system for assigning and assessing liability for environmental damages was not agreed to within the protocol but is reserved for elaboration in a future annex.

COMPETING MANAGEMENT PRINCIPLES

The future management of the Southern Ocean and its resources remains a matter of considerable controversy among states. A group of states not involved in the Antarctic Treaty system has sought to substitute a UN-based universal participation regime for the Antarctic Treaty. The minerals discussions inspired a separate debate that has divided the Antarctic Treaty Consultative Parties over how best to protect the Antarctic and Southern Ocean environments. In all of these debates, proponents of three distinct management principles have offered competing versions of the Antarctic future.

The environmental issues pit preservationists against conservationists. The preservationists would give protection of the environment absolute priority by banning resource exploitation in perpetuity and placing severe constraints on other activities to keep the Antarctic environment as pristine as possible. The conservationists agree that protecting the environment is important, but they do not accord maintaining the Antarctic environment in its current state the same absolute priority over all other considerations. The debate is three-sided, however, because of differences over the proper characterization of use rights and management authority in the Antarctic.

operated under UN auspices. Some of these arguments reach too far; state practice is replete with examples of only limited acceptance even of regimes negotiated under UN auspices. However, the desire for universal acceptance is understandable. Third, they believe that a regime thus freed from the necessity of political compromise would contain clearer and more general rules on such thorny questions as jurisdiction over individuals and groups in Antarctica. Fourth, they believe that such a regime would be able to operate under majoritarian rules and thus not be as subject as the Antarctic Treaty system to delays imposed by the need for unanimity.

Several institutional forms have been advanced. A few analysts have considered the implications of establishing a UN Trusteeship for Antarctica, something the US government briefly contemplated in the late 1940s.[18] This has not proven popular. Many regard Antarctica's lack of native human population as making Trusteeship as defined in the UN Charter inappropriate. In addition, adopting Trusteeship notions would mean entrusting administration to the UN Trusteeship Council, a body of only five members.[19] Others, particularly environmentalist nongovernmental organizations, have argued that preservation would be best assured through the creation of a UN-based Antarctic environmental protection agency, perhaps as part of the broader UN Ecological Agency that was discussed in preparations for the 1992 Conference on Environment and Development. The final resolution of UNCED, however, did not approve such an agency.

A few have argued that including all of Antarctica within the existing system of world parks would suffice. This position fails to pay sufficient attention to the fact that each current world park is administered by the state within whose territory it is found and is declared only upon the request of that state. Such inhibitions would not affect declaring all or part of Antarctica and the Southern Ocean a "biosphere reserve" under UNESCO's Man and the Biosphere Programme, since such a designation can apply to areas outside national jurisdiction (see Keage 1987). However, biosphere reserve designations in other areas have had largely honorary significance and little practical, operational effect.

One major obstacle to such plans is opposition by several Antarctic Treaty parties. Unless all of them accept its supercession, the Antarctic Treaty has the legal weight that attaches to existing treaties under international law.[20] Legally, nonparties could not claim to supercede it with a treaty among themselves. Politically, supercession is likely to be difficult: for every party ready to think about giving it up, there is at least one adamantly opposed. The New Zealand government considered giving up its claim in order to create a world park in 1975 but conditioned this move on parallel moves by other claimants. The Chilean government was ready to accept a world park if it could retain its claim, but other claimants never reacted (Auburn 1982). The political situation is also complicated by the fact that many provisions of the Antarctic Treaty system are actually acceptable to all states; no one seems anxious to disturb scientific cooperation, weaken nonmilitarization, or end Antarctica's relative isolation from global and regional politicomilitary competition.

Opting for a nature preserve or other full protection alternative would be consistent with the Antarctic Treaty system. The Antarctic Treaty itself selects no principle for resource management; part of the compromise on territorial claims involved not raising such issues in the treaty. The enumeration of Consultative Party concerns in Article IX, paragraph 1, makes no reference either to "the environment" or to "resources"; only "the preservation and conservation of living resources" is mentioned. Over time, however, the Consultative Parties developed rules relating to the environment and came to accept the obligation to protect the Antarctic environment as an additional norm of their regime. This obligation began as ancillary to the promotion of scientific cooperation; it was quickly recognized that making too many messes or interfering too much in the natural life patterns of plants and animals would reduce the value of scientific observations.

As the international community has come to place greater value on protecting the environment, the Antarctic Treaty system has adopted more and more rules and has referred in them to the direct value of environmental protection as well as to its value for preserving the "scientific laboratory." The preamble of Recommendation XV-1 of 1989 refers in consecutive paragraphs to "the vulnerability to human interference of the Antarctic environment," "the unique opportunities Antarctica offers for scientific research on processes of global as well as regional importance," and "international concern for the environment and the importance of Antarctica to the global environment" (*Polar Record* 1990a), as reflected in the preamble to the 1991 Protocol on Environmental Protection. In Article 3 of the protocol, the signatories agree that

> the protection of the Antarctic environment and dependent and associated ecosystems and the intrinsic value of Antarctica, including its wilderness and aesthetic values and its value as an area for the conduct of scientific research, in particular research essential to understanding the global environment, shall be fundamental considerations in the planning and conduct of all activities in the Antarctic Treaty area.[21]

plankton abundance that some have attributed to increased ultraviolet radiation penetrating the water because of ozone depletion. This reduction of phytoplankton would reduce absorption of carbon dioxide (contributing to global warming) and decrease the abundance of other Southern Ocean marine life by reducing food supply at the lower end of the food web.[23] The Southern Ocean ecosystem has not been affected significantly by extraregional pollutants. Samples of snow taken from the continent indicate that heavy metals occur in concentrations comparable to the levels found in soils and other strata lying on the surface fifty years ago in other regions.[24] Yet the ecosystem has suffered from various forms of more local pollution and disturbance, and it could suffer from more if activity increases.

The Southern Ocean has special characteristics that must be considered in any management effort. Coastal and shallow sea areas are vulnerable to human disturbance. Penguins can be so disturbed by close human presence that their metabolism accelerates and forces them to seek food before the end of their spell of incubating an egg. The egg is abandoned and freezes before the mate returns from its own feeding to take its turn keeping the egg warm (Heap and Holdgate 1986). Less is known about the resilience of the animal life in shallow seas, but it, too, may be vulnerable to human disturbance (Heap and Holdgate 1986). It is also vulnerable to pollution from spilled oil or other sources. The broad expanse of the Southern Ocean and polar ice pack, although cold, can bear greater human influence without much effect because the winds and currents assure a vigorous mixing. Yet even experts convinced of this resilience conclude that people should take great care to avoid pollution.

The relationship among marine species also requires careful policy, as reflected in the ecosystem approach envisaged by CCAMLR. A single species, krill, accounts for about half the total biomass of Southern Ocean zooplankton. Krill are so central to the Southern Ocean food web, being eaten by nearly all larger marine species, that they are rightly regarded as a "keystone species." The full implications of this depend on whether krill form one or several distinct stocks. If krill are one stock and the Southern Ocean is a single ecosystem, then serious overfishing could cause a general collapse of krill populations and with it severe reductions of predator populations. Predator populations might not collapse as far as the krill populations because some predators eat a mixed diet. If krill are actually several stocks, the effects of overfishing would be more localized. One particular stock might collapse, with deleterious effects on predators in the area, but other stocks, and their predators, might remain numerous. The relations among species have already been disturbed by general overexploitation of whales and local overexploitation of finned species, complicating efforts to arrive at sound policy.

Until recently, Antarctica has been a "frontier region" outside the area of permanent human settlement. Until the International Geophysical Year in 1957–58, humans visited Antarctica or the Southern Ocean in small groups for relatively short periods. Even the sealers active in the nineteenth century and the whalers active in the twentieth were visitors, carrying out their activity in the summer season and heading back home or into other areas during the long Antarctic winter. People came, took what resources or information they could carry away, and left. Few had any regard for the long term: all acted as if they believed something not taken today was more likely to be taken by someone else than left in place.

In sealing, whaling, and fishing, all of the negative aspects of the frontier mentality soon appeared. Each individual or enterprise took as much as it could, made money while it could, and reinvested the profits in new ventures elsewhere once the Southern Ocean resource was too depleted to exploit. Scientists' desire for data rather than physical resources and the organization of their activity along lines of authority permitting some monitoring and mutual accountability served to modify the frontier pattern. Yet even with them some frontier behavior remained. Until station site cleanliness became an issue in the 1980s, some stations earned reputations for poor sitekeeping.

Much remains to be done, even though governments and others are shedding frontier attitudes and treating Antarctica and the Southern Ocean as environments to be managed for the long term (UN General Assembly 1992). First, the actual hazards to environmental security must be identified and their likely effects assessed. Second, robust institutions and procedures for making rules, monitoring activity, and assuring compliance must be created. The first task involves learning more about the Southern Ocean environment, the likelihood that various activity-related risks will materialize, and the likely damage that will result if risk becomes actuality. The second task involves translating that knowledge into a basis for effective institutions.

The Hazards of Offshore Oil and Gas Exploitation

The prospects for offshore oil and gas development in the Southern Ocean are quite remote, but international attention has already focused on the potential institutions under which it might take place and the environmental risks it would pose. The risks of offshore oil and gas operations are well known from experience in other areas. The Arctic in particular provides a reasonably good baseline for extrapolating estimates of Southern Ocean hazards and determining how to minimize their likelihood or impact. The combination of low temperatures and the nonexistence of indigenous hydrocarbon-degrading bacteria have led analysts to believe that any oil trapped in coastal areas would persist for as much as a century (Keys 1984).[25]

Preliminary surveying and prospecting pose the least serious hazards. Seismic surveying, the essential first step in both broad-scale prospecting and the finer surveys used to determine where to drill exploratory wells, is a nonintrusive technique relying on reflected soundwaves that produce no physical damage. Yet even it may disturb marine mammals, which rely on sounds for communication and to locate one another or fish nearby ("Seismic Oil Searches" 1985). Seismic surveying on Alaska pack ice in the spring, for example, is subject to restrictions intended to reduce its effect on the ringed seal populations in the area (*Federal Register* 1988).

The drilling of exploratory wells is far more hazardous. The first hazard is the presence of gas hydrates, solid crystalline mixes of water and gas that form at the top of some hydrocarbon-bearing sediments. They have been identified by seismic work in the Wilkes Sea and may also exist in the Ross Sea. Although they do indicate a hydrocarbon deposit, some geologists believe they should not be disturbed by drilling because they may act as seals which, if ruptured, could lead to seepage of hydrocarbons from substrata to the sea floor (Behrendt 1988). If ever permitted to operate, Southern Ocean well drillers would probably be required to follow rules similar to Arctic practice and drill only during the milder and less ice-infested summer season (Croasdale 1986). Most of the technology for drilling exploratory wells in ice-covered areas exists, having been developed for Arctic operations, but considerable evolution would probably be necessary before Antarctic application is permitted.

The hazards of exploitation are far greater. Producing wells, even with the current safety technology, occasionally entails a blowout in which oil flows uncontrollably. Capping an underwater blowout is particularly difficult, as demonstrated by the blowout at Pemex's Ixtoc I Well in the Gulf of Mexico during 1979–80.[26] Normal blowout hazards are compounded by icebergs or flowing glaciers. Most experts believe that in the Southern Ocean, as in the more distant Arctic locales, new subsea production systems would have to be developed. These would involve such new devices as seabed storage facilities, seabed pipelines to remote processing facilities, or floating storage facilities, depending on the nature of the ice hazard in a particular area (OTA 1985).[27] They would either have to be kept fairly deep[28] or be capable of iceberg avoidance, by use of either towing vessels or rapid disconnect and movement systems. Even these systems, however, would not be able to cope with the hazards posed by ice scouring (scraping of the seabed) caused by the breaking up of pack ice in the spring or by the passage of large icebergs (Deacon 1984). Scour marks have been observed all around the continent at depths of 200 meters or more,[29] and this hazard is taken seriously in all discussions of possible production systems (e.g., OTA 1989).

The whole process of storing, transporting, and processing the crude oil and natural gas entails additional hazards. Storage tanks or pipelines may burst or get broken. Valves might give way under pressure. Oil might spill in the course of being transferred from one storage or transport system to another. The relatively ineffective boom and skimmer technology currently available for spill recovery is utterly useless in the rough sea conditions usually encountered in the Southern Ocean (Deacon 1984). Extremely cold Southern Ocean waters make oil recovery even more difficult because certain crude oils are impossible to pump at those temperatures (OTA 1990).

The actual extent of pollution damage would depend on where spills occurred. Damage might not be very long lasting in the deep ocean. About one-third of the oil lost in spills evaporates, and the rest gets broken down into simpler compounds. However, both processes are slower in cold water, and more oil is thus likely to mix with the water (OTA 1990).[30] Krill stocks would suffer because of the creature's small size, but fish might not because few dwell near the surface (Laws 1989). Penguins (which can die within five minutes if their thermal insulation is disturbed), walruses, and baleen whales, which feed near the surface (and so are vulnerable to getting their baleen plates covered with oil), would all die fairly quickly (Barney 1982). Oil slicks adhering to the edge of the pack ice could prevent the passage of light and oxygen needed to sustain the spring algal blooms that begin the annual cycle of food production (Deacon 1984). The direct physical damage to plant and animal life would be more severe (Sakshang and Skjdal 1989). In fall, the expanding pack ice would serve as an additional barrier to the spread of oil,[31] but melting in the spring would release remaining fractions (Centre for Cold Ocean Resource Engineering 1983). Near-shore spills would have all the negative effects on coasts, intertidal limpids, kelp, sponges, and marine animal life in the affected areas that were observed in the *Exxon Valdez*[32] and *Bahia Paraiso*[33] groundings. The extreme cold means that oil would not quickly degrade but would continue to hug the edges of ice or coast for many years.

Clearly, the actual risk of pollution from oil and gas exploitation would depend in large part on the extent of the activity. Quite apart from the Antarctic Treaty System's fifty-year prohibition, near-term economic prospects are essentially nonexistent. The very long-term potential is harder to predict, especially if allowance is made for politically motivated government subsidization.

The Hazards of Deep Seabed Mining

The prospect of deep seabed mining anywhere in the world is remote today given the development of better land-based extractive techniques capable of exploiting lower-grade ores, greater use of substitutes for many metals, and low prices reflecting the reduced demand for many minerals. The Southern Ocean is one of the least attractive mining sites because the richness of the main resource, polymetallic (or "manganese") nodules, varies roughly inversely with distance from the equator. The richest nodules are all found in the tropics, particularly in the mid-Pacific. The poorest nodules are found in Arctic and Antarctic seas. It is fair to conclude, then, that there will be no deep seabed mining in the Southern Ocean within the next half century, if ever.

The Hazards of Land-based Marine Pollution

Although the exploitation of land-based resources has the greatest effects on land areas near the mine or well, it could affect the marine environment by direct runoff or atmospheric transportation of pollutants. Dredging to recover near-shore seabed minerals or to build wellhead and transportation facilities would cause local pollution as dredged material was redeposited elsewhere, increasing water column turbidity temporarily and destroying benthic fauna. The process of recolonization could be much slower in the Southern Ocean than in temperate waters (Zumberge 1979).

Just exploring Antarctica is expensive (OTA 1989; Zumberge 1982). The US Geological Survey estimates that it would take ten seasons to evaluate properly Antarctica's resource potential, with the onshore portions of the work alone costing about $50 million (Cooper 1989).[34] At present, only gold and diamonds located in the Antarctic Peninsula seem to be economically viable candidates for exploitation within the next several decades, and the likelihood of finding a rich enough source of diamonds seems remote (OTA 1989). The Environmental Protection Protocol may well have banned a set of activities that would not have occurred even without it.

The Hazards of Ordinary Navigation and Waste Disposal

All navigation, whether it involves transporting scientists and their supplies, bringing tourists in, or transporting resources out to other parts of the world, carries certain environmental hazards. If the ship founders, the cargo may escape and pollute the immediate area. Even when there is no collision or break-up at sea, a ship may leak oil from bilges or machinery, or disposal of crew and passenger trash and wastes may be done carelessly. The danger of oil leakage has been reduced by successive amendments to the rules on oil discharge. In November 1990, the whole Southern Ocean was designated a "special area" under MARPOL 73/78.[35] The designation took effect on March 16, 1992, and it requires that ships operating in the Southern Ocean be capable of retaining oil residues and garbage until they reach an approved port reception facility. Since there are no large Antarctic ports, the Consultative Parties have concluded that this will require obligating states whose ports serve as departure or arrival points to ensure the provision of adequate port reception facilities.[36] Ships will have to be designed to retain residues and wastes from both legs of a round trip to the Southern Ocean. At-sea disposal of incineration ash and certain other wastes is banned by the 1972 Convention on the Prevention of Marine Pollution by Dumping of Wastes and Other Matter (the London Convention). The Southern Ocean has never been a favored place for this activity because of its relatively great distance from heavily populated areas. The convention rule helps assure that it will not be considered as newer national and international restrictions limit such activity in other oceans. The restrictions spelled out in MARPOL 73/78 and the 1972 London Convention are reiterated and reinforced in the Antarctic Treaty area by the 1991 Environmental Protection Protocol.

The hazards of accidents were demonstrated dramatically when the Chilean ship *Bahia Paraiso* ran aground just off the US Palmer Station on the west side of the Antarctic Peninsula in January 1989. Proximity to shore meant that the passengers and crew were rescued without great difficulty, but roughly 170,000 gallons of aviation fuel and 250,000 gallons of diesel fuel were spilled into the water. The heavier substances came ashore, killing krill and hundreds of newly hatched skuas and penguins. The destruction of habitat and animals was sufficient to destroy the scientific value of a quarter-century's work by scientists at the US Palmer Station (Barinaga 1990). This incident was all the more troubling because the Chilean operators had been warned by American personnel that the waters were shallower than indicated on the Chileans' charts (Barnes 1989).

The Hazards of Exploitation of Living Marine Resources

The main environmental hazard is the possibility of overexploiting particular stocks or species. The effects of overexploitation can include near extinction of the target species or a change in the relative numbers of different species.[37] Such changes might involve reduction in their numbers as less food becomes available or an increase in abundance as competitors for food are taken away. Some biologists fear that local krill taking has reduced the adelie and chinstrap penguin populations on King George Island (Anderson 1991). There is broad consensus that the sharp reduction of whale populations has permitted other krill-eaters, most notably seals and penguins, to increase.

The past record of Southern Ocean marine resource exploitation offers grounds for great concern. Seal populations were reduced to the point of commercial (although not physical) extinction in a few decades. Whale populations still have not recovered from the heavy taking between 1925 and 1970.[38] Four species of finned fish were seriously depleted off South Georgia and the Kerguelens between 1969 and 1974. Soviet fleets were the most active participants in an effort that featured use of coordinated trawling and factory ship operations. Around South Georgia, several stocks were reduced to 10 percent of their original size, and the balance among bottom-dwelling species was altered ("Overfishing in Antarctica" 1984). During the 1992–93 season, the total reported finfish catch was 5,800 metric tons, less than one-tenth of the 1991–92 catch, and only one species was taken, the Patagonian toothfish (*Dissostichus eleginoides*) (Marine Mammal Commission 1994). Current krill fishing does not threaten the overall population, but the concentration of 20 to 40 percent of effort in the Scotia Sea has raised concerns about local overexploitation. What now appear as obvious errors of omission (lack of management) or commission (poor decisions or low compliance with good ones) form part of the basis for environmentalist critiques of Southern Ocean resource management, efforts to secure an end to commercial whaling, and proposals to ban the taking of all or some other marine creatures.

Some environmentalists believe that whale numbers are so reduced that certain species could become extinct. Most observers believe, however, that whale populations could recover if whales found sufficient food to support larger populations. All agree that any large increase in Southern Ocean whale populations will be slowed, or perhaps made impossible, by the increased populations of seals, penguins, and other krill-eating species. Their populations grew when the drastic reduction of whale populations removed many of their greatest competitors (including competitors capable of eating them as well as the common food). This effect is far stronger in the Southern Ocean than elsewhere because of the unique place of krill in the food web. Reduction of one species provides opportunities for nearly all the others.

This shift in the balance among marine species poses some very difficult issues for anyone attempting either ecosystems management as specified in CCAMLR or preservation of the environment as advocated by many environmentalist groups. The problem of defining the environment to be protected raises serious issues. If it is to be defined as the Southern Ocean ecosystem in as near as possible to its original, non-human-disturbed state, that could well require the culling of seals, penguins, and other species whose populations have grown with the reduction of competition for food and direct predation by whales. Such culling would probably meet with considerable opposition. The exploitation of marine mammals is seriously opposed in many countries, and environmental groups have attempted to use their influence in these countries to win international bans on taking those creatures. This has caused serious disputes among governments, as those that do not accept the environmentalists' ethics arguments find themselves threatened with trade and other sanctions if they do not force their nationals to quit the activity. From their perspective, others who are not bearing any costs (because they have quit or never seriously participated in taking) are imposing their preferences without due acknowledgment of the costs inflicted. Any effort to define the baseline environment that is to be attained and then protected is also likely to generate considerable opposition among the ecologically minded.

The closer integration of Southern Ocean finfish and krill taking also poses problems for management. The largest fleets, particularly the Russian fleet, contain trawlers and processing ships that can switch between krill and other species. This will have to be taken into account in issuing permits and monitoring operations. Present krill nets are rather unselective, catching not only mature krill but also larvae and even roe (Voronina 1990). Krill catches have generally increased since 1986, but the total catch for 1991–92 (289,000 metric tons) represented a 19 percent decline from the previous year, and the 1992–93 catch (88,000 metric tons) fell another 70 percent, largely because of reduced effort by Russia and the Ukraine (Marine Mammal Commission 1994). Despite the unfavorable economic conditions in the former Soviet Union, Russia had previously taken the largest catches of krill, followed by Japan and the Ukraine (ASOC 1993). Marketing problems still inhibit greater development of the krill fishery, but these might be overcome if current experiments in using chitin and chitosan for manufacture of recyclable plastics prove successful (Nicol 1989, 1991b).

Effective ecosystems management also requires that some major unknowns be resolved. The CCAMLR was drafted on the assumption that all Antarctic krill form one population. Catching might have to be prohibited in particular areas to protect other marine species, but the current regulations assume that catching in one part of the open ocean is the same as catching in another. If, however, krill in different areas form distinct stocks, then it will be necessary to write more finely tuned regulations establishing stock-by-stock catch limits. There is also controversy about krill lifespans. Until recently, most scientists believed they lived only for a year or two, meaning that the population was renewed very quickly. Now many believe that krill live for six to ten years, which necessitates a reduction in estimates of sustainable yield. Aspects of krill interactions with other species also remain unknown. Several models of what would happen if krill populations were seriously reduced have been developed, but there is as yet no way to determine their relative predictive value.[39]

Experience with the CCAMLR regime thus far shows that ecosystem management of fisheries is still elusive. The convention did not specify the sorts of data fishers should report, for the good reason that no one was quite sure what would be needed. This had the unfortunate effect of permitting several years' wrangling as fishing states sought minimal requirements while others sought maximal ones (Howard 1989; Gulland 1988b). Debate was eventually tempered by realization on both sides that too few data would impede management but too many data would be too great a burden on both fishers and the scientists. These disagreements did not prevent adoption of all management measures. Regulators could adopt provisional rules of thumb[40] and impose local quotas or bans when overfishing was obvious.

Agreement on data has not solved the next problem, that of determining what should be concluded from the data collected. The CCAMLR Commission and its Scientific Committee are using standard and well-developed models of population dynamics to analyze the effect of fishing on each identifiable stock of fish. When it comes to analyzing interactions, they, like all other fisheries biologists, are exploring new domains. Very little work has been done on this subject, even in places like the North Sea, where it is well known that population changes in one species have major effects on the population of another (Bailey and Steele 1992). The CCAMLR Commission is thus operating on the frontiers of science—an uncomfortable place for policymakers who must be able to explain the rationale of their decisions to skeptical addressees. In 1985, the Scientific Committee identified a number of species as deserving careful monitoring because changes in their population serve as indicators of changes in populations of other species that cannot be assessed very easily.

Many observers, particularly the environmentalist organizations that follow Antarctic affairs, were disappointed by the CCAMLR Commission's early decisions. Its failure to impose any reduction in fishing near South Georgia before 1986 has been contrasted to the far-tougher French management of fisheries off Kerguelen Island. The French closed their 200-mile EEZ around the island to all fishing in the 1978–79 season and have since allowed only a limited number of vessels to fish (Gulland 1988b). Many environmentalists believed from the start that CCAMLR will never effectively manage the fisheries (e.g., Barnes and Beck 1988). Their views were only reinforced when the long controversy about procedure in the Scientific Committee was resolved in favor of consensus (Howard 1989; Vicuna 1991). Though the committee's reports are to cover all views expressed, many environmentalists saw this rule as another triumph of economic over ecological considerations. Others thought CCAMLR could be effective if the nonfishing states were prepared to impose restraint on the major fishing states through trade sanctions.[41]

The experience of the 1980s does demonstrate that a consensus rule yields many opportunities for delay by the least conservation-minded participants (Howard 1989). They can first urge delay until scientists provide specific unanimous advice that particular measures are essential to the continued health of stocks (e.g., Gulland 1988a, 1988b; Nicol 1991a). They can slow this process by failure to provide data needed for making such assessments. They can even continue to resist in the face of agreed advice by pointing to uncertainty in the scientists' predictions.

This problem is not unique to CCAMLR; it affects every effort at fisheries management around the world and will be overcome only when governments adopt two attitudes. First, they will need to accept that uncertainty is an endemic feature of fisheries management and cannot be entirely eliminated by more scientific work. Second, they will have to learn to accept cautious rather than optimistic estimates as the basis of policy rather than fall back on optimism to defer hard decisions.[42]

Uncertainty abounds in current knowledge of Southern Ocean marine life. A few examples should indicate the problems involved. In the mid-1980s, Polish researchers came to the CCAMLR Scientific Committee with data indicating that krill fishing produced large by-catches of young fish. This finding, which went against the grain of previous research, was discussed at length, with the Scientific Committee concluding that in general by-catch was actually much lower than the Polish data indicated.[43] Only in 1985–86 was it shown that krill can survive the winter by feeding on algae they find under or within cracks of the ice (Sahrhage 1988), as well as by shrinking at each molting to compensate for lower food supplies.[44] This provided the basis for current controversies about krill life-span (fig. 7.1). Controversy about the size of krill populations continues. In 1988–89, researchers reported results indicating that stocks were much smaller than had been thought (e.g., Miller and Hampton 1989). These were challenged on grounds the researchers were underestimating the value of target strength in interpreting data from their acoustic surveys (Everson et al. 1990). Uncertainty remains high. In the early 1970s, biomass estimates ranged from 750 million to 7,500 million metric tons (Peterson 1986). Current estimates are even less precise, ranging from 55 million to 7,000 million metric tons (Miller and Hampton 1989). Formulation of management measures is also hindered by the fact that it is extremely difficult to determine the age of individual krill, although most researchers now believe they live between six and ten years (Nicol 1990). Recent work has also shown that the size of most Antarctic fish stocks, even those not suffering overexploitation, are smaller than thought and that the fish grow more slowly than previously believed (Sarhage 1988). Some stocks, particularly the *Notothenia rossi* around South Georgia, are known to have been seriously depleted by overfishing.

Both conclusions mean that fishing must be more severely limited than was anticipated in the 1970s. CCAMLR is beginning to take hold, though unevenly. Krill fishing declined sharply in the early 1980s because fishers found few markets for their products.[45] Although recovering generally before the most recent decline, it has never returned to the peak levels.[46] However, the concentrated krill fishing around South Georgia may need to be limited more severely. While the main overfishing of Antarctic finned species occurred before CCAMLR existed, concentrated fishing in certain areas has inspired calls for regulation. The commission expanded the range and depth of its fishing

Fig. 7.1. Although krill account for roughly half the total biomass of Southern Ocean zooplankton and are eaten by nearly every larger marine species, scientists have not determined whether Antarctic krill constitute one or more distinct stocks, how long they live, and how other species are affected by substantial fluctuations in the krill population.

rules between 1984 and 1989. Before 1984 there were no specific regulations on fishing. After the 1989 meeting there were mesh size regulations to prevent taking juveniles; temporary bans on taking the most exploited species near South Georgia, the South Orkneys, and the Antarctic Peninsula; limits on total allowable catches for all major finfish target species; by-catch limits; limitations on bottom trawling; and a newly improved data reporting system (Heap 1991). The commission had also asked the Scientific Committee to formulate recommended limits for krill taking, particularly in the Scotia Sea (Heap 1991). Yet enforcement remains a serious problem. The commission has only begun acting on the parties' commitment in Article XXIV, paragraph 1, to establish a joint observation and inspection scheme. The current system is not sufficiently strong to deter violation, and the high cost of an adequate system makes one unlikely to materialize (Puissochet 1991).

The Hazards of Research Activity

Most scientific research in Antarctica is environmentally benign (National Research Council 1993b). The same cannot be said, however, of the logistical activity that supports science. Sustained human activity in the Antarctic requires construction of lodging (from temporary tents to permanent stations) and provision of food, heating, medical care, equipment, and transportation. The pressure for an increased number of stations, an artifact of the Consultative Parties' political decision to require a full-time research program for admission to their ranks, has led to local overcrowding at certain accessible ice-free sites (most notably King George Island). Any resource exploration would also be preceded by an increase in broad geological surveying, bringing additional scientists and their research vessels into the Southern Ocean.

In recent years the housekeeping of Antarctic research stations or the construction of certain facilities has provoked charges of carelessness and worse (see Kriwoken 1991; Harris 1991). The longest and most publicized controversy involved French construction of a new rock airstrip near Dumont d'Urville Station on Pointe Géologie. Greenpeace and other environmental organizations charged that the construction disturbed penguin rookeries and caused the deaths of many birds and that its operation will make it very difficult for penguins to get from the sea to their favorite breeding site because the airstrip lies along their migration route.[47] The French acknowledged that some birds had died during blasting, but they denied that breeding had been seriously disrupted during construction or would be afterward, as the penguins had several routes to and from the sea ("French Biologists Reply to Greenpeace Report" 1990).

Even normal station operations were criticized for laxity in the disposal of wastes. Concerns were great enough for the Scientific Committee on Antarctic Research to commission a study in the early 1980s (Benninghoff 1985). Response came, though environmentalists considered it too slow. The Consultative Parties urged better practices in Recommendations XIV-2 of 1987 and XV-3 of 1989 (*Polar Record* 1988, 1990b). Annex III of the Protocol on Environmental Protection, dealing with waste disposal, requires that the generators of hazardous and nondegradable wastes must dispose of them outside Antarctica, bans open air burning of wastes, limits incinerator burning, specifies other limits on waste disposal, and meets environmentalist demands that each Consultative Party apply in Antarctica the same regulations that cover waste disposal at home (SCAR 1991; National Research Council 1993b).

This new attention to station-generated pollution has quickly acquired political implications. At least one Antarctic scientist regards it as an attack on Antarctic science generally disguised as concern for the environment (Laws 1991). Third World critics of the Antarctic Treaty System have used the issue to reinforce their calls for substituting a UN-based regime or at least starting international research stations to end the duplication of facilities and improve coordination of research efforts (Beck 1991b).

The Hazards of Tourism and Recreation

Tourism activities in Antarctica have grown rapidly in recent years, but tourists still represent a far smaller presence there than do scientific personnel (National Research Council 1993b). Tourist visits totaled fewer than 7,000 in 1992–93. Because most tourism is ship based, it has been estimated that, in the aggregate, tourists' time on the continent is less than 1 percent that of scientific personnel (National Research Council 1993b). Recent incidents, however, have underscored the hazards posed by tourism and other recreational activity. The *Bahia Paraiso* was carrying a tourist party when it ran aground. Some tourist parties landing along the coast in zodiacs (large rubber rafts with outboard motors that can reach almost any shore area) have caused damage reminiscent of the activity of early explorers and scientists by disturbing nesting birds and leaving scars on local plant life that disappear very slowly because the plants grow so little each year (Laws 1989). Visits by tourist parties not only interrupt the schedules of stations, but also increase the litter and waste problems that plague many of them. Some people have proposed limiting tourism through a licensing system, but that would probably raise the territorial issue, inasmuch as Argentina and Chile are the most active promoters of Antarctic tourism.

At the same time, Antarctic tourism has increased awareness of the area's ecology among the general public in Western nations. Those who go return impressed with the stark beauty of Antarctica and the Southern Ocean. The larger tour operators provide on-board lectures by naturalists well versed in Antarctic science. They are also careful to brief passengers beforehand, provide competent guides for all onshore visits, and follow the suggestions for avoiding damage and disruption that have been developed by the Antarctic Treaty Consultative Parties.[48] Popular articles, books, and TV programs based on such trips bring vivid Antarctic images to the many unable to afford their own Antarctic voyage.[49]

WHICH REGIME?

Current debates can be confusing to those who have not been closely following Antarctic affairs because they deal with two logically independent questions at the same time. The main political issue is who will manage Antarctic and Southern Ocean affairs in the future. The contenders are the Antarctic Treaty states, particularly the Consultative Parties, on one side and the potential UN agency advocated by members of the so-called Malaysia group on the other.[50] The main environmental and economic issue is whether Antarctica will be treated indefinitely as a wilderness preserve off limits to mineral and fuel resource exploitation. Although some people are worried about the effects of overfishing in the Southern Ocean, there is much less support for a ban on fishing among environmentalist groups and virtually no support for it among governments. The Southern Ocean is included, however, in the international whale sanctuary declared by the IWC in 1994.

Beginning in the early 1980s, several transnational environmental organizations engaged in vigorous campaigning for the ban on mineral and fuel resource activity. The Convention on the Regulation of Antarctic Mineral Resource Activities may be considered to have contained among the strongest environmental standards and safeguards ever negotiated for an international commons regime (Joyner 1992). CRAMRA was opened for signature in Wellington, New Zealand, in November 1988, but it faced stern opposition from some quarters because of important omissions in its provisions. In April 1989 the French and Australian governments announced that they would not sign or ratify the agreement. First individually and then jointly, these two governments pressed for replacing CRAMRA with a comprehensive agreement on Antarctic environmental protection and a permanent ban on mineral and fuel resource exploitation. Their opposition was sufficient to create a stalemate because each, as a claimant, is among the countries whose ratification is necessary for CRAMRA's entry into force.[51] By the time the 15th Consultative Meeting assembled in Paris in October 1989, additional governments, including those of Austria, Belgium, Greece, and Italy, were expressing support for the Franco-Australian position (Coles 1989). But others opposed simply dropping CRAMRA. The New Zealand government advanced the most persuasive arguments for this view: without an agreed set of rules, any resource find that sparked near-term interest in exploitation would trigger severe political contention among the Consultative Parties, and dropping CRAMRA would leave Antarctica and its continental margin completely unprotected.

The Paris meeting ended in a compromise: CRAMRA was put on hold, although the discussions to elaborate rules on liability would proceed. The interim moratorium on resource exploitation was left in effect. At the same time, discussions of a regime for "comprehensive environmental protection" were taken up. Submitted papers and discussions at the Paris meeting revealed areas of agreement, as well as certain points of continuing disagreement, among the participants.[52]

The Consultative Parties took up the challenge of resolving their differences during the sessions of the 11th Special Consultative Meeting in November-December 1990, April 1991, and July 1991. The dispute between those wanting a permanent resource activity ban (in particular the governments of Australia, Belgium, France, Italy, New Zealand, and Sweden) and those opposing such a ban (most earnestly the governments of Japan, the United Kingdom, and the United States)[53] continued through the spring.[54] The shape of a compromise emerged when Germany and then, more surprisingly, Japan, shifted to some form of ban. Norway and Spain then proposed a ban that could be lifted after fifty years if 75 percent of the then–Consultative Parties, including all twenty-six of the current ones, agreed ("Hands Off Antarctica" 1991).

This became the basis of the new draft for a Protocol to the Antarctic Treaty on Environmental Protection.[55] Serious negotiations on the terms of the protocol had begun in November-December 1990, in Vina del Mar, Chile, and were continued in Madrid during April, June, and October of 1991. The United States proved a late holdout, throwing the process into disarray in July by insisting on lifting the ban after fifty years if any Consultative Party wished. The US position sparked intense controversy among the negotiators and at home, where the Bush administration was subjected to withering criticism. In the end, international and domestic pressures forced the administration to accept an agreement under which the ban could be lifted after fifty years only if two-thirds of the Consultative Parties agreed. The final version of the protocol was adopted in Madrid in October 1991 but awaits ratification by all of the Consultative Parties before entering into force. President Bush transmitted the protocol to the US Senate in February 1992 for its consent to ratification, which was granted in October 1992. US ratification, however, will await the enactment of domestic implementing legislation, as is the case with Russia and another eighteen of the Consultative Parties.

Although this agreement provides great protection for the Antarctic environment, it may have been attractive to some parties for other than ecological reasons. Many Australian officials fully accept the ecological case for preventing mineral and fuel resource exploitation, but some others, most notably Treasurer Paul Keating (Minister of Finance), have argued that treating Antarctica as a wilderness reserve would be less damaging to the continued validity of the Australian claim than would participating in CRAMRA.[56] In CRAMRA, Australia would have the right to sit on any regulatory committee established to deal with operators desiring to work within the Australian claim, but it would share decision making with other parties. Traditionally, the regulation of mineral and fuel resource exploitation on national territory has been the sole prerogative of that country's government, a point reinforced in contemporary thinking by wide acceptance of the doctrine of "permanent sovereignty over natural resources."[57]

Any such motivations cannot be expressed openly very often, however, because they go against the grain of joint management under the Antarctic Treaty. Nonclaimant governments could accept ecological reasons for imposing a ban without weakening their legal position, but explicitly or tacitly accepting this second argument would amount to granting claimants the very "special privileges" nonclaimants resisted so long in the CRAMRA negotiations. Several nonclaimants, particularly the Japanese (whose claimant status is blocked, or at least complicated, by the terms of the 1951 Peace Treaty), were reluctant to make the concessions to claimant desires for special rights incorporated in CRAMRA (Friedheim and Akaha 1989).

The whole CRAMRA decision process was, in any case, weighted against a rapid start of resource activites. This creates a situation opposite to that pertaining under CCAMLR. In CCAMLR, restrictions are adopted only if all parties agree. Under its negative trigger system,[58] activity proceeds unless a decision to ban or limit the activity is made. In sharp contrast, CRAMRA would have established a positive trigger decision mechanism, under which activity other than broad surveys and prospecting could occur only if and when authorized.[59] This does not cure all of the ambiguities and other flaws pointed out by critics, but the rules on when authorization is appropriate are framed cautiously enough to give advantage to the more environmentally oriented parties.

The intergovernmental debates were made more difficult to understand by efforts to blur certain distinctions. The choice between a world park or nature preserve, on one side, and CRAMRA on the other was often posed as one between environmental protection and immediate unregulated resource exploitation leading to serious environmental harm. It is true that a resource activity ban would make associated environmental harm impossible, but it is not necessarily true that resource activity, even under CRAMRA, would provoke immediate harm. Earlier sections detailed possible harms, but none of the possibilities holds a strong prospect of rapid disaster. Nor, as things stand, is much mining and hydrocarbon activity, the two greatest hazards, likely to occur soon, even without a moratorium. Perhaps the greatest assurance of the region's environmental security is the absence of economic incentives to threaten it (Joyner 1984).

Adding environmental damage to the cost side of the calculations increases the barrier to Antarctic economic competitiveness with other resources (Wright and Williams 1974). For preservationists, of course, the barrier is quite high; the pristine Antarctic ecosystem is given such high value that essentially no resource activities can be justified. For conservationists, the barriers are not as high but are still formidable compared to those that characterized the older frontier mentality.

A second confusion is sometimes fostered by advocates of UN management. Until 1989, most of their claims on behalf of universal participation focused on the need to have all countries participating in resource decisions if all were to be able to benefit from resource activity. Their arguments were cast in terms of the need to democratize the management of all common spaces, including Antarctica. In 1989 and 1990, a new argument found its way into the General Assembly debates. A number of delegations expressed a belief that effective environmental protection is possible only under a universal participation regime.[60] The precise reasoning underlying this claim has not been clearly stated, but, in an era when openness and mutual cooperation in the solution of global problems are regarded as essential to meeting the challenges facing humanity, the greater moral weight attaching to decisions made by all can be important. Certainly a broader basis would remove the possibility that nonparties would feel free to ignore any restrictions established in the Antarctic Treaty or associated agreements simply because there would be no nonparties in a universal participation regime. A universal participation body also would not be prey to the possibility of a nonmember's seeking to bully its way in by threatening to cause major damage to the environment.

Universal participation institutions provide superior environmental regulation only under certain circumstances, however. First, of course, the quality of environmental protection depends heavily on the attitudes of participating governments. No international institution—universal or small group, majoritarian or unanimity bound—provided very effective protection before the late 1960s because environmental concerns did not loom large in any government's calculation. Even when all agree that environmental concerns merit attention, the mere fact of participation does not guarantee that every party will be equally energetic in living up to the obligations they acknowledge. In a world where some governments are more concerned about such matters than are others, the results will depend very much on the play of politics and the procedures followed by various international institutions.

It is generally assumed, if only for reasons of workability, that a universal participation organization will operate under a simple or qualified majority rule. If restrictions on resources or other activities can be imposed only by a positive trigger, a majority vote rule is more likely to produce restrictions than is a unanimous consent rule because they will not be subject to veto by a single government. This is why conservationist governments sought to have majority voting in CCAMLR while fishing states pushed hard for retaining the unanimity rule even in the Scientific Committee (Howard 1989). Clearly it would be easier to raise a restrictionist majority under a simple rather than a two-thirds or three-quarters majority rule, but a qualified majority rule might be desirable for other reasons. Under a negative trigger system, however, the opposite is true. One environmentally concerned party can keep the rest from acting until it is satisfied. It may come under considerable pressure (as would a single exploitation-minded holdout in a positive trigger system), but it would have every right to hold its ground. A small group—fewer than the one-third that could block action—would mutually reinforce efforts to hold out. Thus, the effect of the voting rule depends partly on whether decisions regulate ongoing activity or are required before any activity begins.

The play of coalition formation would also be very important. Many international institutions produce only least-common-denominator decisions. On environmental issues, this means adopting the amount of regulation that is acceptable to the least enthusiastic or least concerned party, which gives environmental laggards the preponderant influence over policy. This pattern can be overcome, but only under certain conditions. International institutions can escape the least-common-denominator pattern when they permit a conjunction of intergovernmental and domestic pressures to be brought to bear.[61] The environmental NGOs that actively monitor Antarctic affairs are attempting to create such conditions by lobbying Western governments and presenting position papers to both the Consultative Parties and the Malaysia group. The Malaysia group has found the assistance helpful. The Consultative Parties have mixed reactions with respect to the various organizations involved, some of which have acquired reputations as unappeasable critics, while others are regarded as valuable observers. When there are domestic as well as international political rewards for environmentally sound policies or domestic as well as international losses for environmentally unsound ones, governments will pay more sustained attention.

Either a UN-based or an Antarctic Treaty-based regime would depend on member governments for the implementation of decisions. The current scale of Antarctic resource activity would not support an extensive international regulatory body. If a "full protection" option were adopted, there would be virtually no revenue for this purpose. It is highly unlikely, therefore, that any Antarctic management agency would acquire the supranational authority and staffing level needed to enforce decisions on member states or other actors. This, too, means that international institutions provide as much environmental protection as governments believe they need.

Though often ignored by those who follow Antarctic affairs, there remains a third possibility: territorial division. Despite the formidable difficulties involved, a number of claimant governments still seem to prefer division over UN management. They firmly believe in the legitimacy of their claims and are unimpressed by Malaysia group arguments that Antarctic claims are a relic of past colonialist eras with no legitimacy in the contemporary world.

A straightforward acceptance of pre-1961 claims would pose serious problems, however. First, the Argentine-British-Chilean overlap (see map 7.1 and source) would have to be resolved. Second, those governments now active in the Antarctic that had no claims in 1961 would have to be accommodated somehow. Both the United States and Russia reserve the right to assert a claim should the need arise. Germany seemed ready to raise claims in the mid-1930s but has not pursued them since (Beck 1986). Japan was required to abandon all previous claims it might have had in the 1951 Peace Treaty, but some Japanese legal experts have argued that this would not preclude new claims based on post-1951 activity should the Antarctic Treaty collapse. Accommodating any or all of these countries would require a redrawing of the Antarctic political map. The presently unclaimed area would not satisfy any of these states (not even the United States, for which the area was tacitly left by the current claimants) because its resource potential is relatively poor. Some claimants, particularly those with extensive claims, such as Australia, New Zealand, and Norway, would have to yield up territory. This would make a redivision far less appealing to them. Whether it would push them all the way to supporting a UN-based agency is another question; they might prefer continuing the Antarctic Treaty system under which they at least have the protection of the consensus rule. The large Third World Consultative Parties cannot be ignored, either. Participating in a division might prove uncomfortable for Third World states after having denounced colonialism for as long as they have, but they could argue that it would not be colonialism in the usual sense of the term because no indigenous human population is involved. However, their standing in the nonaligned movement might suffer enough to make them reluctant to participate. As Third World unity fades, even they might find the Antarctic Treaty consensus rule preferable to the majoritarian rules of a UN body.

Thus, it is unclear who will manage Antarctica and, by extension, the Southern Ocean. It is clear, however, that environmental concerns are receiving much greater attention and that management has moved in more conservationist directions than anyone would have expected in 1959. This movement is unlikely to satisfy the most preservationist-oriented environmentalists, who have maintained a steady barrage of criticism since the mid-1970s. Nor will the most use-oriented governments and interest groups be happy. Fishing and navigation will continue, but deep seabed mining is very unlikely even fifty years or more from now. Land-based mining and offshore hydrocarbon exploitation are scarcely more likely in that time but, in any case, have been banned for fifty years. The precise balance to be drawn in the future between resource activity and environmental protection will be determined by how the managers of Antarctica and the Southern Ocean—whoever they are—establish the balance between exploitation and conservation that is implicit in the already accepted principle of "rational use."

CHAPTER 8

The Law of the Sea

INTERNATIONAL ENVIRONMENTAL LAW is an essential component of environmental security. As a system of legal norms and principles governing relations among nations in protecting the planet's natural environment, it provides necessary means for the protection against threats to national well-being or the common interests of the international community associated with environmental damage (Broadus and Vartanov 1991). Although international law lacks a police force to ensure compliance, it fulfills two important functions: it provides a structure for international cooperation and assistance to prevent environmental harm, and it offers a framework for avoiding disputes by clarifying the rights and duties of states.[1]

Among the currently underutilized tools for maintaining environmental security is the 1982 United Nations Convention on the Law of the Sea. As described by the World Commission on Environment and Development, adherence to the convention is "the most significant initial action that nations can take in the interests of the oceans' threatened life-support system" (WCED 1987). Similarly, the Oceans Chapter of Agenda 21 (Chapter 17), adopted at the 1992 United Nations Conference on Environment and Development, recognizes the Law of the Sea Convention as the "international basis upon which to pursue the protection and sustainable development of the marine and coastal environment and its resources." But, until the convention is in force with widespread participation, its full contribution to the maintenance of environmental security will not be realized.[2]

Nevertheless, the Law of the Sea Convention remains the only global agreement that provides comprehensive coverage of all aspects of the various uses, abuses, and resources of the world's oceans. Its basic objective is to establish "a legal order for the seas and oceans which will facilitate international communication, and will promote the peaceful uses of the seas and oceans, the equitable and efficient utilization of their resources, the conservation of their living resources, and the study, protection and preservation of the marine environment" (Preamble, para. 4).

The Secretary General of the United Nations is constantly making efforts aimed at achieving universal participation in the convention. As the convention's deep seabed mining regime (Part XI) has made the United States and other developed nations reluctant to ratify the convention, the focus of these discussions has been on finding a satisfactory compromise in this area (UN Secretary General 1991).[3]

Some environmentalists, including some from the Law of the Sea community, believe that the current discussions on seabed mining constitute an opportunity to reconsider the environmental provisions of the convention (Part XII) as well. Otherwise, they observe, ten years will have to pass before a State Party may propose a substantive amendment (Article 312).

However, there exists another point of view shared by many scholars, who believe that reconsideration of Part XII at this juncture is unnecessary and could even lead to the collapse of negotiations. The nonacceptance of Part XI by a large number of countries already constitutes a danger for the viability of the Law of the Sea Convention as a comprehensive compromise package of regulative norms concerning ocean activities. To introduce the possibility of reconsidering Part XII (or any other part) at the same time runs the risk of undermining this compromise and thus the interests of the world community. Adherents to this view are convinced that even environmental interests would be better served by an imperfect convention than by a return to the situation where each nation establishes its own environmental norms.

Certainly there are elements of the Law of the Sea Convention that will have to be added or changed over time, to keep pace with changing circumstances. To a certain extent, events have outpaced the convention already.[4] The fact that some important issues are currently omitted from the convention does not mean, however, that they cannot be adopted and implemented through other legal instruments or perhaps even through implementation under the convention itself.

THE SOURCES OF INTERNATIONAL ENVIRONMENTAL LAW

"International law, like all living law, is in a process of continuous growth and adaptation to new needs and circumstances" (Henkin et al. 1980). It is generally acknowledged to have developed from three sources: treaties, custom based on state practice evidencing a sense of legal obligations, and general principles of law.[5] The growing significance of multilateral international declarations and resolutions in this respect is described below.

Treaties and other legal agreements serve both to codify existing customary law and to establish new norms and principles to govern nations' behavior. For example, while much of the Law of the Sea Convention is a restatement or codification of existing law, some of its most significant elements go beyond customary law. In the environmental area, it introduced the concept of a comprehensive environmental law of the sea based on the obligation of all states to protect and preserve the marine environment and to control all sources of pollution.

Customary law also evolves. Even before the Law of the Sea Convention's entry into force, many of its innovative provisions that went beyond customary law had prompted common acceptance as to the development of new customary law. For example, it introduced the concept of the exclusive economic zone, which extends coastal state jurisdiction for certain purposes such as the protection and preservation of the marine environment from the traditional 12 nautical miles out to 200 nautical miles. The 200-mile EEZ has already acquired customary status.[6]

General principles of law are the third primary source of international law. Courts and arbitral tribunals, when asked to decide upon questions of international law, often rely on equitable notions common to most national legal systems to fill in the gaps left by customary and treaty law.[7] Such principles are often of a general nature, based on such notions as "elementary considerations of humanity" and "every State's obligation not to allow knowingly its territory to be used for acts contrary to the rights of other states."[8] This second principle forms the basis for the duty of states to prevent or abate significant environmental harm to areas outside their jurisdiction.[9]

Some scholars are beginning to recognize a fourth source of law, based on resolutions and declarations of the UN General Assembly and other international bodies (see, e.g., Birnie 1988; Kiss and Shelton 1991). This source of law (called *soft law*) is distinguished from the first three in that the declarations and resolutions are not per se legally binding.[10] Instead, they promote changes in conduct by providing visionary goals and norms of behavior. Hard law, on the other hand, represents the rules that states have generally agreed to be bound by and can be held responsible for breaching.

As more and more nations comply with and accept the principles set forth in global declarations, the lines between soft and hard law grow vague. Such principles often form the basis for new legal agreements and are thereby transformed into hard law. But they may also receive such widespread approval that they become accepted as customary law. For example, certain principles adopted by the Stockholm Declaration, issued by the 1972 United Nations Conference on the Human Environment, formed the basis for the environmental provisions of the Law of the Sea Convention (UN Secretary General 1989).[11] These principles have since met with such widespread acceptance that, notwithstanding the delayed entry into force of the Law of the Sea Convention, most states now accept them as reflective of customary international law.[12]

An Overview of the Law of the Sea Convention, Part XII

The environmental provisions contained in Part XII of the Law of the Sea Convention (together with associated provisions) embody the first attempt to develop a public international law framework to respond to marine pollution and degradation of the marine environment (UN Secretary General 1989).[13] Most significantly, the convention contains the most concise and absolute formula of the principle that states have the obligation to preserve the environment in areas inside and outside their territorial jurisdiction (Article 192) (Kiss and Shelton 1991). The obligation of states to protect and preserve the marine environment extends to all maritime zones, from internal waters and coastal ports to the high seas. This is made clear by Article 193, which recognizes the sovereign right of states to exploit their natural resources but requires that this be done "in accordance with their duty to protect and preserve the marine environment."

Part XII does not contain concrete marine pollution standards, however, nor does it purport to substitute for special agreements. Rather, it aims at resolving the principal issues of determining the major rights and duties of nation states in this regard. Its main objective is to delimit states' competence pertaining to (1) the establishment of concrete national and international rules and standards to prevent, reduce, and control pollution of the marine environment and (2) the enforcement of these rules and standards. In addition, it aims to determine the interrelationship between international norms and standards and national environmental legislation. An important feature of the Law of the Sea Convention is that states' competence to establish and enforce marine environmental protection norms and standards is closely linked to the specific conditions established by the various international legal regimes that relate to different categories of ocean spaces—the high seas, EEZs, contiguous zones, archipelagic waters, straits used for international navigation, territorial seas, and the internal waters of coastal states.

Thus, Part XII is expressly designed to operate as an umbrella for future global and, particularly, regional action. It sets forth rights and obligations of states with respect to the marine environment and obliges states to implement and elaborate upon this framework both through global conventions addressing particular forms of pollution and through regional agreements tailored to the requirements of discrete sea areas.[14] States are obliged to cooperate not only in the creation of new legal instruments, but also in many other areas. The obligation to cooperate includes notifying other countries of imminent threats of pollution, cooperating in eliminating the effects of such pollution and preventing or minimizing the damage, and jointly developing contingency plans for responding to pollution incidents in the marine environment. It also includes undertaking research programs, exchanging information and data, and establishing appropriate scientific criteria. Cooperation must be carried out directly or through competent international organizations (Articles 197, 200, and 201).

From this jurisdictional scheme arises one significant but essentially unavoidable drawback of the convention: the scope of a coastal state's authority, which depends on the regime of ocean spaces, does not always reflect the dynamics of large marine ecosystems, which may straddle two or more jurisdictional zones (French 1990; Thorne-Miller and Catena 1991). Despite the convention's recognition of the duty to protect and preserve the marine environment, this obligation is at times thus made subservient to other states' interests in unhampered commercial and military navigation (see Section III.A.1.(d)). Such, however, is the nature of the compromise (among various ocean users) approach that the convention represents.

Another significant drawback is that the convention does not adequately address the most pressing problem of marine pollution, that from land-based sources, including airborne pollution (see Section III.A.1.(b) and (c)).[15] As a result, the environmental regime established by the Law of the Sea Convention is not entirely satisfactory from a precautionary point of view.

Despite its limitations and the fact that it is not yet in force, the Law of the Sea Convention has effectively served as a catalyst for a whole series of international agreements of both global and regional character.[16] Nevertheless, bringing the convention into force with widespread participation remains a necessary starting point for universal action to protect the marine environment.

POLLUTION CONTROL AND ECOSYSTEM PROTECTION

The convention's provisions for pollution control are quite general. They delineate states' rights and obligations with respect to protection of the marine environment and call on states to take further action. The convention calls for states to take measures individually, as well as jointly, to prevent, reduce, and control pollution of the marine environment from any source.[17] Although the convention provides few specific recommendations, it does provide some general goals (Article 194). Measures taken by states are to include those designed to "minimize to the fullest possible extent"

- the release of toxic, harmful or noxious substances, especially those that are persistent, from land-based sources, from or through the atmosphere or by dumping;
- pollution from vessels, in particular measures for preventing accidents and dealing with emergencies, ensuring the safety of operations at sea, preventing intentional and unintentional discharges, and regulating the design, construction, equipment, operation and manning of vessels;
- pollution from installations and devices used in exploration or exploitation of the natural resources of the seabed and subsoil; [and]
- pollution from other installations and devices operating in the marine environment. (Article 194.3)

Rather than set forth the rules that states are to follow to achieve these ends, the convention clarifies states' rights and obligations with respect to particular sources of pollution.

Pollution by Dumping

For control of pollution by dumping (i.e., the transport and disposal of wastes from ships and offshore structures at sea), the convention calls for all states to adopt laws and regulations and to take other necessary measures (Article 210.1-2). National laws are to be "no less effective" than rules adopted at the global level (Article 210.6). Coastal states have the right to control dumping activities by other states within their waters (including the EEZ) and on their continental shelf; no dumping may occur without their prior express approval (Article 210.5).

Ocean dumping is regulated at the international level by the 1972 Convention on the Prevention of Pollution by Dumping of Wastes and Other Matter (the London Convention, previously known as the London Dumping Convention). The London Convention was already in existence when the Law of the Sea Convention was negotiated and does not fulfill all requirements set forth in the latter. With membership limited to sixty-three countries, the London Convention lacks universal applicability.[18] Consequently, it does little to control dumping off nonmember coastal states, whether by those states themselves or by others. The coming into force of the Law of the Sea Convention will provide at least three added benefits. It will (1) clarify the rights of coastal states to prohibit dumping off their shores, (2) require states to enact measures domestically that are at least as effective as the global requirements, and (3) encourage more states to participate in the international regime.

Pollution from Land-based Sources

By all accounts, land-based pollution (including airborne pollution originating from land) is responsible for almost 80 percent of all marine pollution, yet it has received the least attention at the international level.[19] The Law of the Sea Convention contributes little in this regard because it fails to direct states to take any specific action at the international level to fulfill their obligation to protect the marine environment from land-based pollution (Boyle 1992).

The convention expresses a preference for dealing with land-based pollution at the national and regional levels. States are to adopt laws and regulations and take other necessary measures to prevent, reduce, and control land-based marine pollution (Article 207.1). They are to endeavor to harmonize their policies "at the appropriate regional level" (Article 207.3). Several regions have adopted comprehensive programs for the elimination and control of land-based marine pollution.[20] Before the conclusion of negotiations for the Law of the Sea Convention, agreements had already been adopted for the Baltic Sea, the North-East Atlantic, and the Mediterranean.[21] Agreements have since been adopted for the South-East Pacific, the Persian Gulf region, and the Black Sea.[22] Work is currently under way on similar protocols for the Wider Caribbean region and Western and Central Africa.

Many coastal areas do not fall naturally into regions, however, and their concerns remain unaddressed. Nor is there any mechanism to coordinate the various regional programs that do exist, to prevent duplication of efforts, or to stimulate the transfer of experience, technology, and financial assistance from developed to developing countries or from one region to another. The Baltic Sea and North-East Atlantic programs are not part of UNEP's Regional Seas Programme, under whose auspices the other land-based agreements have been negotiated. Bringing the Law of the Sea Convention into effect will enhance international coordination and cooperative efforts, but it can do little to strengthen national commitment to international action.

At the international level, the convention prescribes only vague, nonbinding actions for regulating land-based pollution. States have only to "endeavor" to establish global and regional rules, standards, and recommended practices and procedures and "to take into account" such measures in domestic pollution control laws (Article 207). This is in contrast to the obligations regarding ocean dumping, which make the London Convention the obligatory minimum standard for state practice. International agreements on land-based pollution are further to take into account characteristic regional features, the economic capacity of developing states, and their need for economic development (Article 207.4). As a result, Article 207 lacks both precise content and any means for direct enforcement (Boyle 1992).

Pollution from or through the Atmosphere

Airborne pollution of the sea has received relatively little attention at the international level. Initially thought to be only a minor problem, it is now recognized as a major source of marine pollution, responsible for more than 33 percent of the total load (GESAMP 1990). It can also travel vast distances, making it difficult to deal with effectively at the regional level alone. For example, pesticides applied in West Africa are a suspected source of the high DDT, chlordane, and heptachlor levels found in coral tissue off the coast of South Florida (Keating 1991).

As with land-based sources, the Law of the Sea Convention imposes no specific requirements at the international level for the control of airborne marine pollution. Instead, it requires states to adopt domestic legislation that is merely to "take into account" internationally agreed rules, standards, and recommended practices and procedures (Article 212). States are also to "endeavor" to establish global and regional rules, standards, and recommended practices and procedures.

Several regional agreements do specifically address the topic of airborne marine pollution, including the Paris Convention, the Helsinki Convention, and the 1976 Barcelona Convention for the Protection of the Mediterranean Sea against Pollution (through an annex to its land-based pollution protocol). Other regional seas treaties echo the Law of the Sea Convention in that they contain general language requiring states to take all appropriate measures to prevent, abate, and control this form of pollution (Kiss and Shelton 1991). A convention involving Northern Hemisphere countries, the Geneva Convention on Long Range Transboundary Air Pollution (1979), while not directly aimed at reducing pollution of the marine environment, will have indirect benefits (Boyle 1992).[23] There is no coordinated international effort, however, to stem the release of toxic, harmful, or noxious substances from airborne sources.

Vessel-Source Pollution

The need to reconcile competing ocean uses and interests is recognized by the preamble to the Law of the Sea Convention, which states that "the problems of ocean space are closely interrelated and need to be considered as a whole." The detailed jurisdiction and enforcement rules set forth by the convention with respect to vessel-source pollution have struck an important balance between the interests of international navigation and the concerns of coastal states for environmental protection (see also chapter 4). The concept of the exclusive economic zone introduced by the convention fostered the enclosure of vast areas of ocean space that were formerly high seas and made them subject to national jurisdiction and control. States with strong maritime interests took great care to ensure that such expansion of coastal state jurisdiction did not interfere with vessels' traditional rights of innocent passage and freedom of navigation. Although the resulting jurisdictional rules expand the powers of coastal and port states compared with prior customary norms,[24] these states' jurisdictional competence remains subject to substantial restriction (Bodansky 1992), as necessary to protect the interests of international navigation from possible abuses by states other than the flag state.

The primary authority to apply and enforce national and international standards regarding vessel-source pollution rests with the state where the vessel is registered (the flag state). International standards are to be established by states acting through the "competent international organization" (i.e., the International Maritime Organization) or general diplomatic conference (Article 211.2). Port states may establish particular requirements for pollution prevention as a condition of entry for foreign vessels into their ports or internal waters (Article 211.3).

In contrast to flag states and port states, coastal states have limited authority to regulate passing ships: in their territorial seas, they may adopt national laws that are stricter than international standards and apply them to foreign vessels—provided, however, that such rules do not hamper innocent passage (Article 211.5).[25] For vessels passing through the EEZ, a coastal state may, as a general rule, apply only "generally accepted international rules and standards" and not its own, potentially stricter ones (Article 211.5).[26] Countries bordering narrow straits whose territorial seas are commonly used for international navigation are even more restricted in their standard-setting capacities: they may not impose norms beyond those passed at the international level (Article 42).

No matter how restrictive these convention limitations might seem from the coastal states' point of view, they were a part of a difficult compromise reached between coastal and maritime interests, and they are a prerequisite for ensuring unhampered international navigation. If these limitations were absent from the convention, they could entail either an inordinate increase in vessels' construction costs or, more likely, the impossibility of entering or transiting the offshore zones where disparate design, construction, or manning standards would be established under the pretext of environmental regulation.[27] That would inevitably force foreign vessels to bypass these zones and incur a substantial lengthening of their routes and an increase in freight costs. Moreover, if such standards would be established by any of the states bordering the international straits, the straits might become effectively closed for international navigation.

The convention also invests coastal and port states with broad powers to investigate and enforce national pollution prevention norms and rules with regard to foreign vessels. However, it defines the admissible limits of interference into foreign navigation with the aim of pollution prevention on the side of a coastal state in the areas under its jurisdiction and control (Articles 218 to 220, 233, and others). Thus, the convention considerably restricts the right of a coastal state to stop a foreign vessel and to undertake subsequent measures (inspection, detention, and so forth) when the vessel violated national norms and rules related to pollution prevention, not only in the EEZ but in the international straits covered by the territorial sea as well.[28] At the same time, the convention significantly broadens port state enforcement powers compared to prior customary norms. It permits port states to institute proceedings against vessels voluntarily within port for violations occurring within its EEZ (Article 220.1), as well as for discharges on the high seas in violation of international standards (Article 218.1)1.[29]

Still, certain provisions of Part XII, drafted as it was at the height of Cold War tensions, are unnecessarily biased toward the protection of maritime, and particularly naval, interests. This helps explain why governmental vessels in noncommercial service are made exempt from all environmental provisions.[30] This concession to military and other governmental vessels is incompatible with the usual principles of immunity, which provide only for exemption from enforcement procedures, not from applicability of the law. As Kiss and Shelton (1991) stated: "Logically and according to general international law, there is no reason why government ships and aircraft should not be governed by marine pollution rules, even though they may be immune from otherwise applicable enforcement procedures."

Pollution from Activities in the International Area

The convention distinguishes between seabed activities in areas subject to national jurisdiction and activities conducted on the high seas, or in the seabed "Area" beyond national jurisdiction. Seabed activities in the Area are subject to a special regime created by the convention to ensure that the resources of the deep seabed, deemed part of the common heritage of humankind, are equitably shared.[31] Such activities are to be subject to special environmental rules as provided for in Part XI, not in the section devoted to protection of the marine environment (Part XII).

Part XI of the convention establishes an International Seabed Authority responsible for overseeing the development of seabed mineral resources and the protection of the marine environment in the Area. The convention confers on the Authority the duty to assure protection of the ocean environment: the Authority is authorized to adopt appropriate rules to prevent pollution or other hazards to the marine environment, including interference with its ecological balance (Article 145).[32]

The Authority is given unique responsiblities regarding environmental protection in the Area outside national jurisdiction. No similar institution exists for the protection of the high-seas environment above the seabed. In this regard, Part XI sets a good example for the creation of an international body charged with protection of an area traditionally considered part of the global commons.[33] If accepted by a wide international community, its example may some day be applied to the creation of an international regime for protection of the marine environment and conservation of the natural resources of the high seas.

Ecosystem Protection

In addition to pollution-oriented directives, the convention contains several significant provisions directed at the protection of marine ecosystems. Some scholars have gone so far as to say that the convention mandates ecosystem-based management (Belsky 1985).

The convention directs states to take measures to protect and preserve rare or fragile ecosystems as well as the habitat of depleted, threatened, or endangered species and other forms of marine life (Article 194.5). As this provision falls under the general section on marine pollution, it requires states to protect these special ecosystems and habitats from the effects of pollution originating from all sources, in addition to other, more general conservation measures.

In addition, states are to take all measures necessary to prevent, reduce, and control pollution of the marine environment resulting from "the use of technologies" under their use or control (Article 196). This provision fails to specify what technologies are envisaged, but it could be read to mean biotechnology or, more broadly, any technology that pollutes or otherwise threatens marine ecosystems. The rest of the sentence refers more specifically to biological pollution. It requires that states "take all measures necessary to prevent, reduce and control pollution of the marine environment resulting from the use of technologies under their jurisdiction or control, or the intentional or accidental introduction of species, alien or new, to a particular part of the marine environment, which may cause significant and harmful changes thereto" (Article 196).

The convention's ability to protect ecosystems is further bolstered by requirements for monitoring the risks or effects of pollution. As described in greater detail in the next section, the convention requires states to monitor continuously their ongoing activities and to assess the potential effects of planned activities to determine whether they are likely to pollute the marine environment or cause significant and harmful changes (Article 204).[34]

Other important ecosystem-protection provisions are found outside Part XII under the sections for conservation of the living resources of the EEZ and the high seas. For example, Part V, on the EEZ, obligates coastal states to ensure through proper conservation and management measures that the maintenance of the living resources in the EEZ is not endangered by overexploitation (Article 61.2). In taking measures to determine the allowable catch of the living resources, states "shall take into consideration the effects on species associated with or dependent upon harvested species with a view to maintaining or restoring populations of such associated or dependent species above levels at which their reproduction may become seriously threatened" (Article 61(4)). The same requirement is applicable on the high seas as well (Article 119.1(b)). Such a provision, however, aims only to maintain the viability of such species, not to protect their role within the food web or the functioning of the marine ecosystem as a whole.

THE LAW OF THE SEA CONVENTION AS A FRAMEWORK FOR ASSURANCE BUILDING

Agreements regulating various forms of pollution as detailed above are not enough to ensure environmental security. To follow through on the initial commitment, nations must see a structure that promotes national implementation and builds confidence that other nations are also fulfilling their obligations. There must also be a mechanism for reconciling conflicting interpretations of treaty obligations. The Law of the Sea Convention facilitates the achievement of environmental security by establishing such mechanisms.

In particular, the convention requires states to commit to three long-term activities:

1. to monitor, assess, and publish reports on environmental risks of certain activities under state control that may cause substantial pollution or significant and harmful changes to the marine environment;
2. to provide scientific and technical assistance to developing states to build national capacities to implement international legal obligations; and
3. to submit to compulsory dispute resolution mechanisms for disagreements regarding the interpretation or application of rights and obligations under the convention.

Monitoring, Environmental Assessment, and Reporting

The monitoring and environmental assessment provisions of the Law of the Sea Convention (Articles 204, 205, and 206) require states to take four steps:

1. to monitor continuously the effects of any activities they permit or in which they engage to determine whether these activities are likely to pollute the marine environment;
2. to publish the results of such monitoring in reports, which are to be made available to all states, and to provide such reports to the competent international organizations;
3. to assess the potential effects of activities when states have "reasonable grounds for believing that planned activities under their jurisdiction or control may cause substantial pollution of or significant and harmful changes to the marine environment"; and
4. to report on the results of such assessments to the competent international organizations.

The obligation to maintain surveillance on the effects of ongoing activities goes beyond the current requirements in any of the regional seas agreements. Regional agreements generally call for monitoring, but the emphasis is on determining the existing levels of contamination. Some of the regional agreements also call for monitoring to determine the effectiveness of the measures adopted (e.g., the Paris Convention on land-based sources of marine pollution, the Athens Protocol), and the recently signed revision of the Helsinki Convention for the Baltic Sea region calls for monitoring of compliance with emissions limitations for point sources of pollution (1992 Helsinki Convention). The 1992 Paris Convention for the Protection of the Marine Environment of the North-East Atlantic requires in Article 6 that the parties undertake and publish joint assessments on a regular basis, including "evaluation of the effectiveness of measures taken and planned for the protection of the marine environment and the identification of priorities for action." In no agreement, however, is the obligation to maintain surveillance of the state of the marine environment so clearly expressed as in the Law of the Sea Convention.

The duty to assess environmental impacts of planned activities originated from the principle that a state has a duty to prevent transfrontier environmental harm (Meng 1987; see "Sources of International Environmental Law" above). The convention broadens that requirement. In addition to activities that threaten the interests of other states, the convention requires environmental assessments for any activities that may threaten the marine environment. Significantly, this includes areas subject to national jurisdiction and areas beyond all nations' jurisdiction (the high seas). Similar provisions are now included in most regional seas agreements, but most are applicable only to the specific area covered by the agreement. As a result, the high seas may be left out of consideration.

Dissemination of the results of environmental impact assessments as mandated by the convention allows countries to react to the potential threat to the marine environment and possibly to enter into an agreement to avert such harm. Consultation with other nations before commencing potentially harmful activities is not, however, explicitly required. At the regional level, only some of the agreements contain such a requirement (e.g., the 1983 Cartagena Convention and the 1992 Helsinki Convention do; the 1976 Barcelona Convention and the 1978 Kuwait Convention do not). It could be argued that the duty to consult with potentially affected states is implicit in the Law of the Sea Convention because, if a state proceeds with an activity that it knows to be contrary to its duty to protect the marine environment, the state would be in breach of the convention's primary obligation. Another state party could then utilize the convention's mechanisms for peaceful resolution of disputes (see below). It would be more useful, however, if such a requirement were made explicit and broadly applicable throughout the marine environment, not just within the regional seas.

Similar requirements have been incorporated into the UNEP regional seas agreements, but some of these go further and require that states parties develop procedures for the dissemination of information and, as appropriate, consult with and seek comments from potentially affected states. The Rio Declaration broadens this obligation by calling on states "to provide prior and timely notification and relevant information to potentially affected States on activities that may have a significant adverse transboundary environmental effect and shall consult with those States at an early stage and in good faith" (Principle 19).

The lack of reporting of relevant information remains a major stumbling block to the achievement of environmental security. International environmental agreements are not well monitored (GAO 1992), and the convention does little to help in this regard. It establishes no institutional structure to receive, process, and distribute reports. Instead, the convention assumes that "competent international organizations" will fill in the gap. However, as demonstrated above, many gaps in coverage exist at the regional level. Moreover, certain activities may have more than regional effects (e.g., airborne pesticide residue) or may affect the high seas.

Scientific and Technical Assistance

Because developing countries may need assistance in obtaining pollution-control technology and access to relevant expertise to protect the marine environment, the convention's technical assistance provisions are of vital importance. States are required to promote in developing countries programs of a scientific, educational, or technical nature[35] and to provide these countries with assistance in responding to major accidents or other incidents of pollution and in preparing environmental assessments (Article 202). International organizations are also to grant preference to developing countries in the allocation of funds and technical assistance and in access to their specialized services (Article 203). Without such assistance and incentives, the convention remains an empty shell because the obligations of developing countries are conditioned upon what can be achieved through the use of the "best practicable means at their disposal and in accordance with their capabilities" (Article 194).

Great emphasis is placed in the convention on the development and transfer of marine science and technology. The duty to cooperate in the development and transfer of specifically marine technology is an innovative feature of the convention, where an entire Part (XIV), consisting of thirteen articles, is devoted to its promotion. Article 266 requires states to cooperate to promote the development and transfer of marine science and technology on fair and reasonable terms and conditions. Particular attention is paid to measures to promote the scientific and technological capacity of developing countries, which includes the establishment of national and regional centers to stimulate and advance marine scientific and technological research. Functions of regional centers identified by the convention include (in addition to basic scientific research and data acquisition) training in the conservation and management of living resources and programs related to protection of the marine environment and prevention of pollution (Article 277).

The duty to assist developing countries has been recognized as a general obligation of international law since the Stockholm Conference of 1972. This duty is also referred to in the Athens Protocol and the Rio Declaration. However, none of these statements appears in a binding legal document of universal application. The coming into force of the Law of the Sea will give this requirement added status and stimulate the transfer of needed technology to developing countries, thereby enhancing environmental security.

Dispute Settlement

The convention was the first international agreement to gain consensus on compulsory binding dispute settlement procedures for questions regarding the implementation of marine environmental protection requirements, among others (Kimball 1989). It provides for a variety of dispute settlement arenas. When a dispute arises over the interpretation or application of the convention, parties are first encouraged to pursue exchange of views, conciliation, or other peaceful means of settlement (Article 284). When such attempts fail, any party to the dispute may request that the dispute be referred to the International Court of Justice or a tribunal authorized by the convention (Article 286). Authorized tribunals include the International Tribunal for the Law of the Sea, an arbitral tribunal selected from a list of general maritime experts, or a special arbitral tribunal selected from a list of experts in particular fields (fisheries, protection and preservation of the marine environment, marine scientific research, and navigation, including pollution from vessels and dumping) (Article 287).

Such mechanisms would enable one state to require another to submit to compulsory conciliation of any dispute over alleged failure to ensure that the living resources of the EEZ are not seriously endangered. Similarly, if a state violates the convention regarding its obligations to prevent marine pollution, a coastal state may submit the dispute for a binding decision.

As creatures of the convention, none of these tribunals will come into being until after the convention enters into force.

CONCLUSIONS

This chapter has explored the significance of the Law of the Sea Convention for the maintenance of environmental security. Compared to the preexisting legal regime governing marine environmental protection, the convention is an invaluable addition. It contains many requirements that go beyond existing agreements. For example,

> it establishes basic treaty obligations to protect and preserve the marine environment from all sources applicable to all states—developed and developing. It goes beyond traditional notions of commercial fishery conservation: it expressly permits stricter protection for marine mammals, and requires protection of the food chain as well as rare or endangered species and their habitats. . . . The Convention establishes a duty to enforce detailed international regulations to protect the environment from all marine sources of pollution [which] . . . applies to all parties to the Convention, even if they do not participate in the adoption of particular international environmental standards. The Conven-

tion also contains sophisticated detailed provisions harmonizing interests in free navigation with coastal state interests in control of pollution from ships. (Panel on the Law of Ocean Uses 1993)

Where the convention's provisions have been incorporated into regional or international agreements, it still provides the harmonizing and coordinating element. It is also applicable in areas not covered by regional seas agreements or where international agreements lack universal participation.

Most significantly, the convention contains the foundation for ecosystem-based management aimed at predicting and preventing environmental harm. It requires states to monitor the effects of ongoing human activities and to assess the potential effects of new activities on the quality of the marine environment. States are further required to minimize to the fullest possible extent the release of toxic, harmful, or noxious substances and to take special measures to protect rare and fragile ecosystems and the habitats of depleted, threatened, or endangered species. Fisheries management is to take into account the effects of harvest on species associated with or dependent on the targeted species. Many of these requirements go beyond those found in regional agreements and are more universally applicable.

The convention also has requirements aimed at building the capacity of developing countries to implement these broad-based requirements. It requires countries to provide assistance to developing countries by promoting programs of a scientific, technical, and educational nature. Assistance is to include help in preparing environmental impact assessments and in responding to major accidents. The convention is the only agreement to place great emphasis on the development and transfer of marine science and technology. The coming into force of the Law of the Sea Convention will stimulate the transfer of needed technology to developing countries.

The convention's institutional elements, such as dispute resolution mechanisms, can make important contributions to the clarification of the rights and duties of states and the peaceful settlement of disputes. Such mechanisms, however, will not come into effect until the convention enters into force, and their practical usefulness will depend on the extent to which all states participate in the convention through ratification or accession. As of November 1993, sixty states had ratified or acceded to the convention, including only one developed economy (Iceland) (table 8.1).

Despite the convention's many important features, there are some shortcomings in its approach to marine environmental protection. As noted earlier, the convention does not permit amendments for ten years after it has come into force. Particularly for this reason, the opportunity to update the environmental provisions should not be ignored as nations contemplate, especially through consultations initiated by the UN Secretary General, means to encourage other nations to ratify the convention.

The convention is weak with respect to land-based and airborne pollution. Although almost 80 percent of marine pollution is generated on land, existing regional agreements may be only partially effective in stemming the flow of pollutants into the sea. They generally lack mechanisms to stimulate the transfer of technology and expertise between regions and to developing countries, and they often fail to address the multiple sources of land-based pollution—from discharges, disposal, seepage, or drainage or through the atmosphere. The call in UNCED's Agenda 21 for the UNEP Governing Council to convene, as soon as practicable, an intergovernmental meeting on protection from land-based sources of marine pollution is a step in the right direction.

If the Law of the Sea Convention is to succeed in providing a stable regime for protecting the marine environment, further work is needed to improve its provisions relating to vessel-source pollution. With the Cold War receding into the past, maritime states should now be willing to redress the existing imbalance between military navigational interests and environmental protection, and the exemption of military vessels from the convention's marine environmental protection provisions should be removed. Furthermore, the international community may decide to rectify the definition of what constitutes innocent passage by excluding actions threatening environmental security, including intentional pollution and reckless conduct, to bring the existing rules in line with two emerging, but widely accepted, norms of international law: the precautionary approach and the "polluter-pays" principle.

Although the convention already conveys the obligation to protect and preserve the marine environment, this duty is at times relegated to secondary importance after short-term commercial and military concerns. The precautionary approach should be recognized as a guiding principle for all activities that may adversely affect the marine environment. As expressed in Principle 15 of UNCED, "Where there are threats of serious or irreversible damage, lack of full scientific certainty shall not be used as a reason for postponing cost-effect measures to prevent environmental degradation." Similarly, the requirement for ecosystem-based management should be made explicit.

Table 8.1
The United Nations Convention on the Law of the Sea[a]: Ratifications and Accessions as of November 16, 1993

1. 10 December 1982 Fiji
2. 7 March 1983 Zambia
3. 18 March 1983 Mexico
4. 21 March 1983 Jamaica
5. 18 April 1983 Namibia
6. 7 June 1983 Ghana
7. 29 July 1983 Bahamas
8. 13 August 1983 Belize
9. 26 August 1983 Egypt
10. 26 March 1984 Côte d'Ivoire
11. 8 May 1984 Philippines
12. 22 May 1984 Gambia
13. 15 August 1984 Cuba
14. 25 October 1984 Senegal
15. 23 January 1985 Sudan
16. 27 March 1985 Saint Lucia
17. 16 April 1985 Togo
31. 25 August 1986 Guinea-Bissau
32. 26 September 1986 Paraguay
33. 21 July 1987 Yemen
34. 10 August 1987 Cape Verde
35. 3 November 1987 Sao Tome & Principe
36. 12 December 1988 Cyprus
37. 22 December 1988 Brazil
38. 2 February 1989 Antigua & Barbuda
39. 17 February 1989 Zaire
40. 2 March 1989 Kenya
41. 24 July 1989 Somalia
42. 17 August 1989 Oman
43. 2 May 1990 Botswana
44. 9 November 1990 Uganda
45. 5 December 1990 Angola
46. 25 April 199 Grenada
47. 29 April 1991 Micronesia[b]

18.	24 April 1985	Tunisia	48.	9 August 1991 Marshall Islands[b]
19.	30 May 1985	Bahrain	49.	16 September 1991 Seychelles
20.	21 June 1985	Iceland	50.	8 October 1991 Djibouti
21.	16 July 1985	Mali	51.	24 October 1991 Dominica
22.	30 July 1985	Iraq	52.	21 September 1992 Costa Rica
23.	6 September 1985	Guinea	53.	10 December 1992 Uruguay
24.	30 September 1985	Tanzania	54.	7 January 1993 Saint Kitts & Nevis
25.	19 November 1985	Cameroon	55.	24 February 1993 Zimbabwe
26.	3 February 1986	Indonesia	56.	20 May 1993 Malta
27.	25 April 1986	Trinidad/Tobago	57.	1 October 1993 St. Vincent/Grenadines
28.	2 May 1986	Kuwait	58.	5 October 1993 Honduras
29.	5 May 1986	Yugoslavia	59.	12 October 1993 Barbados
30.	14 August 1986	Nigeria	60.	16 November 1993 Guyana

[a]The Law of the Sea Convention will enter into force on November 16, 1994.
[b]Indicates that a state has acceded to the convention.

SOURCES: "Ratifications of the Law of the Sea Convention" 1993; Higginson, pers. comm., 1993; "Sixtieth Ratification of the Law of the Sea Convention Deposited" 1993.

Environmental security is dependent on all nations' taking action to anticipate and prevent harm, often in the face of scientific uncertainty. Given continuing deterioration in the health of the world's oceans and threats to the integrity of marine ecosystems, remedial responses on an ad hoc basis are no longer enough.

CHAPTER 9

Conclusions

ENVIRONMENTAL SECURITY IS both a goal and a way of thinking about environmental affairs. It is the reasonable assurance of protection against threats to national well-being or the common interests of the international community associated with environmental damage. Some problems involving the world's oceans, and of high mutual interest to the United States and Russia, call this assurance into question. These are the problems examined in this work, from a shared Russian-American perspective and with a combined emphasis on economics and law.

Viewing the problem from a shared Russian-American perspective has value because of the unique history and position of the two countries in international ocean and security affairs, their similarly linked roles in the world's tumultuous geopolitical transformation, their experience with radically different economic and environmental management systems, and the breadth of the intellectual contribution afforded by their scholars' grounding in different theoretical and analytical traditions (which are most divergent in economics, more similar in international law, and most similar in the natural sciences). During the nearly half-century that the two countries faced each other as hostile nuclear superpowers, the oceans, with their transport lanes and importance to submarine and other naval operations, occupied a special place in the two countries' security interests. Throughout the on-again, off-again efforts at bilateral confidence building and détente, the oceans were a common focal point for cooperative safety arrangements and scientific collaboration ("Report of the First Meeting of the U.S.-Russian Joint Committee" 1993; WHOI 1991).

Now, in the midst of the hopeful but perilous new political realities initiated in Russia, the two countries remain the world's foremost maritime presences, conducting most of the naval operations that might degrade the marine environment, controlling the bulk of the world's maritime shipping capacity, governing two of the planet's longest coastlines, and managing the two largest zones of coastal-state jurisdiction. Both are among the world's leaders in offshore minerals extraction and fisheries exploitation, and together they may well possess more scientific assets needed for improved understanding and use of ocean processes than the other nations of the world combined (including over 80 percent of the world's scientific research fleet and most of its satellite remote sensing capability).

Significantly, the two countries also exercise a preeminent role in setting the pace and direction of post–Cold War military conversion. The radical changes in the Soviet Union in the late 1980s and its final collapse in 1991 exerted an immense, if still only partially realized, influence on contemporary ocean activities and marine policy at the national and international levels. The changes prompted reexamination of priorities in the expenditure of the nation's financial, industrial, scientific, and human resources on the development of strategic naval capabilities. This created an unprecedented opportunity to reduce the nuclear militarization of the world's ocean, as has in fact been begun. Military conversion thus became a practical priority on national and international agendas.

The changes in the former Soviet Union also extended the prospects for international marine scientific research by opening areas for collaborative investigation (such as large areas of the Arctic Ocean, the Sea of Okhostsk, etc.) that previously were effectively closed to the international scientific research community. Russia is among the countries most actively polluting the oceans, and its deep socioeconomic and political crisis has created serious practical obstacles for environmental protection, clean-up, and scientific research projects. The government is still the primary if not the only source of funding for such projects, and government support is sorely limited. Using Russian scientific and technical capabilities in understanding and addressing ocean environmental problems has become a practical prospect, but only with international assistance in support of Russia's marine science facilities and personnel (WHOI 1991).

Human behavior that creates environmental hazards is driven in large measure by economic motivations. Measures to alter that behavior are likely to be more effective to the extent that they marshal market forces and shape economic incentives. Moreover, for every type of oceanic hazard, the marginal cost of reducing its threat through environmental protection measures will eventually exceed, at some level of expenditure, the marginal benefit gained from further reduction. It is in society's interest, therefore, to determine the most cost-effective balance between marine environmental protection and the activities that jeopardize it, taking into account the full range of costs and benefits associated with each. Doing so, of course, is not easy. Problems of environmental security arise precisely because of differences over or disregard for that balance.

Markets and other economic institutions are notoriously imperfect; externalities or spillovers, such as environmental damages, distort allocative decisions and lead to excessive risks. Waste, inefficiency, and suffering abound. Even if economic behavior in general were able to locate an efficient balance between environmental risk and protection, that still would not assure that the outcome would be equitable or would lead to a fair distribution of benefits and costs. International law and international organizations can facilitate efforts to overcome such economic shortcomings, to foster fairness, and to condition the institutional framework within which such activities take place. International legal issues pervade the cases examined in this book, and unresolved legal questions or incomplete legal institutions contribute to problems. Improvements in the legal context can promote solutions to many environmental difficulties. The principles articulated in the Rio Declaration of 1992 provide guidance for the contributions of law to environmental security, but their practical effect will be limited until they are widely expressed in actual state practice and specific legal instruments. The 1982 Law of the Sea Convention promises to be the single most influential and effective instrument of environmental security for ocean areas and activities, but that promise cannot be fully realized without the entry into force of the Law of the Sea Convention with widespread participation.

CRITICAL PROBLEMS OF INTERNATIONAL ENVIRONMENTAL SECURITY

Recalling the criteria for *critical problems* of international environmental security—problems likely to destabilize normal relations between nations and to provoke international countermeasures—the cases we have examined include a critical problem for the long term, that of radioactive materials management. However limited the actual risk imposed by three decades of Soviet radioactive dumping in the Arctic and Pacific and by Russia's continuing dumping to cope with a legacy of neglect and urgent backlogs, perceptions often govern responses and threaten to destabilize normal relations. The responses from Iceland, Norway, and Japan to Russia's recent dumping of radioactive wastes reflect greater concern about threats that their seafood will be perceived as contaminated than about the scope of the actual environmental damage.

If misperceptions contribute to the problem, then broadened *glasnost* may contribute to the solution. The record of secrecy, deception, and violation of international agreements and declarations certainly helped intensify the problem to a critical level. Russia's increasing openness, information gathering and exchange, cooperative research, notification, and consultation represent important improvements. These developments, coupled with continuing understanding and forbearance by its neighbors and greatly accelerated and expanded technical and financial assistance to deal with the backlog, may well resolve the problem, secure normal relations, and remove the likelihood of countermeasures. All evidence suggests, however, that the worst of Russia's radioactive disposal problems have yet to be faced (WHOI 1993). Consequently, serious international attention may be required to respond to this critical problem of environmental security.

The other oceanic threats to environmental security examined here remain at sub-critical levels, although problems of North Pacific living resources management have been critical and could flare up again. Despite the North Pacific's chronic history of bitter resource management disputes, impaired relationships between "victim" and "culprit" states, and international countermeasures, recent multilateral measures have eased the problem and opened the way to long-term solutions. Notable measures include the UN General Assembly's moratorium on large-scale driftnet fishing on the Pacific high seas; the 1994 Convention for the Conservation and Management of Pollock Resources in the Central Bering Sea; the 1993 UN Conference on straddling stocks and highly migratory stocks; agreements embodied in the 1992 Convention for the Conservation of Anadromous Stocks in the North Pacific Ocean, with establishment of the North Pacific Anadromous Fish Commission; and creation of the cooperative North Pacific Marine Science Organization. Many of these measures, however, are only tentative or temporary. Bitter differences remain between American environmentalists and traditional Asian fishing interests, in particular. Stocks remain limited and species linked through ecosystem relationships. Thus, the North Pacific's complex of marine living resource problems remains far from an environmentally secure solution.

The potential for critical problems is always present, too, in the other cases examined. This is most apparent for hazardous materials transport and land-based marine pollution. The uneasy and not entirely secured balance between coastal state powers and navigational freedoms is quite vulnerable to being upset by a single catastrophic accident. The situation will be especially unstable if the Law of the Sea Convention enters into force with almost exclusively developing coastal states as parties. Under those circumstances, the only states enjoying access to the convention's conflict resolution mechanisms would be the least relevant states. Similarly, unresolved international standards for responsibility, liability, and compensation in the event of a marine-pollution incident heighten the potential for destabilized relations, unilateral actions, and countermeasures.

Environmental problems in the Arctic and in the Southern Ocean and those associated with the Law of the Sea Convention seem farther removed from critical status. Nonetheless, deep international differences over Antarctic management policies, increasingly vocal and unified political expressions by and on behalf of northern indigenous peoples, and temptations for unilateral breakaway actions (most likely intended to strengthen coastal state security) threaten the balance achieved in the convention and remind us of the potential for critical problems in these areas as well.

CROSS-CUTTING THEMES

In our case review of environmental security problems, nine cross-cutting themes become apparent (see table 9.1). These not only describe important aspects of the problems examined, they also help guide analysis of the problems and narrow the continuing search for elusive remedies.

From both an economic and a legal perspective, a significant gap in the existing environmental security system is the absence of clearly specified and widely recognized criteria for the assignment of *responsibility, liability, and compensation* in cases of environmental damage. This deficiency arises in virtually all of the cases examined, and it has hampered progress in the areas of transboundary land-based marine pollution, hazardous materials transport, and radioactive contamination. The UN International Law Commission has been working for over a decade to propose rules concerning injurious consequences of "acts not contrary to international law" and state responsibility for "internationally wrongful acts" (Tinker 1990). In the absence of such general guidelines, few treaties specify standards of liability for environmental damage or whether the standards are keyed to negligence, due diligence, or strict liability.

When clear international standards of responsibility, liability, and compensation are acknowledged, economically efficient and equitable

Table 9.1
CROSS-CUTTING THEMES IN OCEAN ENVIRONMENTAL SECURITY

• Responsibility, liability, and compensation	• International cooperation and coordination
• Global perspectives	• International organizations
• Population growth	• Technology and technology transfer
• Criteria and standards	
• Monitoring, data, and information	• Military conversion

outcomes will be fostered by shifting the risks of environmental damages onto those responsible, causing them to take precautionary measures to minimize their exposure, allowing them to reduce the costs of those risks through insurance, and expediting environmental repair and compensation for those most affected. Even a system based on clear standards, however, will result in economic inefficiencies and inequity unless it accounts for all damages, including lost environmental benefits in terms of both use and non-use values. That innovation will require substantial progress in economic methodology as well as in institutional design.

The cases examined here also make clear the value of viewing each problem from a *global perspective*. The concept of environmental security is global in spirit, positing something desirable for all everywhere, if not yet fully recognized as a basic human right to a healthful environment (Gormley 1976; Tinker 1990). The problem of land-based marine pollution is felt not just through its extent in the planet's coastal waters and its transboundary transport via interconnected environmental processes and media (atmosphere, rivers, runoff, and currents). It is also transported internationally through such human mediation as seafood exporting, flows of capital investments, tourism, and heartfelt concern. Widespread caring and commitment to a healthy environment also give a global reach to all of the other cases reviewed in this volume, although the hazard of radioactivity may be global in the most physical sense, by virtue of the major powers' nuclear naval operations. Certainly the problems of hazardous materials transport arise in the context of a global transportation system and touch the interests of all the world's coastal states. Human institutions like markets and transport systems introduce other global connections, such as the pressures exerted on remote Arctic sites or North Pacific ecosystems by distant demand for products. In recent years, nongovernmental organizations have made important contributions in articulating and acting on such global interests.

A related cross-cutting theme, which expresses itself more directly in some cases than in others, is the influence of *population growth*. There is a strong correlation between coastal or drainage basin population growth and land-based marine pollution; in the absence of remedial measures, increased human population in coastal areas puts increased pressures on coastal waters and other environmental resources. The increased density of coastal populations also leaves more humans exposed to harm from land-based pollution and, for that matter, from hazardous cargo spills. More generally, population growth increases the demand for goods and services, stimulating supply responses in the form of greater volumes and varieties of hazardous materials shipments and increased extractive pressure on North Pacific and Arctic ecosystems. Similar connections can be identified for all the cases.

Criteria and standards (for environmental policy, for acceptable exposure, for safe practice, etc.) comprise another set of issues that run through all of the cases. At a fundamental level, sound environmental policy will be determined by the criterion used to govern potentially damaging activities. A strict criterion of no harm to the environment is unrealistic, and even undesirable, for it would rule out all potentially harmful activities no matter how great their offsetting benefits. The precautionary principle, a kind of minimum-worst-consequence criterion, attempts to offer guidance on this score by counseling caution on the side of the environment and calling for preventive measures even when there is no certainty of hazard. The principle does not help, however, in determining what kinds or levels of preventive measures should be taken or how realistic the risk of hazard must be before they are invoked.

Even where there is certainty of some harm, it may be worthwhile accepting that harm as part of the cost of some greater benefits. Thus, the policy criterion implicit throughout most of our case analyses has been a net benefits criterion, in which a goal of policy is to identify and encourage those steps likely to result in the greatest aggregate net benefits. This criterion entails its own problems of measurement, aggregation, and distribution, but it does imply some softening of certain rigid, and environmentally troubling, standards. Where coastal state preventive measures merely impair innocent passage, for example, they might be accepted as compatible with general navigational freedoms if the environmental and economic benefits they afford are sufficiently large. Similarly, flexibility and sensitivity to national differences must be considered in the design and promulgation of all international environmental standards if they are to avoid inefficiency, or even irrelevance.

The establishment and enforcement of environmental standards inevitably raise issues of *monitoring, data, and information management*. These issues, too, pervade our cases. In every case, monitoring is needed to reveal environmental status and trends, to track relevant human activities and their effects, and to gauge the level of compliance with regulations and international obligations. Such basic information as the degree of jurisdictional straddling by fish stocks in the Bering Sea, the level of North Pacific fishing effort, the exact composition and source terms of the radioactive wastes dumped in the arctic seas, and casualty rates in hazardous materials shipping and their correlates remains elusive. The success of monitoring in waters around the Sellafield nuclear reprocessing plant stands out as an exemplary exception, as NOAA's coastal status and trends program promises to be for US land-based marine pollution. Monitoring generates data, of course, and it is important that data from different places be comparable, well managed, and easily accessible. That is a major international challenge, and effective information management requires a high level of international cooperation and coordination. Tracking compliance with international environmental agreements is a priority, and most evidence indicates that the record is poor (GAO 1992).

The importance of *international cooperation and coordination* is a common theme in all of the cases. Cooperation is necessary because of the many mutual interdependencies involved in environmental affairs and because effective responses to problems will be those devised and implemented by sovereign states acting through their domestic authority. Coordination is necessary because of states' differences in capability, endowment, situation, motivation, stakes, priorities, and custom.

International organizations are both an expression of and a vehicle for cooperation and coordination. They take many forms, most generally inter-governmental or nongovernmental. The UN system, with its family of specialized agencies and organizations, is the most important of these, and its involvement and influence pervades our cases. For enhanced effectiveness, the intrinsic limitations of the United Nations, as an instrument of sovereign states rather than a sovereign actor in its own right, must be kept clearly in view (Lefever 1993). The United Nations could be the single most important and influential organization in maintaining ocean environmental security by concentrating on its areas of greatest effectiveness: information gathering and exchange and the promotion and development of international environmental law (Tinker 1990). Issues of environmental security are best handled separately from the UN Security Council's special responsibilities (Tinker 1992).

Many other organizations, of course, have important roles to play. Prominent examples we have seen include CSCE, the Black Sea Commission established by the 1992 Bucharest Convention, the North Pacific Marine Science Organization, the North Pacific Anadromous Fish Commission, maritime classification societies, marine insurance organizations, the South Pacific Forum, the Northern Forum, and the Antarctic Treaty Consultative Parties. A vital role has been assumed by nongovernmental organizations, such as environmental advocacy groups, although some of the wealthier ones may have acquired quasi-governmental influence with little accountability. NGOs serve a constructive purpose in many ways: as independent gadflies, monitors, whistle-blowers, information clearinghouses, educators, catalysts, and windows of transparency into otherwise opaque private or governmental processes. Some NGOs, however, can mislead and distort public perceptions on scientific questions to serve their own organizational self-interests or because of honest misconceptions or incompetence. (Of course, governmental organizations have also been known to exhibit these faults.) Greater clarity and consistency in the standards and procedures through which NGOs are granted standing in intergovernmental affairs would be useful (Tinker 1993).

Another theme running through the cases is the importance of *technology and technology transfer*. Virtually all of the problems reviewed—from the depletion of North Pacific fisheries and wasteful by-catch to the proliferation of hazardous chemicals and their extensive seaborne trade and the improper or unresolved disposal of radioactive wastes—are regrettable spillover effects of advancing technology. Solving these problems will also require the application of technology. Many proposals for the control of land-based marine pollution call for the use of the "best available technology," and advanced electronic monitoring and detection networks are evolving rapidly for deployment in coastal waters (Gaines 1992). Numerous technologies show promise for improvements in hazardous materials transport, including advanced navigational systems and vessel traffic control technologies. Soft technologies, such as information systems, know-how, and best practice and procedures, show promise for useful application in all of the cases. A noteworthy potential for technology's contribution to environmental security is in the verification of compliance with international agreements and obligations (Ausubel and Victor 1992), though institutional innovations may be needed to realize the full potential. In any case, for technologies to be widely used, they must be widely available and accessible. Technology transfer, whether through free-market exchange, international assistance, or some institutional combination that provides incentives for suppliers and users alike, is essential.

Technological advances and technology transfer are among the outstanding potential benefits for environmental security associated with post–Cold War *military conversion* efforts. Some of the greatest scientific and technological skills in both the United States and Russia have been bottled up in their military-industrial establishments. Active efforts are now under way in both countries, through such programs as the US Advanced Research Projects Agency's (ARPA) Technology Reinvestment Project, to redirect some of those advanced capabilities to environmental purposes. Defense-related remote-sensing and data communication technologies point to breakthroughs in environmental monitoring, and defense navigational databases could accelerate progress in the adoption and diffusion of electronic charts and vessel traffic control systems.

As seen clearly in the cases of radioactive and hazardous materials, however, post–Cold War military transitions also raise some new threats to environmental security by hastening the decommissioning of whole classes of weapons, creating new or enlarged disposal problems, and altering the flow of hazardous materials. Recognition of these new threats, together with the lessening of bipolar tensions, enhances the rationale and opportunity for making certain that all military activities meet all generally accepted international standards and for ensuring compliance without prejudicing national security interests. Both countries are already moving to upgrade the environmental practices of their military establishments: the US Navy has developed and is pursuing an "Environmental Strategic Plan," and the US Congress is requiring full Navy compliance with Annex V of MARPOL under the 1987 Marine Plastic Pollution Research and Control Act, even though international law offers immunity to government vessels from enforcement of those standards. In Russia, despite political and economic distractions, the military for the first time sought approval from the Ministry of Environment before proceeding with its 1993 emergency disposal of low-level liquid radioactive wastes in the Sea of Japan (Lystsov, pers. comm., 1993). The world's new military situation also affords the chance, with America and Russia leading the way, to strengthen and extend the currently weak constraints on actions to imperil the environment in the course of armed conflict.

Conclusions   255

The concept of environmental security enhances our understanding of ocean environmental problems by focusing attention on the strategic interdependence of nations in environmental affairs, on the potential for destabilized relations and conflict over resource use and environmental practices, and on the sorts of international measures and institutions that might help solve these problems. The concept begins with an analogy to traditional security affairs, which opens the way for applying the intellectual apparatus of that field to environmental problems, even as it spotlights the relationship of traditional defense activities to environmental affairs.

In that regard, the custom of separate treatment of military facilities and operations from international environmental agreements stands out as an awkward situation. Self-interested actions and agreements by sovereign states must provide the basis for environmental security, as for international security more generally, and sovereign states may reasonably be expected to protect the essentials of their sovereign interests. Greater efforts to conform military activities to general standards and practices established by international law, however, are warranted.

In any event, it is clear that armed conflict ranks among the most serious threats to the ocean environment and to environmental security more generally. The maintenance of peace and international comity is therefore a first priority for the oceans and environmental security.

Notes

CHAPTER 2. LAND-BASED MARINE POLLUTION: THE GULF OF MEXICO AND THE BLACK SEA

1. Two parties that continue to favor a legally binding global convention are the Greenpeace organization and the government of Iceland.
2. The Mexican border states are Baja California, Sonora, Chihuahua, Coahuila, Nuevo Leon, and Tamaulipas. In the United States, California, Arizona, New Mexico, and Texas have jurisdiction in the border area.
3. A closure/advisory is a single beach for which a closure advisory has been issued for a single day (Abdelghani et al. 1993).

CHAPTER 3. LIVING RESOURCE PROBLEMS: THE NORTH PACIFIC

1. The region includes several natural complexes: ecosystems of the Bering Sea, ecosystems of other bordering areas, and the North Pacific Ocean proper. See *Biologischeskie resursi Tikhogo Okeana* 1986; *Issledovanie ekosystemi Beringova Morya* 1973, 1990; *Tikhii okean* 1982 (all in Russian); and *The Management of Marine Regions: The North Pacific* 1982.
2. In particular, there are overlapping distributions of Sakhalin, Kurile, and Hokkaido chum salmon and of North American and Kamchatka chum salmon and a zone of overlapping Asian and American sockeye salmon spanning a distance of 1,500 km from east to west. See Birman 1985 (pp. 5, 55) and Birman 1980 (pp. 17–18) (both in Russian) and Larkin 1988.
3. See Fredin (1987) for detailed discussions of the regulatory history of Alaskan groundfish fisheries.
4. Similar problems occur in an area of high seas in the Sea of Okhotsk that is about one-fifth the size of the Doughnut Hole and has been dubbed the *Peanut Hole*. Unregulated fishing in the Peanut Hole seems to have brought stocks of pollock to the verge of collapse and has fueled tensions between Russia, which recently called for a three-year moratorium on pollock fishing in the area, and Poland, South Korea, and China, which thus far have agreed only to reduce their fishing activities by 25 percent (Pitt 1993).
5. The principles embodied in these unsuccessful proposals included (1) that high-seas fishing must be conducted with no adverse effect on stocks found within coastal state jurisdiction; (2) that a coastal state has priority in

determining the relative interests of states fishing such stocks in the adjacent high seas; (3) that high-seas fishing states have a responsibility to comply with previously prescribed international measures and to ensure compliance by their flag vessels; (4) that such international measures concerning straddling stocks should provide for consistency with the measures applied by coastal states within ther EEZs; and (5) that UN agencies should prepare guidelines for application of the precautionary approach in areas within national jurisdiction. Also advanced but ultimately omitted was a proposal to recognize the principle of large marine ecosystems as a management concept, which had both adherents and opponents among coastal states and high-seas fishing states alike (Burke 1993).

6. Burke, Freeberg, and Miles (1993) argue that the US position on driftnets has from the beginning "been dominated by politics, irrespective of input from those knowledgeable about the fisheries and ecosystems involved," noting that this position stands in stark contrast to the US position in Law of the Sea negotiations (see chapter 8), that "resort to an international agency is inconsistent with the freedom to engage in deep seabed mining."

7. This problem is not confined to the fisheries of the Bering Sea and the adjacent North Pacific. In September 1992, the Food and Agriculture Organization held Technical Consultations on High Seas Fisheries in Rome, where discussion centered on the need for "accurate and complete statistical reporting on fisheries in all waters, particularly on high seas fisheries, both for monitoring and evaluating high seas catches and as an essential instrument for research and fisheries management" (UN Secretary General 1992).

8. This principle was recognized in May 1992 by the International Conference on Responsible Fishing, which adopted the Cancun Declaration and a comprehensive concept of responsible fishing as "encompassing sustainable utilization of fisheries resources in harmony with the environment; the use of practices not harmful to ecosystems, resources or their quality; the incorporation of added value to fish products processed according to required standards; and the conduct of commercial practices providing high quality products" (UN Secretary General 1992).

CHAPTER 4. HAZARDOUS MATERIALS TRANSPORT

1. These include the sagas of the *Khian Sea/Felicia, Karin B, Zanobia, Deepsea Carrier, Mare Equatoriale,* and three *Isola* ships.

2. Provisions for radioactive materials are based on recommendations issued by the International Atomic Energy Agency.

3. A brief but insightful discussion of this situation as it applies to marine underwriters can be found in "Hazardous Cargoes—An Insurer's View," in *Fairplay,* 11 October 1990, p. 46.

4. Although it publishes no tonnage estimates, the United Nations Environment Programme has calculated that 20 percent of the millions of tons of wastes that annually cross international borders move from industrialized to developing nations.

5. The International Code for the Construction and Equipment of Ships Carrying Dangerous Chemicals in Bulk (IBC Code) applies to chemical tankers constructed after July 1, 1986. The Code for the Construction and Equipment of Ships Carrying Dangerous Chemicals in Bulk (BCH) applies to tankers constructed before that date. Requirements in the IBC and BCH Codes cover the design, construction, equipment, and operation of tankers. The IGC Code is the International Code for the Construction and Equipment of Ships Carrying Liquified Gases in Bulk.

6. The Joint Group of Experts on the Scientific Aspects of Marine Pollution (GESAMP) is an advisory scientific body of experts sponsored by the United Nations, IMO, FAO, UNESCO, WMO, IAEA, UNEP, and IOC.

7. Under Article 92 of the Law of the Sea Convention, such exceptional circumstances are those "expressly provided for in international treaties or in this Convention" (para. 1). Article 92, para. 1 further provides that "a ship which sails under the flags of two or more States, using them according to convenience, may not claim any of the nationalities in question with respect to any other State, and may be assimilated to a ship without nationality."

8. Unitary chemical weapons contain just one chemical agent whose molecular composition and structure make it highly toxic or lethal. By contrast, binary chemical weapons, which are also present in the US stockpile, consist of two nonlethal chemicals that, when mixed, form a lethal chemical agent. The nerve agents known by their Army code designations as GB and VX are both used in unitary weapons, and both are soluble in water—GB especially so. VX, which evaporates very slowly, is designated *persistent* (with respect to the environment). GB presents primarily a vapor hazard, whereas VX presents primarily a contact hazard (National Research Council 1993a).

9. An arguable exception, in the case of the CWC, is the Middle East, where Syria, Egypt, and Iraq have indicated their intention to comply with the terms of the convention but not a commitment to sign it, because of their concern over the potential for Israel, a signatory, to use chemical weapons against them (Staples, pers. comm., 1994).

10. The International Maritime Satellite (INMARSAT) Organization, based in London, operates geostationary communications satellites for maritime use.

CHAPTER 5. RADIOACTIVITY IN THE OCEANS

1. Nuclear weapons tests first occurred at Bikini Atoll in 1946. Atmospheric tests were conducted in the United States and in the Pacific between 1954 and 1962. Underground tests continue to the present. Some 360 to 400 test detonations have occurred worldwide.

2. Transuranic elements are radionuclides that are heavier than uranium—for example, americium, curium, neptunium, and plutonium. As radioactive waste products, they are generated from nuclear reprocessing and

fuel fabrication activities. Wastes containing transuranics are considered to be more dangerous than those that do not, because transuranic elements are characterized by long half-lives.

3. Reprocessing is a chemical process in which usable plutonium and uranium products can be recovered from spent reactor fuel. Low- and moderate-level radioactive wastes in both liquid and solid form are generated during fuel reprocessing.

4. At-sea dumping is considered a radionuclide dispersal strategy rather than a long-term containment strategy, the latter being considered necessary for safe management of high-level radioactive waste.

5. Risk is defined in this context as entailing exposure to seawater or seafood containing these radionuclides, which could lead to the development of a fatal cancer or hereditary defects.

6. In the case of the Windscale (1957), Three Mile Island (1979), and Chernobyl (1986) accidents, much of the radioactive exposure and contamination was concentrated on land.

7. A source cited in Berkhout, Suzuki, and Walker (1990) calculated that this amount equals the total quantity of plutonium found in the US and Russian nuclear weapon arsenals combined.

8. Note deleted in proof.

9. Encompassing safety, training, and emergency preparedness and response, for example.

10. This encompasses fissionable material as cargo, as fuel used to power nuclear vessels and satellites, and as a component of weapons.

11. Hypothetically, the fissionable material associated with reactor fuel in some lost submarines and vessels could lie undisturbed on the seabed. This would be dependent on the events surrounding the loss and the physical condition of the wreck. Although no attempt has been made to confirm or update this statement, a published report on the fates of the US submarines *Thresher* and *Scorpion* (which sank in 1963 and 1968, respectively) stated that the nuclear fuel contained in their reactor vessels continued (as of 1982) to hold fission products that had not been exposed to seawater (see US Navy 1982). As noted previously, fuel is leaking, however, from a Soviet submarine that sank in the mid-Atlantic in 1986 (Broad 1994).

12. As early as 1974, nuclear contamination controversies plagued the government of Japan and its nuclear program. In that year, the nuclear-powered cargo vessel *Mutsu* developed a reactor leak while on sea trials and was forced to drift for weeks before being allowed to dock. Local fishermen, fearing contamination of their shellfish beds, organized a boycott of the vessel that lasted until the government agreed to provide compensation for lost income.

13. For further discussion, see Berkhout, Suzuki, and Walker 1990.

14. Japan's demand for plutonium in the future is based on a growing reactor R&D program and on possible use of mixed-oxide fuels in existing thermal reactors.

15. Under the terms of the "Agreement for Cooperation Concerning the Peaceful Uses of Atomic Energy," the United States exercises substantial extra-territorial control over Japan's nuclear programs. This extends to the nuclear fuel and technologies supplied by the United States, as well as to any by-products and materials ultimately produced through their use in the nuclear fuel cycle.

16. The nuclear complex at Sellafield encompasses 600 acres and includes commercial reactors, reprocessing facilities, and storage repositories. An oceanographic perspective on Sellafield can be found in Hain (1986).

17. In the case of the various pathways, annual Sellafield dosage levels calculated for members of critical groups have ranged between 10 and 50 percent of the ICRP's total annual maximum recommended exposure limit.

18. In an effort to gain hard currency, Russian shipping officials have instituted new programs to ferry tour groups to the North Pole and commercialize shipping operations across the Arctic Basin.

19. As for the Soviets' dumping of high-level wastes in their own shallow coastal waters, the most appropriate notions of liability may be those that apply in the case of land-based sources of marine pollution, where transboundary damages typically are overshadowed by the far more immediate and more severe damage done to the polluter's own coastal and territorial waters. In such cases, liability for culpable negligence is typically determined and enforced by national institutions of justice (Barsegov 1993).

19a. Russia abstained from the November 1993 vote to ban low-level radioactive dumping under the LC (as did Belgium, China, France, and the United Kingdom) and provided formal notice to the contracting parties of its nonacceptance and inability to comply with the ban, which entered into force one hundred days after the vote. (The United States supported the ban and played a primary role in blocking the inclusion of an opt-out mechanism, much to the surprise and dismay of France and the United Kingdom.) B. Wood Thomas. Office of International Activities, EPA. Telephone communication, May 1994.

20. Technical information related to marine radioactivity that has been published by Greenpeace, for example, has been discredited or is frequently attacked by the scientific community. The American and Russian media, in the case of nuclear accidents or naval operations involving nuclear weapons, often engage in speculation when confronted with a lack of information or with official attempts to suppress information. This tends to feed public anxiety about nuclear contamination in the oceans.

21. A comprehensive review of research on this alternative, including a list of seminal source documents, can be found in *Seabed Disposal of High-Level Radioactive Waste: A Status Report on the NEA Coordinated Research Programme*, published in 1984 by the Seabed Working Group, Nuclear Energy Agency, Organization for Economic Cooperation and Development.

22. The seminal research of this alternative is summarized in Hollister, Anderson, and Heath 1981.

23. For the most part, these have evolved over time from the negotiating frameworks and preparatory work surrounding international conventions such as the Law of the Sea Convention, the London Dumping Convention, the MARPOL treaties, the Montreal Protocol, the Basel Convention, and the recent UNCED summit. Diplomatic conferences in support of regional treaties have also contributed some of the theoretical foundation behind these general principles.

24. The first acknowledged dumping activity was recorded in 1946 and was conducted by the United States.

25. Under the terms of the London Convention, dumping is defined as (1) any deliberate disposal at sea of wastes and other matter from vessels, aircraft, platforms, or other manmade structures or (2) any deliberate disposal at sea of vessels, aircraft, platforms, or other manmade structures.

26. Three other important regional conventions with detailed provisions pertaining to radioactive discharges—the Paris Convention, the Helsinki Convention, and the Athens Protocol—pre-date the Law of the Sea Convention, which was adopted in 1982. Of these, only the Athens Protocol falls under the aegis of UNEP's Regional Seas Programme.

27. As described more fully in chapter 4, the Salvage Convention recognizes the right to a reward for all salvage operations having a useful result; no payment is due for operations that have no useful result. Criteria for fixing the reward due a salvor include the following: (1) value of the property; (2) skill and efforts of the salvor in preventing or minimizing damage to the environment; (3) measures of success; (4) time, expenses, and losses incurred; (5) risk of liability run by the salvor; (6) promptness of the salvor; (7) availability and use of vessels or equipment; (8) readiness and efficiency of the salvor's equipment; and (9) nature and degree of danger.

28. Note deleted in proof.

29. Scientific knowledge about these characteristics, for example, can be utilized to determine the risk of transboundary pollution associated with an operational discharge or the long-term health implications posed by a nuclear accident, so that appropriate control measures can be developed (see Finn 1983).

30. It would be impossible to provide actual profiles of radioactivity concentrations for large expanses of the oceans and the seabed because of the extreme variations both vertically and horizontally, in the water column and in bottom sediments.

31. The ICRP's radioactive release limitation controls have been recommended by the IAEA for use by national regulatory agencies.

32. These collective dose estimates are made of very small annual doses to a large number of individuals over long periods.

33. Michael Veksler (1990) noted, for example, that the procedures promulgated by the IAEA under the Rules for the Safe Transport of Radioactive Materials in 1985 call for containers carrying spent nuclear fuel to maintain a hermetic seal during a shipboard fire of 800°C for one hour, although the

average shipboard fire lasts twenty hours and can attain a maximum temperature of 2,400°C.

34. The British television network ITV aired a documentary in 1983 entitled "Windscale—the Nuclear Laundry," which precipitated a series of government inquiries into the health effects on local coastal communities of effluent discharges.

35. Annual radiological surveillance and monitoring at Sellafield has continued since 1977. In that year, the NEA instituted a multilateral framework to coordinate dumping of packaged low-level radioactive waste. Before that, dumping operations by European countries were not subject to any supranational regulation (see Calmet and Bewers 1991).

36. The same is true for the monitoring programs that have been undertaken by the OECD's Nuclear Energy Agency (NEA) at the Northeast Atlantic waste dumpsite that was active until 1982.

CHAPTER 6. ENVIRONMENTAL PROTECTION FOR THE ARCTIC OCEAN

1. Perhaps the most striking case in point involves the prospect of global warming associated with worldwide emissions of carbon dioxide and other greenhouse gases. Most scenarios of global climate change suggest that the polar regions are expected to experience by far the greatest warming—from 2.0 to 2.4 times greater than the global annual average (i.e., over 10°C at maximum). That may lead to significant change in sea-ice extent and thickness, in land-ice mass, and in the permafrost regime, with very serious if yet unclear consequences for global climatic conditions and with implications for human activities in the Arctic. See Maxwell and Barrie (1989).

2. Concentrations of heavy metals (particularly cadmium) in top-level arctic predators are much greater than in other regions. For example, some Canadian studies have reported cadmium levels in narwhal kidneys that were among the highest ever reported in marine mammals. See "Arctic Environmental Protection Strategy" (1991).

3. For example, in the Canadian maritime Arctic, the impact of oil and gas production and municipal and industrial developments seems to have been restricted to the immediate vicinity of development projects, with the exception of the widespread distribution of organochlorine contaminants brought to the area via long-range transport mechanisms. See Packman and Shearer (1989).

4. Between 1975 and 1985, there were already 195 accidental spills of hydrocarbons across the Arctic involving a total volume of 1.5–1.8 million liters, comprising diesel oil, jet fuel, gasoline, and crude oil. See Packman and Shearer (1989). The average yearly volume of oil spills in the whole of the Arctic Ocean and its seas is estimated at roughly 400 tons. See Futsaeter et al. (1991).

5. Moreover, it has been estimated that the environmental damage costs of one barrel of oil extracted in Tyumen exceed the official wholesale price per barrel by an order of magnitude. See Katasonov (1990).

6. Production from the Alaskan North Slope is expected to decline in the late 1990s and to be replaced by production from fields in coastal areas and on the OCS.

7. It has been estimated that the Bent Horn field on Cameron Island may contain as much as 350 million barrels of oil.

8. The US Department of the Interior's mean case estimate of economically recoverable oil resources in the ANWR is 3.2 billion barrels. The marginal probability of oil reserves for this area is 0.19.

9. In 1990, nearly 20 percent of Russian-produced oil was exported. It is believed that this figure grew dramatically in 1991 as a result of falling production.

10. The largest of these is Shtokmanovskoye, located 600 km off the Kola peninsula. A consortium of Norwegian, Finnish, and US interests won the right to develop this gas condensate deposit. See Ebel (1992).

11. Taken from Soviet data that are reported in the West in *Polar Record* 23:146 (1987), 24:149 (1988), and 25:154 (1989).

12. The annual growth rate for oil consumption in Asian countries is 4.5 percent, as compared to 1.6 percent worldwide.

13. The feasibility of overcoming the many obstacles and objections to arctic commercial shipping is under study by the five-year International Northern Sea Route Programme. As explained by Willie Østreng, director of the program, its principal objective is "the building up of an applicable knowledge base as the foundation for rational decisionmaking" (Østreng 1993).

14. The impetus provided by the World Conservation Strategy has already given rise to two separate efforts to articulate regional conservation strategies for the Arctic, namely the Inuit Regional Conservation Strategy, sponsored by the Inuit Circumpolar Conference, and the Arctic Marine Conservation Strategy, promoted by the Canadian Department of Fisheries and Oceans. See Griffiths and Young (1990).

15. It is not insignificant that, in addition to the Arctic Eight, nine other observers have participated in the activities of the Finnish Initiative: the governments of Germany, Poland, and the United Kingdom; the Inuit Circumpolar Conference; the Nordic Saami Council; the USSR Association of Small Peoples of the North; the United Nations Economic Commission for Europe; the United Nations Environment Programme; and the IASC.

16. In addition to a four-point implementation strategy, the signatories also agreed to assess cooperatively the present state of the arctic environment through scientific research, to determine the potential effects of development activities, and to report routinely on changes related to these matters.

17. A follow-up meeting of the signatory parties was held in Nuuk, Greenland, in September 1993, and additional meetings have been scheduled since.

18. On November 28, 1990, US Secretary of State for External Affairs Joe Clark announced his intention to propose the establishment of an Arctic Council at the June 1991 ministerial meeting that concluded with the signing

of the Rovaniemi Environmental Initiative. As of August 1993, discussions were continuing on the formation of an Arctic Council.

19. Representing the state and provincial governments of Alaska, British Columbia, Alberta, Yukon, Northwest Territories, Quebec, Home Rule Government of Greenland, Lapland, Trondelag, Vasterbotten, Russian Republic, Magadan, Jewish Autonomous Region, Sakhalin, Chukhotka, Heilogjiang, and Hokkaido.

20. For example, the six priority fields approved by the parties to the Mediterranean Action Plan are aquaculture, water resources management, renewable sources of energy, human settlements, soil conservation, and tourism. See Saliba (1989).

21. Arctic-wide (or Arctic-wide plus major shipping states) management of shipping in the Arctic (including uniform design, crewing, and discharge standards) is warranted, for the day may come when a single ice-strengthened vessel travels through the Northwest Passage, the North Pole route, and the Northeast passage on a multipurpose shipping assignment. See Vanderzwaag and Lamson (1986). Establishment of a multilateral body with both advisory and regulatory authority (e.g., issuing Arctic safety and pollution prevention certificates), composed of personnel from the relevant government agencies, may provide a solution to this problem. For a similar proposal concerning a bilateral US-Canadian commission, see Westermeyer and Goyal (1986).

22. Of course, an arctic coastal state cannot exclude international navigation altogether. Article 234 implies that some navigation must be allowed and that regulations cannot have the effect of restricting innocent passage of foreign ships. In addition, Article 236 exempts military vessels from compliance with the environmental protection provisions of the convention.

23. The dumping of wastes can be prohibited on arctic islands and the continent as well, in cases where these wastes may reach arctic waters.

24. This source of pollution, and especially offshore oil and gas development, seems to be in urgent need of regional pollution control measures. No adequate universal conventions or agreements exist, and the effects of a large blowout could be disastrous for arctic marine ecosystems. As in the case of ship-generated pollution, particular care should be taken of preventive measures regarding safe design and construction of rigs and platforms as well as adequate training of personnel. Additional emphasis must be placed on appropriate joint contingency planning, training for emergency situations, and, possibly, the formal establishment of an intergovernmental emergency group to supervise or take control of offshore accidents. See Harders (1987).

25. The first instance of such cooperation was the agreement reached at the US-Soviet Summit meeting in June 1990 on the establishment of an international park in the Beringia region on both coasts of the Bering strait, where similar nature protection rules would apply. The most likely candidates for future designation as protected areas are large, recurrent polynyas, like the

North Water in the northern part of Baffin Bay, and thermal oases, like Polar Bear Pass in the Canadian High Arctic. The emerging IASC could become an important player in the identification of specially protected areas in the Arctic. See Griffiths and Young (1990).

26. For instance, for periodic checks of the status of oil spill contingency plans.

27. To their credit, the parties to the Rovaniemi process are increasingly aware of the need to incorporate an aboriginal perspective. Indigenous peoples were thus included at Yellowknife in the delegations of Canada, Denmark, Finland, Norway, the Soviet Union, and Sweden. Also, the Inuit Circumpolar Conference was granted and made effective use of observer status at the April 1990 meeting. See Griffiths and Young (1990).

28. For the most part, transfers of technology into Siberia and the Russian Arctic have involved capital equipment for oil and gas development, icebreaking capacity, and a small number of turnkey projects. Primary sources have included firms in Finland, Norway, Japan, and the United States. Compensation for goods and equipment is often made in the form of raw materials or surplus production.

29. On January 10, 1989, the Minister of the Merchant Marine of the USSR issued an order assigning concrete environmental protection functions to enterprises, organizations, and institutions of the Ministry of Marine Transportation. Special emphasis has also been placed on providing conditions for Russian vessels to observe all provisions of MARPOL-73/78 presently in force through the release in 1991 by the Department of Navigation and Oceanography, Ministry of Defense, of "Regulations for Navigation on the Sea-ways of the Northern Sea Route."

30. Present technology for oil spill containment and clean-up in the Arctic is even more limited in its effectiveness. Most technologies are not effective in offshore areas with seas of more than 3 feet or in winds of 15 knots or greater—conditions that are present much of the time in the High Arctic.

31. One of Environment Canada's (ministry of the environment) largest research and development partners in the US firm, Marine Spill Response Corporation, which was formed in the aftermath of the *Exxon Valdez* spill.

32. One exception was the Oil and Hazardous Materials Simulated Test Tank (OHMSETT) program, in which four US agencies (Minerals Management Service, Environmental Protection Agency, Navy, and Coast Guard), along with Environment Canada, participated in joint simulation research toward improving arctic oil spill response capabilities (see Smith 1986). Until the program was disbanded in 1988, several R&D projects were funded each year. The OHMSETT facility, located in Leonardo, NJ, was recently reopened by the Minerals Management Service.

33. The Lena, Yenisey, and Ob river systems in Siberia; the Yukon in Alaska; and the Mackenzie in Canada's Northwest Territories are important contributors.

34. Denmark, Norway, and the United States take explicit exception to the sector theory, maintaining that the ice and waters within the arctic rim are subject to traditional international law and the law of the sea (Larson 1989).

CHAPTER 7. THE SOUTHERN OCEAN

1. The term *Southern Ocean* is used here with full awareness that it is not universally accepted. Many experts seeking to avoid the political controversies that accompany use of the term use less direct expressions such as *Antarctic circumpolar ocean*. See discussion in Deacon 1984.

2. Though not universally shared (see Sulden 1989).

3. Argentina, Australia, Belgium, Brazil, Chile, China, Ecuador, Finland, France, Germany, India, Italy, Japan, the Republic of Korea, the Netherlands, New Zealand, Norway, Peru, Poland, Russia, South Africa, Spain, Sweden, the United Kingdom, the United States, and Uruguay.

4. Text of the Stockholm Declaration in UN Doc. A/CONF.48/14 and Corr.1 (16 June 1972), reprinted in *International Legal Materials* 11:1416 (1972). Discussed in Elzinga and Bohlin (1989). Text of the Rio Declaration in UN Doc. A/CONF.151/51 Rev. 1 (13 June 1992), reprinted in *International Legal Materials* 31:874 (1992).

5. This can occur only under rather special circumstances. See the judgment of the International Court of Justice in North Sea Continental Shelf Cases (Federal Republic of Germany v. Denmark and Federal Republic of Germany v. Netherlands, I.C.J. Reports, 1969, paras. 61–78).

6. E.g., Guyer (1983). Brazilian statement of views in the Secretary General's Report on Antarctica, UN Doc. A/39/583, part II, p. 10 (2 November 1984); and a closely related claim that the international community has by nonobjection given tacit consent in the Chilean statement of views (UN 1984, pp. 30, 32). Although the notion of an objective regime was discussed at length by the International Law Commission, the concept was not included in the Vienna Convention on the Law of Treaties (1969). The whole question is explored at length in Simma (1986).

7. The CCAMLR approximates the Antarctic Polar Front with a line connecting a series of points described by latitude and longitude in Article I, para. 4. The greatest deviation from the Convergence occurs off Tierra del Fuego, where the line was drawn to avoid the 200-mile EEZs off the Argentine and Chilean mainlands.

8. The lengthiest treatment remains Birnie's (1985). Scarff (1977) provided a more ecologically centered discussion.

9. See "Report of the 1988 Meeting to Review Operation of the Convention for Conservation of Antarctic Seals" (1991).

10. Some exceptions are allowed for scientific research and subsistence hunting, and the continued effectiveness of the moratorium, and even of the IWC itself, has been called into question by some whaling nations that have

announced they will resume whaling despite IWC quotas. (See Marine Mammal Commission 1994.)

11. The French government, which has sovereignty over the Crozet and Kerguelen islands, where fishing was already extensive in the late 1970s, pressed hardest for acceptance of this rule (Rowland 1988).

12. Argentina, Australia, Chile, France, New Zealand, Norway, and the United Kingdom.

13. See discussions in Vicuna 1988 and Beck 1991b.

14. Area of application as defined in Article 5.

15. The idea of towing icebergs to temperate zones for melting was discussed in the mid-1970s (see Lundquist 1977).

16. See terms of CRAMRA, Article V, paras. 2 and 3.

17. For example, Resolution of the 16th IUCN General Assembly (1984, noted in Barnes and Beck 1988).

18. See documents in *Foreign Relations of the United States* (1948, Vol. 1, pp. 948–1001). The US draft of an agreement to create the Trusteeship also appears in Bush 1991a.

19. Amending the charter to alter the membership would be a laborious process, involving ratification by two-thirds of the UN membership, including all five permanent members of the Security Council. Creating a new UN organ would not be any more difficult.

20. For a particularly strong exposition of this view, see Kindt 1988.

21. Draft approved at Madrid, June 1991 (text in Bush 1991b).

22. Although the term *Southern Ocean* suggests one ecosystem, marine scientists are divided on whether there is in fact a single ecosystem or a set of ecosystems.

23. Schneider (1991), reporting on testimony given in a US Senate hearing on the effects of ozone depletion. According to Schneider, Dr. C. Susan Weiler of Whitman College reported 6 to 12 percent reductions in phytoplankton abundance relative to 1987.

24. See reports in *Atmos-Environ* 16(10), 1982.

25. According to the Office of Technology Assessment (OTA) (1990), a NOAA study of the effects of the 1974 grounding of the supertanker *Metula* in the Straits of Magellan concluded that heavy oil deposits in sheltered areas will persist 100 years.

26. OTA (1989) quotes estimates of oil loss ranging from 139 million to 428 million gallons.

27. OTA (1989, appendix A) discusses three hypothetical systems.

28. Deacon (1984) said that, to avoid icebergs in certain parts of the Southern Ocean, oceanographers must be sure that the top float of any water column instrument array is at least 400 meters from the surface.

29. Deacon (1984) said that scientists are not sure whether the marks found by Norwegian researchers at 350 meters in the Weddell Sea are modern or date to earlier geological eras when the sea level in the area was lower.

30. Izmaylov (1988) noted that the difference is years, not decades, as previously believed.

31. Hempel (1991) summarized the seasonal patterns, which were revealed by a series of year-round research cruises in 1984–88.

32. The *Exxon Valdez* ran aground in Prince William Sound, Alaska, on March 24, 1989, eventually spilling about 10.8 million gallons (35,000 tons) of oil. For a description of the damage to coasts, plants, and animal life, see Hodgson 1990.

33. But see Barinaga (1990) for debate about whether oil pollution alone or one of the regular periodic collapses of juvenile population for lack of enough food was the main culprit in bird deaths in this instance.

34. The total program, which includes considerable offshore components, would cost about $250 million. The estimate of $100 million for the land portion attributes half of the aircraft support costs to land operations and the other half to offshore ones.

35. Text of the 1973 Convention and Annexes in *International Legal Materials* 1973; text of the 1978 Protocol in *International Legal Materials* 1978.

36. Such provisions had already been written into the Consultative Parties' agreements. See Protocol to the Antarctic Treaty on Environmental Protection 1991.

37. Fisheries management specialists have long debated whether overfishing will actually cause extinction. Most believe that it becomes uneconomic to pursue a particular species well before the species becomes physically extinct. They see a commercial extinction at which numbers are too low to fish but still sufficient to maintain the species. A few question this view, at least for species with long ranges, long periods to sexual maturity, and gregarious habits. Scarff (1977) noted arguments that whales, although not yet hunted to extinction, may well become extinct because there are not enough whales still alive to ensure timely pairing off and breeding. This is less likely with fish species that are quicker to mature sexually and that spawn rather than carry offspring singly or in small broods, as mammals do.

38. *The Arctic and Southern Oceans* (1985) estimated that the current population of blue whales is 1 percent of its original size, with corresponding estimates of 3 percent for humpback whales and 20 percent for fin whales.

39. For a discussion of the possibilities, see Voronina 1990.

40. For instance, D.S. Butterworth's (1986) suggestion that the total allowable catch be set such that stock size remains at 75 percent of the estimated preexploitation size. However, even this approach could overestimate prudent catch levels unless based on the more conservative estimates of current population size.

41. Two US environmental organizations, the Antarctica Project and the Environmental Defense Fund, advocated imposition of sanctions in the mid-1980s (MacKenzie and Rootes 1985). At the time, US law had provision for two sorts of sanctions against states judged by the US executive branch to be violating an international conservation agreement—exclusion of its nationals

from fishing in the US EEZ or embargo on imports of fish products originating in that country. Later US decisions to phase out virtually all foreign fishing in its EEZ mean that trade embargoes are the only sanctions available today. These trade sanctions are, in turn, under attack internationally as violative of US legal commitments under the General Agreement on Tariffs and Trade (GATT).

42. See, for example, Heap (1987), who argued that the difference between majority and consensus voting is insignificant while optimism reigns.

43. Noted in remarks of Sahrhage (1988).

44. Nicol (1990) noted that in lab experiments krill were able to endure periods of up to 200 days with no food by shrinking as they went through their normal molting process.

45. Few Americans have even seen krill products. More former Soviet citizens have, but they initially resisted them so strongly that Soviet enterprises gave up attempting to sell them for human consumption (see Quigg 1983). More recently, Russian and Japanese processors have been able to produce more attractive products; they have high expectations of doing even better in the near future. Both co-authors of this chapter (Peterson and Barsegov) ate Soviet-processed krill in 1991 and found it acceptable, but no delicacy.

46. Krill catches were 39,931 metric tons in 1975–76, reached 530,003 in 1981–82, fell to 128,218 in 1983–84, rose to 445,673 in 1985–86, and were 395,470 in 1989–90 (*FAO Yearbook of Fisheries Statistics*, Vol 1: Catches and Landings, 1978, 1982, 1986, and 1988, table B-46; Nicol 1990).

47. For environmentalist views, see ASOC 1985, 1986; Greenpeace 1990.

48. Guidelines first embodied in Recommendation VIII-9, Annex A (actually completed in Recommendation X-8; text in Bush 1991a, b).

49. Including airfare from North America to Argentina or Chile, an Antarctic voyage costs at least $7,500.

50. The Malaysia Group is an informal and shifting coalition of states that favor establishment of a UN agency to manage Antarctic and Southern Ocean affairs. Malaysia has been the most vocal advocate of this approach, which has been steadfastly supported by Antigua and Barbuda as well.

51. CRAMRA, Articles 62 (ratification) and 29 (composition of Regulatory Committees).

52. See the Australian-French, Chilean, New Zealand, Swedish, and United States working papers (reprinted in Bush 1991a).

53. Positions noted in Kimball 1990.

54. These discussions are analyzed and commented upon in Kimball 1990.

55. Revised Negotiating Text of a Protocol to the Antarctic Treaty on Environmental Protection adopted at Madrid, 29 April 1991 (in Bush 1991b).

56. Letter to Foreign Minister Evans, made public during May 1989 parliamentary debates about CRAMRA. For discussion see Blay and Tsamenyi 1990.

57. Its most authoritative expression occurs in Chapter 2, Article 2 of the Charter of Economic Rights and Duties of States, UN General Assembly Resolution 3281 (XXIX) (text in General Assembly Official Records, Twenty-Ninth Session, Supplement 31, p. 52).

58. This trigger terminology was used during negotiation of the 1985 London Ozone Treaty and its 1987 Montreal Protocol. For discussion, see Haas 1992.

59. Lee A. Kimball's (1990) is one of the few NGO reports acknowledging this point. For further discussion see OTA 1989.

60. For example, Kenya (p. 6), Pakistan (p. 11), Nigeria (p. 18) in A/C.1/45/PV. 43 and Philippines in A/C.1/45/PV. 42, p. 33 (20 November 1990).

61. This is one of the central conclusions of Haas, Keohane, and Levy 1993.

CHAPTER 8. THE LAW OF THE SEA

1. The main role of international law in the environmental field, as described by one expert, is "avoiding or minimizing disputes by establishing a framework for the building of a legal order within which activities that might adversely affect the environment take place only within a regulated, monitored regime" (Birnie 1988). By setting forth the rights and obligations of states with respect to the sea, the United Nations Convention on the Law of the Sea (signed at Montego Bay in 1982) provides the basis for such a regime.

2. The convention enters into force on November 16, 1994, twelve months after the date of deposit of the sixtieth instrument of ratification or accession. As of November 16, 1993, sixty states had ratified or acceded to the convention (see table 8.1). Although 119 nations signed the Law of the Sea Convention, only one developed country, Iceland, has since ratified it. The United States refused to sign because of its dissatisfaction with the provisions in Part XI for deep seabed mining in areas beyond national jurisdiction. Many other developed countries have been reluctant to ratify the convention, either for the same reasons or because it has not been ratified by the United States. It took almost ten years (1973–82) to negotiate the Law of the Sea Convention, and more than a decade later it had yet to receive sufficient ratifications to bring it into force. Ongoing efforts at the United Nations to resolve questions surrounding Part XI give hope that the convention may still enter into force with widespread or universal participation, including the United States, Russia, and other developed maritime states (Charney 1992).

3. There is, of course, great reluctance to reopen the negotiations on the Law of the Sea Convention, for many reasons. It was negotiated as a package deal full of compromises by all parties. The convention was to be regarded as an indivisible whole; no reservations are accepted (Article 309).

4. As degradation of the ocean continues, for example, states have begun to recognize in a variety of international fora, including the 1992 UN Confer-

ence on Environment and Development, the need for a more anticipatory and precautionary approach to preventing harm to the marine environment. The introductory paragraph of UNCED's Agenda 21 on Oceans Protection calls for "new approaches to marine and coastal area management and development . . . that are integrated in content, and precautionary and anticipatory in ambit" (Chapter 17, para. 1). The precautionary principle, which represents a shift away from the concept of assimilative capacity toward an emphasis on pollution prevention and clean production technologies, has been adopted in numerous intergovernmental declarations, from the International Conferences on the North Sea to the Governing Council of the United Nations Environment Programme (see Freestone 1991). In its emphasis on preservation and protection of the marine environment, the Law of the Sea Convention is consistent with the precautionary approach.

5. Article 38(1) Statute of the International Court of Justice. Decisions of the World Court and arbitral and other international tribunals and scholarly teachings are referred to as *subsidiary sources of international law*. In practice, such decisions are often considered the affirmation or the revelation of customary international rules (Kiss and Shelton 1991).

6. In the Malta-Libya Continental Shelf Case, the International Court of Justice found that the EEZ had acquired customary status through widespread state practice (Boyle 1992). For a norm to become customary law, unanimity is not always essential. According to one expert, marine pollution norms and standards that have received substantial support from maritime powers can be regarded as "universally accepted" (Hakapaa 1986). In declaring its own EEZ, the United States issued a statement referring to the Law of the Sea Convention as a document reflecting (other than the deep seabed provisions) customary international law (Statement Accompanying US Presidential Proclamation No. 5030 of 10 March 1983). Although the convention may be treated as customary international law by some states, some of its provisions and norms give rise to the question of whether it has achieved widespread state practice.

7. The International Court of Justice Statute Article 38(1) refers to "general principles of law recognized by civilized nations." While the propriety of the term *civilized nation* is today questionable, the underlying concept remains valid (Kiss and Shelton 1991).

8. Corfu Channel Case, International Court of Justice Reports, 9 April 1949.

9. A classic statement of the principle not to cause substantial transboundary environmental harm or interference is found in the Arbitral Award of the Trail Smelter Case (The United States v. Canada): "Under the principles of international law, as well as of the law of the United States, no State has the right to use or permit the use of its territory in such a manner as to cause injury by fumes in or to the territory of another or the properties or persons therein, when the case is of serious consequence and the injury is established

by clear and convincing evidence" (3 Rep. of Int. Arb. Awards, p. 1911 et seq. [1965]).

10. Repeated endorsement and public acceptance of principles by government representatives may give rise to legal obligations by those governments based on the principle of good faith (Freestone 1991).

11. Principles adopted in the Stockholm Declaration include the duty of states to prevent marine pollution, their responsibility to ensure that their activities do not cause damage to the marine environment outside their jurisdiction, and the duty to cooperate to develop further the international law regarding liability and compensation for victims of pollution (Principles 7, 21, and 22).

12. The text of the Oceans portion of Agenda 21 approved at the 1992 UNCED makes continuous reference to the Law of the Sea Convention as the basis for actions by states. It commences with a fairly significant statement: "International law, as reflected in the provisions of the United Nations Convention on the Law of the Sea referred to in this chapter of Agenda 21, sets forth rights and obligations of States and provides the international basis upon which to pursue the protection and sustainable development of the marine and coastal environment and its resources" (Chapter 17 of Agenda 21, as adopted at UNCED, 14 June 1992).

13. The first UN Conference on the Law of the Sea, held in 1958, adopted four separate conventions governing activities in different ocean spaces: the Convention on the Territorial Sea and the Contiguous Zone, the Convention on the Continental Shelf, the Convention on the High Seas, and the Convention on Fishing and Conservation of the Living Resources of the High Seas. None of them dealt particularly with the protection of the marine environment from pollution (Meng 1987). Two of the conventions did, however, contain prohibitions relating to pollution by oil or pipeline as well as by radioactive wastes and wastes resulting from oil drilling on the continental shelf. A second UN Conference on the Law of the Sea in 1960 failed to reach any substantive agreements.

14. "States shall co-operate on a global basis, and as appropriate, on a regional basis, directly or through competent international organizations, in formulating and elaborating international rules, standards and recommended practices and procedures consistent with this Convention, for the protection and preservation of the marine environment, taking into account characteristic regional features" (Article 197).

15. According to GESAMP (1990), water-borne pollution is responsible for about 44 percent of global marine pollution, airborne for 33 percent, ocean dumping for 10 percent, marine transportation for 12 percent, and offshore oil production for 1 percent.

16. Under the auspices of the UNEP's Regional Seas Programme, eleven regions have developed agreements aimed at regulating all forms of marine pollution along the lines envisaged in the Law of the Sea Convention. Regional Sea conventions fostered by UNEP's Programme cover or are being

actively developed to cover the Mediterranean, the Persian Gulf and the Gulf of Oman, the Caribbean, West and Central Africa, the South-East Pacific and North-West Pacific, the Red Sea and Gulf of Aden, the East and South Asian Seas, and the East African Region. Regional conventions have been developed outside of UNEP's auspices for the Baltic, North-East Atlantic, and Black Sea. The conventions contain general obligations with regard to protection and preservation of the marine environment, whereas detailed provisions relating to these obligations are elaborated in separate protocols (French 1990; Schröder 1992).

17. In addition to taking measures at the national level, states are to cooperate at the global and regional levels to establish rules, standards, and recommended practices and procedures to prevent, reduce, and control pollution from various sources. States are directed for such purposes to act especially through competent international organizations or diplomatic conferences. They are furthermore required to reexamine from time to time such rules, standards, and recommended practices and procedures (Articles 207, 208, 209, 210, 211, and 212).

18. See the Annex to the UN Secretary General's Report on the Law of the Sea for 1989 for a list of agreements pertinent to the protection of the marine environment and their participants (UN Secretary General 1989). The gap is only partially filled by existing regional agreements. Only four regions currently have agreements specifically regulating ocean dumping: the North-East Atlantic (1972 Oslo Convention for the Prevention of Marine Pollution by Dumping from Ships and Aircraft, now incorporated into the 1992 Paris Convention for the Protection of the North-East Atlantic); the Baltic (1974 Helsinki Convention on the Protection of the Marine Environment of the Baltic Sea Area and its 1992 successor); the Mediterranean (1976 Protocol for the Prevention of Pollution of the Mediterranean Sea by Dumping from Ships and Aircraft); and the South Pacific (1986 Protocol for the Prevention of Pollution of the South Pacific Region by Dumping).

19. The only universal international instrument currently applicable to land-based sources is the voluntary guidelines prepared by UNEP at Montreal in 1985. The Montreal Guidelines were expected to serve as the basis for the development of bilateral, regional, and multilateral agreements, as well as national legislation.

20. The 1974 Convention for the Prevention of Marine Pollution from Land-based Sources in the North-East Atlantic (the Paris Convention) was the first multilateral agreement to deal exclusively with land-based pollution. (In 1992, the 1974 Paris Convention was combined with the 1972 Oslo Convention preventing dumping from ships and aircraft in the 1992 Convention for the Protection of the North-East Atlantic, also known as the Paris Convention). Unlike the 1974 Paris Convention, the 1974 Convention on the Protection of the Marine Environment of the Baltic Sea Area (Helsinki Convention) was a comprehensive agreement governing all sources of marine pollution. As compared to the later regional agreements negotiated under the auspices

of UNEP, the 1974 Helsinki Convention (and its 1992 successor, also known as the Helsinki Convention) contained extensive and specific provisions regarding land-based pollution. Only three of the regions of the UNEP Regional Seas Programme—the Mediterranean, the South-East Pacific, and the Persian Gulf area—have developed specific agreements to deal with land-based pollution, although developing "a concerted approach to land-based activities damaging the marine environment" is among the priority activities that the program advises coastal states to undertake (Schröder 1992).

21. The 1974 Helsinki Convention, the 1974 Paris Convention, and the 1980 Protocol for the Protection of the Mediterranean Sea Area against Pollution from Land-based Sources (the Athens Protocol). In 1992, the 1974 Paris Convention was combined with the 1972 Convention for the Prevention of Marine Pollution by Dumping from Ships and Aircraft (Oslo Convention) in the 1992 Convention for the Protection of the Marine Environment of the North-East Atlantic (also known as the Paris Convention).

22. The 1983 Protocol for the Protection of the South-East Pacific against Pollution from Land-based Sources (the Quito Protocol), the 1990 Protocol on the Protection of the Marine Environment against Pollution from Land-based Sources (the Kuwait Protocol), and the 1992 Protocol on the Prevention of Pollution from Land-based Sources (the Bucharest Protocol).

23. Two protocols to the Geneva Convention on Long Range Transboundary Air Pollution have been negotiated, one on sulfur dioxide (1985) and a second on nitrogen (1988). Both have only limited participation, with twenty and eighteen parties, respectively (Boyle 1992).

24. Before the Convention, a port state could inspect vessels that are voluntarily in port and bring proceedings for violations occurring within its territorial sea and internal waters. With respect to other violations, however, the port state was limited to reporting the violation to the flag state. Similarly, coastal states had prescriptive and enforcement competence only in their international waters and territorial sea.

25. Passage is considered innocent "so long as it is not prejudicial to the peace, good order or security of the coastal state" (Article 19.1). The next paragraph then specifies which activities make passage no longer innocent. With respect to marine environmental protection, only acts of "wilful and serious pollution contrary to this Convention" are considered not to be innocent (Article 19.2 (h)). Although a single discharge from a passing vessel may not reach the "serious" standard set by the convention, the cumulative effect of many small discharges from passing ships can be severe.

26. Two important exemptions from this limited coastal state jurisdiction are provided in the convention. (1) In ice-covered areas, coastal states have the right to impose unilateral measures beyond generally accepted international standards (including those relating to vessel design, construction, manning, or equipment standards) in the interests of preventing major harm or irreversible disturbance to the ecological balance (Article 234). (2) Outside ice-covered areas, coastal states must submit applications to the competent

international organization (i.e., the International Maritime Organization) establishing why special protective measures are necessary for "recognized technical reasons" within discrete areas of their exclusive economic zones (Article 221(6)(a)). They may then adopt rules that are made applicable for such areas through the international organization. They may also adopt rules relating to discharges or navigational practices, but not those relating to standards of vessel design, construction, manning, or equipment (Article 211.6(c)).

27. Even today, some coastal states (Antigua and Barbuda, Guyana, India, Maldives, Mauritania, Mauritius, Nigeria, Pakistan, Seychelles) have promulgated laws and regulations reserving the right to regulate navigation in their EEZs under the pretext of declaring designated or special areas within them.

28. For example, when a vessel has violated a coastal state's norms and rules within the EEZ, the state's authorities may, as a rule, require the vessel to supply information necessary to establish whether the violation has occurred; in those cases when violation resulted in a "substantial discharge," causing or threatening "significant pollution" of the marine environment, and if the vessel has refused to give information or the information supplied "is manifestly at variance with the evident factual situation," the state may undertake physical inspection of the vessel. Only when "there is clear objective evidence" of such a violation resulting in a discharge causing "major damage" may the coastal state institute proceedings, including detention of the vessel, in accordance with its laws (Article 220).

29. In addition, port states may undertake investigations and bring proceedings for discharge violations in another state's coastal waters if the discharge has or is likely to cause pollution in the port state's coastal waters (Article 218.2). Port states must also, as far as practicable, undertake investigations of discharge violations if requested by the flag state injured by the discharge (Article 218.3, 4) and may institute proceedings at the request of either state (Article 218.2).

30. Article 236 states: "The provisions of the Convention regarding the protection and preservation of the marine environment do not apply to any warship, naval auxiliary, or other vessels or aircraft being used by the government in non-commercial service." States are nevertheless required to act in a manner consistent with the convention under certain circumstances; they are to take "appropriate measures not impairing operations or operational capabilities of such vessels or aircraft owned or operated by it, that such vessels or aircraft act in a manner consistent, so far as is reasonable or practicable, with this Convention." Although conformity with environmental standards by military vessels is expected, enforcement is lacking and allowance for special circumstances is available. Thus, Kiss and Shelton (1991) remarked that "it is difficult to see in this tortured phraseology anything other than a reaffirmation of state sovereignty and an invitation to military or other state actors to ignore applicable environmental norms."

31. For seabed activities subject to national jurisdiction, the convention

requires coastal states to adopt laws and regulations to prevent, reduce, and control pollution and to endeavor to harmonize their policies at the regional level. States are also to act through international organizations or conferences to establish global or regional rules (Article 208). State laws are to be no less effective than international rules.

All of the regional seas treaties reaffirm the Law of the Sea principles in general, but only the Kuwait Convention has adopted a Protocol Concerning Marine Pollution Resulting from Exploration and Exploitation of the Continental Shelf. Agreements to regulate pollution from offshore oil and gas activities are now under consideration in six regions, but only discharges from machinery are regulated internationally (Agenda 21:17.21). UNEP has also elaborated forty-two principles that should govern such operations (Kiss and Shelton 1991). Other forms of seabed activities are not subject to any regional or international agreement.

32. Article 145 is quite comprehensive. It provides

Necessary measures shall be taken in accordance with this Convention with respect to activities in the Area to ensure effective protection for the marine environment from harmful effects which may arise from such activities. To this end the Authority shall adopt appropriate rules, regulations and procedures for *inter alia:*

(a) the prevention, reduction and control of pollution and other hazards to the marine environment, including the coastline, and of interference with the ecological balance of the marine environment, particular attention being paid to the need for protection from harmful effects of such activities as drilling, dredging, excavation, disposal of waste, construction and operation or maintenance of installation, pipelines and other devices related to such activities;

(b) the protection and conservation of the natural resources of the Area and the prevention of damage to the flora and fauna of the marine environment.

33. The regime envisaged by the current Draft Regulations on Protection and Preservation of the Marine Environment from Activities in the Area is nevertheless imperfect; among other shortcomings, it fails to incorporate a precautionary approach (Nollkaemper 1991). Significant tension is also created by the fact that the body charged with environmental protection is the same as that charged with (and funded by) the development of economic activities in the Area (Nollkaemper 1991).

34. The Convention's definition of marine pollution fails, however, to make explicit reference to its effects on marine ecosystems: "'Pollution of the marine environment' means the introduction by man, directly or indirectly, of substances or energy into the marine environment, including estuaries, which results or is likely to result in such deleterious effects as harm to living

resources and marine life, hazards to human health, hindrance to marine activities, including fishing and other legitimate uses of the sea, impairment of quality for use of sea water and reduction of amenities" (Article 1.1.(4)). This deficiency has been remedied in many of the regional seas conventions but is not yet universally applied.

35. Such assistance under Article 202 is to include, *inter alia:*

(i) training their scientific and technical personnel;
(ii) facilitating their participation in relevant international programs;
(iii) supplying them with necessary equipment and facilities;
(iv) enhancing their capacity to manufacture such equipment;
(v) advice and development of facilities for research, monitoring, educational, and other programs.

References

Aarkrog, A. 1988. "The Radiological Impact of the Chernobyl Debris Compared with That from Nuclear Weapons Fallout." *Journal of Environmental Radioactivity* 6:151–162.

———. 1989. Proceedings of a Seminar on the Radiological Exposure of the Population of the European Community from Radioactivity in North European Marine Waters, Project MARINA, 14–16 June, Bruges. Commission of the European Communities, XI/4669/89-EN.

Aarkrog, A., E. Buch, A.J. Chen, et al. 1987. "Environmental Radioactivity in the North Atlantic Region Including the Faroe Islands and Greenland." Risø-R 564. Roskilde, Denmark: Risø National Laboratory.

Aarkrog, A., H. Dahlgaard, L. Hallstadius, and E. Holm. 1983. "Radiocaesium from Sellafield Effluents in Greenland Water." *Nature* 304:49–51.

Abdelghani, A., W.R. Hartley, F.R. Esmundo, and T.F. Harris. 1993. "Biological and Chemical Contaminants in the Gulf of Mexico and the Potential Impact on Public Health: A Characterization Report." Draft. New Orleans: Tulane University School of Public Health and Tropical Medicine.

Ainley, D.G., A.R. DeGange, L.L. Jones, and R.J. Beach. 1981. "Mortality of Seabirds in High-Seas Salmon Gillnets." *Fishery Bulletin* 79:800–806.

Amano, M. 1990. "Outline of Marine Mammal and Sea Bird Bycatch Investigations on Board the *Koei Maru* No. 68 in 1987." In: K. Shimazaki (ed.). *Report of the Bycatch Investigations for the Land-Based Salmon Driftnet Fishery* (in Japanese).

AMS (American Management Systems, Inc.) and G.L. Webster. 1991. *Marine Debris Action Plan for the Gulf of Mexico*. Report prepared for the Gulf of Mexico Program. Stennis Space Center, MS (October).

———. 1992. *Public Action Agenda for the Gulf of Mexico*. Report prepared for the Gulf of Mexico Program. Stennis Space Center, MS (December).

Anderson, C. 1991. "Is Krill Fishing behind Penguins' Decline?" *Washington Post*, 8 April, p. A3.

The Arctic and Southern Oceans. 1985. Moscow: Nauka Publishing House.

"Arctic Environmental Protection Strategy." 1991. Rovaniemi, Finland (14 June).

Arkin, W.M., and J.M. Handler. 1989a. *Neptune Papers No. 3. Naval Accidents 1945–1988*. Washington, DC: Greenpeace Institute for Policy Studies.

———. 1989b. "Nuclear Disasters at Sea, Then and Now." *Bulletin of the Atomic Scientists* 45(6).

Arrow, K., R. Solow, P.R. Portney, E.E. Leamer, R. Radner, and H. Schuman. 1993. *Report of the NOAA Panel on Contingent Valuation*. Washington, DC: Resources for the Future (January 11).

ASOC (Antarctica and Southern Ocean Coalition). 1985. "Background Paper on the French Airfield at Pointe Géologie, Antarctica." Sydney, Australia: Antarctica and Southern Ocean Coalition (1 March).

———. 1986. "Antarctica Briefing No. 9: The French Airstrip—A Breach of Antarctic Treaty Rules?" Sydney, Australia: Antarctica and Southern Ocean Coalition (30 July).

———. 1993. "ASOC Report on the XIth Meeting of the Commission for the Conservation of Antarctic Marine Living Resources." Sydney, Australia: Antarctica and Southern Ocean Coalition.

Associated Press. 1993. "Government Proposes Storage for Plutonium." *Cape Cod Times*, 29 July, p. C-5.

Auburn, F.M. 1982. *Antarctic Law and Politics*. Bloomington: University of Indiana Press.

Ausubel, J.H., and D.G. Victor. 1992. "Verification of International Environmental Agreements." *Annual Review of Energy and the Environment* 17:1–43.

Bailey, R.S., and J.H. Steele. 1992. "North Sea Herring Fluctuations." In: M.H. Glantz (ed.). *Climate Variability, Climage Change, and Fisheries*. Cambridge: Cambridge University Press.

Baker, D.J. 1986. "The Arctic's Role in Climate." *Oceanus* 29(1):41–46.

Barinaga, M. 1990. "Eco-Quandary: What Killed the Skua?" *Science* 249:243.

Barnes, J.N. 1989. "Protecting Antarctica's Environment." Testimony before the Subcommittee on Science, Technology and Space, US Senate, 8 September, p. 103.

Barnes, J.N., and P.J. Beck. 1988. "Legal Aspects of Environmental Protection in Antarctica." In: C. Joyner and S. Chopra (eds.). *The Antarctic Legal Regime*. Dordrecht, Netherlands: Nijhoff.

Barney, B. 1982. *Antarctica: Wilderness at Risk*. San Francisco: Friends of the Earth.

Barsegov, Y. 1991. "The USSR and the International Law of the Sea." *Oceanus* 34(2):35–40.

———. 1993. "Certain Legal Issues Concerning Prevention of Radioactive Pollution of the Arctic Marine Environment from Land-based Sources." Paper presented at the Conference on Radioactivity and Environmental Security in the Oceans: New Research and Policy Priorities in the Arctic and North Atlantic, 7–9 June, Woods Hole Oceanographic Institution, Woods Hole, MA.

Baskaran, M., J.J. Kelley, A.S. Naidu, and D.F. Holleman. 1991. "Environmental Radiocesium in Subarctic and Arctic Alaska following Chernobyl." *Arctic* 44:346–350.

Beck, P.J. 1986. *The International Politics of Antarctica.* New York: St. Martin's.
———. 1991. "Antarctica, Vina del Mar, and the 1990 UN Debate." *Polar Record* 27:213.
Behrendt, J.C. 1988. "Geophysical and Geological Research in Antarctica Related to the Assessment of Petroleum and Mineral Resources and Potential Environmental Hazards." In: R.R. Wolfrum (ed.). *Antarctic Challenge III: Conflicting Interests, Cooperation, Environmental Protection, Economic Development.* Berlin: Dencker and Humblot.
Belsky, M.H. 1985. "Management of Large Marine Ecosystems: Developing a New Rule of Customary International Law." *San Diego Law Review* 22:733–763.
Benninghoff, W.S. 1985. "Man's Impact on the Antarctic Environment: A Procedure for Evaluating Impacts from Scientific and Logistical Activity." Cambridge, UK: Scientific Committee on Antarctic Research.
Berkhout, F., T. Suzuki, and W. Walker. 1990. "Surplus Plutonium in Japan and Europe: An Avoidable Predicament." Working paper for the MIT-Japan program. Cambridge: Massachusetts Institute of Technology.
Bewers, J.M., and C.J.R. Garrett. 1987. "Analysis of the Issues Related to Sea Dumping of Radioactive Wastes." *Marine Policy* 11(2):105–123.
Biologischeskie resursi Tikhogo Okeana. 1986. Moscow: Nauka.
Birman, I. 1985. *Morskoi period jizni i voprosi dinamiki stad Tikhookeanskikh lososei.* Moscow: Agropromizdat.
Birman, I.V. 1980. *Kosnovam prognozirovania zapasov gorbushi, Okeana zapasov promyslovikh ryb i prognozirovania ulovov.* Moscow (Mimeo).
Birnie, P. 1985. *The International Regulation of Whaling.* (2 vols.) New York: Oceana.
———. 1988. "The Role of International Law in Solving Certain Environmental Conflicts." In: J.E. Carroll (ed.). *International Environmental Diplomacy: The Management and Resolution of Transfrontier Environmental Problems.* Cambridge: Cambridge University Press.
Blay, S.K.N., and B.M. Tsamenyi. 1990. "Australia and the Convention for the Regulation of Antarctic Mineral Resource Activities." *Polar Record* 26:200.
"BNFL Gets Okay for One Phase of Thorp Plant Commissioning." 1993. *Nucleonics Week,* 22 July.
Bockstael, N.E., W.M. Hanemann, and C.L. Kling. 1987. "The Value of Water Quality Improvements in a Recreational Demand Framework." *Water Resources Research* 23:951–960.
Bodansky, D. 1990. "New Directions for the Law of the Marine Environment." Unpublished paper. Seattle: University of Washington.
———. 1991. "Scientific Uncertainty and the Precautionary Principle." *Environment* 33(7):4–5, 43–44.
———. 1992. "Protecting the Marine Environment from Vessel-Source Pollution: UNCLOS III and Beyond." *Ecology Law Quarterly* 18:719–777.

Bogoslovskaya, L.S., and L.M. Votrogov. 1981. "Mass Overwinterings of Birds and Whales in Polynyas in the Bering Sea." *Priroda (Nature)* 1:42–43 (in Russian).
Boyle, A.E. 1992. "Land-based Sources of Marine Pollution: Current Legal Regime." *Marine Policy* 16(1):20–35.
Brigham, L.W. (ed.). 1991. *The Soviet Maritime Arctic*. Annapolis: Naval Institute Press; London: Bellhaven Press.
Broad, W.J. 1994. "Sunken Soviet Sub Leaks Radioactivity in Atlantic." *New York Times* February 8, p. A13.
Broadus, J.M. 1991. "The Sea Environment: Good News, Bad News." *Naval Institute Proceedings* 117(10):50–55.
———. 1992. "Creature Feature Too: *Principium precautionarium.*" *Oceanus* 35(1):6–7.
Broadus, J.M., S. Demisch, K. Gjerde, P. Haas, Y. Kaoru, G. Peet, S. Repetto, and A. Roginko. 1993. *Comparative Assessment of Regional International Programs to Control Land Based Marine Pollution: The Baltic, North Sea, and Mediterranean*. Report prepared for the Office of International Activities of the US Environmental Protection Agency. Woods Hole, MA: Marine Policy Center, Woods Hole Oceanographic Institution.
Broadus, J.M., and R.V. Vartanov. 1991. "The Oceans and Environmental Security." *Oceanus* 34(2):14–19.
Brookshire, D.S., M.A. Thayer, W.D. Schulze, and R.C. d'Arge. 1982. "Valuing Public Goods: A Comparison of Survey and Hedonic Approaches." *American Economic Review* 72:165–177.
Burke, W.T. 1991. "Anadromous Species and the New International Law of the Sea." *Ocean Development and International Law* 22:95–131.
———. 1993. "UNCED and the Oceans." *Marine Policy* 17:519–533.
Burke, W.T., M. Freeberg, and E.T. Miles. 1993. *The United Nations Resolutions on Driftnet Fishing: An Unsustainable Precedent for High Seas and Coastal Fisheries*. Seattle: School of Marine Affairs, University of Washington (August).
Busch, B.C. 1985. *The War against the Seals*. Montreal: McGill-Queen's University Press.
Bush, W.R. (ed.). 1991a. *Antarctica and International Law*. Booklet AT6. New York: Oceana.
———. 1991b. *Antarctica and International Law*. Booklet AT91. New York: Oceana.
Butterworth, D.S. 1986. "Antarctic Ecosystems Management." *Polar Record* 23:45.
Caggiano, M.J.J. 1993. "The Legitimacy of Environmental Destruction in Modern Warfare: Customary Substance over Conventional Form." *Boston College Environmental Affairs Law Review* 20(3):479–506.
Calmet, D.P., and J.M. Bewers. 1991. "Radioactive Waste and Ocean Dumping: The Role of the IAEA." *Marine Policy* 15(6):413–430.
Cameron, T.A., and M.D. James. 1987. "Efficient Estimation Methods for 'Closed-Ended' Contingent Valuation Surveys." *Review of Economics and Statistics* 69(2):269–276.

Canada Department of Fisheries and Oceans. 1991. *Scientific Review of North Pacific High Seas Driftnet Fisheries, Sidney, B.C., June 1–14, 1991.* Report for presentation to the United Nations pursuant to Resolutions 44/225 and 45/197. Sidney, B.C.: Institute of Ocean Sciences.

Canfield, J.L. 1993a. "The Independent Baltic States: Maritime Law and Resource Management Implications." *Ocean Development and International Law* 24:1–39.

———. 1993b. "Recent Developments in Bering Sea Fisheries Conservation and Mangement." *Ocean Development and International Law* 24:257–289.

Carnegie Commission on Science, Technology, and Government. 1993. *Facing toward Governments: Nongovernmental Organizations and Scientific Technical Advice.* New York: Carnegie Commission (January).

Caron, D.D. 1993. "Toward an Arctic Environmental Regime." *Ocean Development and International Law* 24:377–392.

Carson, R. 1962. *Silent Spring.* Boston: Houghton-Mifflin.

Carson, R., and P. Navarro. 1988. "Fundamental Issues in Natural Resource Damage Assessment." *Natural Resources Journal* 28:815–836 (Fall).

Centre for Cold Ocean Resource Engineering. 1983. "An Oil Spill in Pack Ice." Saint Johns, New Brunswick: Centre for Cold Ocean Resource Engineering.

Chadwick, W.A. 1984. "Marine Casualties and How To Prevent Them." *Transactions of the Institute of Marine Engineers* 96:12 (Paper 46).

Charney, J.I. 1992. "The United States and the Revision of the 1982 Convention on the Law of the Sea." *Ocean Development and International Law* 23:279–303.

"China Bucks at US over Yinhe Incident." 1993. *Beijing Review,* August 30–September 5, p. 5.

Chircop, A.E. 1988. "The Marine Transportation of Hazardous and Dangerous Goods in the Law of the Sea—An Emerging Regime." *Dalhousie Law Journal* 11(2):612–638.

Chittleborough, G. 1984. "Nature, Extent, and Management of Antarctic Living Resources." In: S. Harris (ed.). *Australia's Antarctic Policy Options.* Melbourne: Centre for Resource and Environmental Studies, Australian National University.

Chopra, S.K., and A. D'Amato. 1991. "Whales: Their Emerging Right to Life." *American Journal of International Law* 85:21–62.

Choucri, N., and R.C. North. 1975. *Nations In Conflict: National Growth and International Violence.* San Francisco: W.S. Freeman.

Church, R.F., and G. Jones. 1988. *Environmental Assessment of Annex III, Regulations for the Prevention of Pollution by Harmful Substances Carried by Sea in Packaged Form, the International Convention for the Prevention of Pollution by Ships, 1973, as Modified by the Protocol of 1978.* Staff Study SS-46-U7-162. Cambridge, MA: US Department of Transportation Systems Center (November).

Clark, C.W. 1976. *Mathematical Bioeconomics: The Optimal Management of Renewable Resources.* New York: Wiley-Interscience.

Clark, J.W. 1991. "Oil and Gas Resources in the Offshore Soviet Arctic." In: L.W. Brigham (ed.). *The Soviet Maritime Arctic*. Annapolis: Naval Institute Press; London: Bellhaven Press.
Clark, R.B. 1989. *Marine Pollution*. New York: Oxford University Press.
Clawson, M., and J.L. Knetsch. 1966. *Economics of Outdoor Recreation*. Baltimore: Johns Hopkins University Press.
Coachman, L.K., K. Aagard, and R.B. Tripp. 1975. *Bering Strait: The Regional Physical Oceanography*. Seattle: University of Washington Press.
Cogema, Inc. (n.d.) "La Hague: The UP 3 Processing Plant." (Promotional materials provided by facsimile.) Cap de la Hague, Normandy, France.
Coles, P. 1989. "France, Australia Left in Cold." *Nature* 341:678.
Comptroller General of the United States. 1981. *Better Monitoring Techniques Are Needed to Assess the Quality of Rivers and Streams*. (2 vols.) Washington, DC: US General Accounting Office.
Conference on Shared Living Resources of the Bering Sea. 1990. Draft transcript of the meeting held 6–8 June 1990, University of Alaska, Fairbanks.
Cooper, A.I. 1989. *Polar Prospects: A Minerals Treaty for Antarctica*. Unpublished paper of the US Geological Survey. Cited in OTA (US Congress) Report OTA-O-428. Washington, DC: US Government Printing Office (September 1989).
Craighead, F.L., and J. Oppenheim. 1985. *Population Estimates and Temporal Trends of Pribilof Island Seabirds*. Final Report No. 30, NOAA, OCSEAP. Washington, DC: US Department of Commerce.
Croasdale, K.R. 1986. "Arctic Offshore Technology and Its Relevance to the Antarctic." In: US Polar Research Board. *Antarctic Treaty System: An Assessment*. Washington, DC: National Academy Press.
Cummings, R.G., D.S. Brookshire, and W.D. Schulze (eds.). 1986. *Valuing Environmental Goods: An Assessment of the Contingent Valuation Method*. Totowa, NJ: Rowman and Allenheld.
Curtis, C. 1993. "The London Convention and Radioactive Waste Dumping at Sea: A Global Treaty Regime in Transition." Paper presented at the Conference on Radioactivity and Environmental Security in the Oceans: New Research and Policy Priorities in the Arctic and North Atlantic, 7–9 June, Woods Hole Oceanographic Institution, Woods Hole, MA.
Dahlgaard, H. 1993. "Anthropogenic Radioactivity in the Arctic Seas: Time Trends and Present Levels." Paper presented at the Conference on Radioactivity and Environmental Security in the Oceans: New Research and Policy Priorities in the Arctic and North Atlantic, 7–9 June, Woods Hole Oceanographic Institution, Woods Hole, MA.
Dahlgaard, H., A. Aarkgrog, L. Hallstadius, E. Holm, and J. Rioseco. 1986. "Radiocaesium Transport from the Irish Sea via the North Sea and the Norwegian Coastal Current to East Greenland." *Rapports et Procès-Verbaux des Réunions du Conseil International pour l'Exploration de la Mer (ICES)* 186:70–79.

Davis, W.J., and J.M. Van Dyke. 1990. "Dumping of Decommissioned Nuclear Submarines at Sea: A Legal and Technical Analysis." *Marine Policy* 14(6):467–476.

Deacon, G. 1984. *The Antarctic Circumpolar Ocean*. Cambridge: Cambridge University Press.

"Delayed-Action Mines in the Seas." 1993. *Spasenie (Rescue)*, 16 April (in Russian).

Domning, D.P. 1978. "Sirenian Evolution in the North Pacific Ocean." *University of California Publications in Geology* 118:1–176.

Draft Protocol to the Antarctic Treaty on Environmental Protection. 1990. In: W.R. Bush (ed.). *Antarctica and International Law*, Booklet AT 90, p. 82.

Ebel, R.E. 1992. "United States–Soviet Cooperation in Oil." *Geopolitics of Energy* 14(1):6.

"Ecocide in Russia." 1992. *Woods Hole Currents* 2(1):8–9.

El-Hinnawi, E. 1985. *Environmental Refugees*. Nairobi: United Nations Environment Programme.

Elliott, H. (ed.). 1974. *Proceedings of the Second World Conference on National Parks, 1972*. Morges, Switzerland: International Union for the Conservation of Nature and Natural Resources.

Elzinga, A., and I. Bohlin. 1989. "The Politics of Science in Polar Regions." *Ambio* 18:74.

Emerson, P.M., and E.W. Bourbon. 1991. "The Border Environment and Free Trade." Paper prepared for the North American Institute Conference on Harmonizing Economic Competitiveness with Environmental Quality: A North American Challenge, 8–10 November, Santa Fe, NM.

Emerson, P.M., and R.A. Collinge. 1993. "The Environmental Side of North American Free Trade." In: T.L. Anderson (ed.). *NAFTA and the Environment*. San Francisco: Pacific Research Institute for Public Policy.

Environmental Defense Fund. 1990. *EDF Letter* 21(2):8.

EPA (Environmental Protection Agency). 1989a. *Assessing Human Health Risks from Chemically Contaminated Fish and Shellfish*. EPA Office of Marine and Estuarine Protection, Office of Water Regulations and Standards. EPA 503/9-89-002. Washington, DC: US Environmental Protection Agency (September).

———. 1989b. *Marine and Estuarine Protection: Programs and Activities*. EPA Office of Water. EPA 503/9-89-002. Washington, DC: US Environmental Protection Agency (February).

———. 1990a. *Meeting the Environmental Challenge*. EPA 21K-2001. Washington, DC: US Environmental Protection Agency (December).

———. 1990b. *National Water Quality Inventory: 1988 Report to Congress*. EPA Office of Water. EPA 440/4-90-003. Washington, DC: US Environmental Protection Agency (April).

———. 1990c. *Progress in the National Estuary Program: Report to Congress*. EPA 503/9-90-005. Washington, DC: US Environmental Protection Agency (February).

―――. 1990d. *Report on Implementation of the U.S.-USSR Agreement on Cooperation in the Field of Environmental Protection, Feburary 1988–December 1989.* Washington, DC: US Environmental Protection Agency.

―――. 1990e. *Report to Congress on Implementation of Section 403(c) of the Federal Water Pollution Control Act.* EPA Office of Marine and Estuarine Protection. EPA 503/6-90-001. Washington, DC: US Environmental Protection Agency (June).

―――. 1992a. *EMAP Monitoring.* EPA Office of Research and Development, Environmental Monitoring and Assessment Program. EPA 600/M-91-051. Washington, DC: US Environmental Protection Agency.

―――. 1992b. *Managing Nonpoint Source Pollution.* EPA Office of Water. EPA-506-9-90. Washington, DC: US Environmental Protection Agency (January).

―――. 1993. *Virginian Province Demonstration Report. EMAP-Estuaries: 1990.* EPA Office of Research and Development, Environmental Monitoring and Assessment Program. EPA 620/R-93/006. Washington, DC: US Environmental Protection Agency (June).

EPA and NOAA (National Oceanic and Atmospheric Administration). 1992. "Environmental Monitoring and Assessment Program: Near Coastal Component." Program brochure. Washington, DC: US Government Printing Office (June).

Estes, J.A., D.O. Duggins, and G.B. Rathbun. 1989. "The Ecology of Extinctions in Kelp Forest Communities." *Conservation Biology* 3:252–264.

Everson, I., J.L. Watkins, D.G. Bone, and K.L. Foote. 1990. "Implications of a New Acoustic Target Strength for Abundance Estimates of Antarctic Krill." *Nature* 345:338–339.

Fairplay. 1990. 11 October, p. 45.

Falk, R.A. 1971. *This Endangered Planet.* New York: Random House.

FAO (Food and Agriculture Organization). 1989. *Yearbook of Fishery Statistics,* Volume 70. Food and Agriculture Organization of the United Nations.

―――. 1990. *Report of the FAO Expert Consultation on Large Scale Pelagic Driftnet Fishing.* FAO Fishery Report 434, Rome.

"Far East Latest Addition to Russian Offerings." 1993. *Offshore/Oilman* 53(7):3, 6 (July).

Favorite, F., A.J. Dodimead, and K. Nasu. 1976. "Oceanography of the Subarctic Pacific Region, 1960–1971." *Bulletin of the International North Pacific Fishery Commission* 33:1–187.

Fay, F.H., and T.J. Cade. 1959. "An Ecological Analysis of the Avifauna of St. Lawrence Island, Alaska." *University of California Publications in Zoology* 63:73–150.

FBIS (Foreign Broadcast Information Service). 1992a. *Daily Report—Central Eurasia.* FBIS-SOV-92-046.12. Washington, DC: US Government Printing Office (19 March).

―――. 1992b. "Submarine Talks with Sweden Assessed." *Daily Report—Central Eurasia.* FBIS-SOV-92-133. Washington, DC: US Government Printing Office (10 July).

Federal Register. 1988. 53:31918 (22 August).
Finn, D.P. 1983. "International Law and Scientific Consultation on Radioactive Waste Disposal in the Ocean." In: P.K. Park, D.R. Kester, I.W. Duedall, and B.H. Ketchum (eds.). *Wastes in the Ocean,* Volume 3: *Radioactive Wastes and the Ocean.* New York: Wiley-Interscience.
Fluharty, D., and C. Danson. 1979. "Management of Living Resources in the Northeast Pacific and the Unilateral Extension of the 200-mile Fisheries Zone." *Ocean Development and International Law* 6:11.
Foreign Relations of the United States. 1948. 1:948–1001.
Føyn, L. 1993. "Radioactivity in the Barents Sea, Past and Present Status, and Its Impact on Fisheries." Paper presented at the Conference on Radioactivity and Environmental Security in the Oceans: New Research and Policy Priorities in the Arctic and North Atlantic, 7–9 June, Woods Hole Oceanographic Institution, Woods Hole, MA.
Franckx, E. 1993. *Maritime Claims in the Arctic: Canadian and Russian Perspectives.* Dordrecht, Netherlands: Nijhoff.
Fredin, R.A. 1987. *History of Regulation of Alaskan Groundfish Fisheries.* Northwest and Alaska Fisheries Center Processed Report 87-07. National Marine Fisheries Service, National Oceanic and Atmospheric Administration, US Department of Commerce, Seattle, WA.
Freestone, D. 1991. "The Precautionary Principle." In: R. Churchill and D. Freestone (eds.). *International Law and Global Change.* London: Graham and Trotman.
"French Biologists Reply to Greenpeace Report." 1990. *Polar Record* 26:54.
French, G. 1990. "Legal Mechanisms for Protection and Preservation of the Marine Environment: Their Relationship to Particularly Sensitive Areas." In: *Proceedings of the International Seminar on Protection of Sensitive Sea Areas, 25–28 September 1990, Malmo, Sweden.* (Preliminary edition.) London: International Maritime Organization.
Friedheim, R., and T. Akaha. 1989. "Antarctic Resources and International Law: Japan, the United States and the Future of Antarctica." *Ecology Law Quarterly* 16:144–147.
Funke, O. 1992. "Environmental Dimensions of Security: The Former Soviet Bloc Nations." Paper prepared for presentation at the meeting of the International Studies Association, 31 March–4 April, Atlanta, Georgia.
Futsaeter, G., G. Eidnes, G. Halmo, S. Johnsen, H.P. Mannvik, L.K. Sydnes, and U. Witte. 1991. "Report on Oil Pollution." In: *The State of the Arctic Environment: Reports Rovaniemi, 1991.* Rovaniemi: Arctic Centre Publications.
Gaines, A.G. 1992. "Electronics Instrumentation and Environmental Security." *Sea Technology,* August, p. 93.
GAO (General Accounting Office, US Congress). 1992. "International Environment: International Agreements Are Not Well Monitored." GAO Resources, Community, and Economic Development Division. GAO/RCED-92-43. Washington, DC: US General Accounting Office.
Garcia, S.M., and J. Majkowski. 1990. "State of High Seas Resources." Work-

ing Paper for 24th Annual Conference of the Law of the Sea Institute, 24–27 July, Tokyo.
Gardner, P.F. 1988. "Tuna Poaching and Nuclear Testing in the South Pacific." *Orbis* 32(2):249–262.
GESAMP (Group of Experts on the Scientific Aspects of Marine Pollution). 1988. Unpublished draft of Annex IID on radionuclides. UNEP/GESAMP/WG.26/V/12 (Working Group 26). United Nations Environment Programme.
———. 1990. *GESAMP: The State of the Marine Environment*. UNEP Regional Seas Reports and Studies No. 115. United Nations Environment Programme.
Giziakis, K. 1982. "Economic Aspects of Marine Navigation Casualties." *Journal of Navigation* 35:466–478.
Gjerde, K.M. 1992. "The World Ocean and Environmental Security: The Law of the Sea Convention and Protection of the Marine Environment." Unpublished manuscript.
Gjernes, T., S. McKinnell, A. Yatso, S. Hayase, J. Ito, K. Nagao, H. Hatanaka, H. Ogi, M. Dahlberg, L.L. Jones, J. Wetherall, and P.J. Gould. 1990. *Final Report of Squid and Bycatch Observations in the Japanese Squid Driftnet Fishery for Neon Flying Squid* (Ommastrephes bartrami), *June–December, 1989 Observer Program*. Joint Report of Fisheries Agency of Japan, Canadian Department of Fisheries and Oceans, and United States Fish and Wildlife Service.
Godschalk, D.R. 1992. "Implementing Coastal Zone Management: 1972–1990." *Coastal Management* 20(2):93–116.
Gold, E. 1988. "New Directions in Ship-generated Marine Pollution Control: The New Law of the Sea and Developing Countries." In: E.M. Borgese and N. Ginsburg (eds.). *Ocean Yearbook 7*. Chicago: University of Chicago Press.
———. 1990. "Dangerous, Hazardous and Noxious Cargoes: A New Role for Maritime Arbitration?" *Marine Policy* 14:378–384.
Goldman, H. 1990. "The Environmental Crisis Facing Local Government." In: EPA. *Paying for Progress: Perspectives on Financing Environmental Protection*. Washington, DC: US Environmental Protection Agency.
Gong, Y., Y.S. Kim, and S.S. Kim. 1985. "Distribution and Migration of Flying Squid, *Ommastrephes bartrami* (LeSueur), in the North Pacific." *Bulletin of the Korean Fisheries Society* 18:166–179.
Gorbachev, M.S. 1987. In *Pravda*, 2 October.
Gormley, W.P. 1976. *Human Rights and Environment: The Need for International Cooperation*. Leyden, Netherlands: Sijthoff.
Gould, P., and V. Mendenhall. (n.d.) Personal written communication to the authors.
Granberg, A. 1992. "The Northern Sea Route and the Policy of New Russia." *International Challenges* 12(1):5–14.
Grebmeier, J.M., H.M. Feder, and C.P. McRoy. 1989. "Pelagic-Benthic

Coupling on the Shelf of the Northern Bering and Chukchi Seas: II. Benthic Community Structure." *Marine Ecology Progress Series* 51:253–268.
Grebmeier, J.M., and C.P. McRoy. 1989. "Pelagic-Benthic Coupling on the Shelf of the Northern Bering and Chukchi Seas: III. Benthic Food Supply and Carbon Cycling." *Marine Ecology Progress Series* 53:79–91.
Grebmeier, J.M., C.P. McRoy, and H.M. Feder. 1988. "Pelagic-Benthic Coupling on the Shelf of the Northern Bering and Chukchi Seas: I. Food Supply Source and Benthic Biomass." *Marine Ecology Progress Series* 48:57–67.
Greenpeace. 1990. "Report on a Visit to Dumont d'Urville, Antarctica." *Polar Record* 26:26–51.
Griffiths, F., and O.R. Young. 1990. "Protecting the Arctic's Environment: Impressions of the Co-Chairs." *Reports and Papers 1990–1*. Hanover, NH: Working Group on Arctic International Relations.
Grigalunas, T.A., J.J. Opaluch, S.C. Chung, D. French, and M. Reed. 1987. "Adaptation of an Integrated, Ocean Systems/Economics Damage Assessment Model to Korea: Some Preliminary Results." Paper presented at the Third Conference on Asia/Pacific Marine Resources and Development, January 1988, Koahsiung, Taiwan.
Grigalunas, T.A., J.J. Opaluch, D. French, and M. Reed. 1988. "Measuring Damages to Marine Natural Resources from Pollution Incidents under CERCLA: Applications of an Integrated Ocean Systems/Economic Model." *Marine Resources Economics* 5:1–21.
Gulland, J. 1988a. *Fish Population Dynamics*. New York: John Wiley and Sons.
———. 1988b. "The Management Regime for Living Resources." In: C. Joyner and S. Chopra (eds.). *The Antarctic Legal Regime*. Dordrecht, Netherlands: Nijhoff.
Guyer, R.E. 1983. "Antarctica's Role in International Relations." In: F.O. Vicuna (ed.). *Antarctic Resources Policy*. Cambridge: Cambridge University Press.
Haas, P.M. 1992. "Banning Chlorofluorocarbons: Epistemic Community Efforts to Protect Stratospheric Ozone." *International Organization* 46:209 (January).
Haas, P.M., R.O. Keohane, and M. Levy (eds.). 1993. *Institutions for the Earth: Sources of Effective Environmental Protection*. Cambridge: MIT Press.
Hain, J.W.H. 1986. "Low-Level Radioactivity in the Irish Sea." *Oceanus* 29(3):16–26.
Hakapaa, K. 1986. *Marine Pollution in International Law: Material Obligations and Jurisdictions*. Moscow: Progress.
Hamre, J. 1988. *The Aspects of the Interrelation between the Herring in the Norwegian Sea and the Stocks of Capelin and Cod in the Barents Sea*. ICES Report CM 1988/H:42. Copenhagen: International Council for the Exploration of the Sea.
Handler, J. 1993a. "Radioactive Pollution from Accidents Involving Nuclear-powered Vessels: Problems and Solutions." Poster presentation

at the Conference on Radioactivity and Environmental Security in the Oceans: New Research and Policy Priorities in the Arctic and North Atlantic, 7–9 June, Woods Hole Oceanographic Institution, Woods Hole, MA.

———. 1993b. "Trip Report: Greenpeace Visit to Moscow and Russian Far East July–November 1992." Washington, DC: Greenpeace Nuclear Free Seas Campaign (15 February).

Handler, J., and W.M. Arkin. 1990. *Neptune Papers No. 5. Nuclear Warships and Naval Nuclear Weapons 1990: A Complete Inventory.* Washington, DC: Greenpeace (September).

Handler, J., A. Wickenheiser, and W.A. Arkin. 1990. *Neptune Papers No. 4. Naval Safety 1989: The Year of the Accident.* Washington, DC: Greenpeace Institute for Policy Studies (April).

"Hands Off Antarctica." 1991. *Economist,* 4 May, p. 47.

Hanemann, W.M., and R.T. Carson. 1987. *Southcentral Alaska Sport Fishing Economic Study,* chapters 8 and 9. Report prepared for Alaska Department of Fish and Game by Jones and Stokes Associates, Inc., Sacramento, CA (November).

Hanemann, W.M., P.-O. Johansson, B. Kristrom, and L. Mattson. 1992. "Natural Resources Damages from Chernobyl." *Environmental and Resource Economics* 2:523–525.

Hansell, D.A., J.J. Goering, J.J. Walsh, C.P. McRoy, L.K. Coachman, and T.E. Whitledge. 1989. "Summer Phytoplankton Production and Transport along the Shelf Break in the Bering Sea." *Continental Shelf Research* 12:1085–1104.

Hansson, R. 1992. "An Environmental Challenge to Half the Arctic." *International Challenges* 12(1):90–96.

Harders, J.E. 1987. "In Quest of an Arctic Legal Regime: Marine Regionalism—A Concept of International Law Evaluated." *Marine Policy* 12:285–298.

Harris, C.K. 1989. "The Effects of International Treaty Changes on Japan's High Seas Salmon Fisheries, with Emphasis on Their Catches of North American Sockeye Salmon, 1972–1984." Ph.D. thesis. Seattle: University of Washington.

Harris, C.M. 1991. "Environmental Effects of Human Activities on King George Island, South Shetland Islands." *Polar Record* 27:193–204 (July).

Harrison, C.S., K.-S. Choe, H. Fen-qi, N. Litvenenko, and Y. Shibaev. 1990. "A Summary of the Laws and Treaties of Canada, China, Japan, Korea, U.S.A., and U.S.S.R. That Protect Seabirds at Land and Sea." Abstract of paper presented at Pacific Seabird Group Annual Meeting, 22–25 February, Victoria, British Columbia, Canada.

Haushofer, K. 1932a. *Ruckblick und Vorschau auf das geopolitische Kartenwesen mit 18 Abbildungen.* Berlin-Grunewald: K. Vowinckel Verlag.

———. 1932b. *Wehr-Geopolitik geographische Grundlagen einer Wehrkunde.* Berlin: Junker und Dunnhaupt.

Hayase, S., Y. Watanabe, and T. Hatanaka. 1990. "Preliminary Report on the Japanese Fishing Experiments Using Subsurface Gillnets in the South and the North Pacific, 1989-1990." Working Paper for IWC Workshop on Mortality of Cetaceans in Passive Fishing Nets and Traps, 20-25 October, La Jolla, CA.

Heap, J.A. 1987. "Current and Future Problems Arising from Activities in the Antarctic." In: G.D. Triggs (ed.). *The Antarctic Treaty Regime.* Cambridge: Cambridge University Press.

———. 1991. "Has CCAMLR Worked? Management Policies and Ecological Needs." In: A. Jorgensen-Dahl and W. Østreng (eds.). *The Antarctic Treaty System in World Politics.* London: Macmillan.

Heap, J.A., and M.W. Holdgate. 1986. "The Antarctic Environment." In: U.S. Polar Research Board. *Antarctic Treaty System: An Assessment.* Washington, DC: National Academy Press.

Heininen, L. 1990. "Environmental Threats of Military Presence in Northern Waters and the Arctic." In: L. Heininen (ed.). *Arctic Environmental Problems.* Tampere Peace Research Institute Occasional Papers 41:41-69.

Hempel, G. 1991. "Life in the Antarctic Sea Ice Zone." *Polar Record* 27:249-254.

Henkin, L., R.C. Pugh, O. Schachter, and H. Smit. 1980. *International Law: Cases and Materials.* St. Paul, MN: West Publishing.

Herz, John H. 1959. *International Politics in the Atomic Age,* chapter 10. New York: Columbia University Press.

Higginson, C. 1993. Council on Ocean Law, Washington, DC. Telephone communication (8 October).

Highsmith, R.C., and K.O. Coyle. 1990. "High Productivity of Northern Bering Sea Amphipods." *Nature* 344:862-863.

Hodgson, B. 1990. "Alaska's Big Spill." *National Geographic* 177:2-43.

Hollister, C.D., D.R. Anderson, and G.R. Heath. 1981. "Subseabed Disposal of Nuclear Wastes." *Science* 213:1321-1326.

Homer-Dixon, T.F., J.H. Boutwell, and G.W. Rathjens. 1993. "Environmental Change and Violent Conflict." *Scientific American,* February, pp. 38-45.

Honjo, S., S.J. Maganini, and G. Wefer. 1988. "Annual Particle Flux and a Winter Outburst Sedimentation in the Northern Norwegian Sea." *Deep-Sea Research* 35:1223-1234.

Horgan, M. 1993. Personal telephone communication (13 October).

Howard, M. 1989. "The Convention on the Conservation of Antarctic Marine Living Resources: A Five Year Review." *International and Comparative Law Quarterly* 38:124-129.

Hume, H.R., I. Buist, D. Betts, and R. Goodman. 1983. "Arctic Marine Oil Spill Research." In: L. Rey (ed.). *Arctic Energy Resources.* New York: Elsevier.

Huntington, E. 1924. *Civilization and Climate.* New Haven: Yale University Press.

IAEA (International Atomic Energy Agency). 1991. *Inventory of Radioactive*

Material Entering the Marine Environment: Sea Disposal of Radioactive Waste. IAEA-TECDOC-588. Vienna: International Atomic Energy Agency.

Iakimets, V. 1993. "Arctic Radioactive Contamination: Past Legacy and Future." Paper presented at the Conference on Radioactivity and Environmental Security in the Oceans: New Research and Policy Priorities in the Arctic and North Atlantic, 7–9 June, Woods Hole Oceanographic Institution, Woods Hole, MA.

"Iceland and the Environment." 1991. Document of the Icelandic Ministry for the Environment (4 June). Washington, DC: Embassy of Iceland.

"Iceland Takes Initiative on Marine Conservation." 1992. *News from Iceland*, April, p. 1.

"Iceland's Security and Defense." 1993. Report of the Special Committee on Iceland's Security and Defense. Reykjavik: Icelandic Ministry of Foregin Affairs (March).

Imperial, M.T., D. Robadue, Jr., and T.M. Hennessey. 1992. "An Evolutionary Perspective on the Development and Assessment of the National Estuary Program." *Coastal Management* 20(4):311–341.

Intergovernmental Task Force on Monitoring Water Quality, Interagency Advisory Committee on Water Data, and Water Information Coordination Program. 1992. *Ambient Water Quality Monitoring in the United States: First Year Review, Evaluation, and Recommendations*. Washington, DC: Office of Management and Budget (December).

International Challenges. 1992. Special issue on the Northern Sea Route 12(1).

International Legal Materials. 1972. 11:251–288.

———. 1973. 12:1319–1344.

———. 1978. 17:546–578.

———. 1979. 18:1434–1441.

———. 1980. 19:841–859.

———. 1982. 21:1293–1308.

———. 1988. 27:868–900.

———. 1991a. 30:846 (UN Sec. Res. 687, passed April 3).

———. 1991b. 30:864 (UN Sec. Res. 692, passed May 20).

———. 1993. 32:1101–1133.

Issledovanie ekosystemi Beringova Morya. 1973. Volume 1. (Also Volume 2, 1990.) Leningrad: Gidrometizdat.

IUCN (International Union for the Conservation of Nature and Natural Resources). 1980. *World Conservation Strategy: Living Resources Conservation for Sustainable Development.* Gland, Switzerland: International Union for the Conservation of Nature and Natural Resources.

Izmaylov, V.V. 1986. "Vozmozhnjye ekologicheskiye posledstviya neftyanogo zagryazneniya ledyanogo pokrova polyarnykh basseinov." In: *Metodologiya prognozirovaniya zagryazneniya okeanov i morej*. Moscow: Gidrometeoizdat (in Russian).

———. 1988. *Transformation of Oil Films within the Ocean-Atmosphere System.* Leningrad: Soviet Ministry of Hydrometeorology (in Russian).

Jervis, R. 1978. "Cooperation under the Security Dilemma." *World Politics* 30:167–186.
Jin, D., H. Kite-Powell, and J.M. Broadus. 1993. "Dynamic Economic Analysis of Marine Pollution Prevention Technologies: An Application to Double Hulls and Electronic Charts." Draft. Woods Hole, MA: Marine Policy Center, Woods Hole Oceanographic Institution.
Johnston, D.M. 1990. "The Driftnetting Problem in the Pacific Ocean: Legal Considerations and Diplomatic Options." *Ocean Development and International Law* 21:5–39.
———. 1993. "Vulnerable Coastal and Marine Areas: A Framework for the Planning of Environmental Security Zones in the Ocean." *Ocean Development and International Law* 24:63–79.
Joint Resolution of the Fifth Bering Sea Conference. 1992. Press Release of the Joint Resolution of the Fifth Conference on the Conservation and Management of the Living Marine Resources of the Central Bering Sea (14 August).
Jones, L., M. Dahlberg, and S. Fitzgerald. 1990. "High Seas Driftnet Fisheries of the North Pacific Ocean." Working Paper for IWC Workshop on Mortality of Cetaceans in Passive Fishing Nets and Traps, 20–25 October, La Jolla, California.
Jones, L.L., and A.R. DeGange. 1989. "Interactions between Seabirds and Fisheries in the North Pacific Ocean." In: J. Burger (ed.). *Seabirds and Other Marine Vertebrates: Competition, Predation and Other Interactions.* New York: Columbia University Press.
Joyner, C.C. 1984. "Oceanic Pollution and the Southern Ocean: Rethinking the International Legal Implications for Antarctica." *Natural Resources Journal* 24:1–40 (January).
———. 1992. *Antarctica and the Law of the Sea.* Dordrecht, Netherlands: Nijhoff.
"Judge Orders U.S. Sub Out of Brazilian Waters." 1990. *Honolulu Advertiser,* 5 November, p. D1.
Kaitala, V., and G.R. Munro. 1993. "The Management of High Seas Fisheries." *Marine Resource Economics* 8:313–329.
Kasoulides, G.C. 1989. "The 1986 United Nations Convention on the Conditions for Registration of Vessels and the Question of Open Registry." *Ocean Development and International Law* 20:543–576.
Katasonov, V. 1990. "Tyumenskiy apokalipsis." *Zeolyony Mir* 4:6.
Keage, P.A. 1987. *Antarctic Protected Areas: Future Options.* Hobart: University of Tasmania.
Keating, D. 1991. "Pesticides in Reefs Scrutinized." *Miami Herald,* 13 May, p. 1A.
Kelso, Adm. F.B. 1991. Speech by the US Chief of Naval Operations at the International Seapower Symposium, 7 October, Newport, RI.
Kennan, G. 1970. "To Prevent a World Wasteland." *Foreign Affairs* 48:404–413.

Keondzhyan, V.P., A.M. Kudina, and U.V. Terehena (eds.). 1990. *Practical Ecology of Marine Regions: The Black Sea*. Kiev: Naukova Dumka.

Keys, J.R. 1984. *Antarctic Marine Environments and Offshore Oil*. Wellington, New Zealand: Commission for the Environment.

Khen, G.V. 1989. "Oceanological Conditions of the Bering Sea Biological Productivity." In: *Proceedings of the International Science Symposium on Bering Sea Fisheries, Sitka, Alaska, 19–21 July 1988*. NOAA Technical Document NMPS F/NWC, Volume 163.

Kimball, L.A. 1989. "International Law and Institutions: The Oceans and Beyond." *Ocean Development and International Law* 20(2):147–165.

———. 1990. "Southern Exposure: Deciding Antarctica's Future." Washington, DC: World Resources Institute (November).

Kindt, J.W. 1988. "A Regime for Ice-covered Areas." In: C. Joyner and S. Chopra (eds.). *The Antarctic Legal Regime*. Dordrecht, Netherlands: Nijhoff.

Kiss, A., and D. Shelton. 1991. *International Environmental Law*. New York: Transnational.

Kite-Powell, H. 1992. "Rationale, Processes, and Effects of Technological Standards in Marine Electronics." Ph.D. thesis, Joint degree program of the Massachusetts Institute of Technology, Department of Ocean Engineering, Cambridge, MA, and the Marine Policy Center, Woods Hole Oceanographic Institution, Woods Hole, MA.

Knopman, D.S., and R.A. Smith. 1993. "20 Years of the Clean Water Act." *Environment* 35(1):17–20, 34–41.

Kolmogorov, D., and I. Sinelnikov. 1984. "Mery regulirovania promysla dalnevostochnykh lososei." *Rybnoe khoziaistvo*, No. 4 (in Russian).

Kopp, R.J., and V.K. Smith. 1989. "Benefit Estimation Goes to Court: The Case of Natural Resource Damage Assessments." *Journal of Policy Analysis and Management* 8(4):593–612.

Kraska, J.C. 1993. "The U.S. Navy and the Marine Pollution Convention." *American Bar Association National Security Law Report* 15(9):1–2, 5–6.

Kraus, G.F., Jr. 1992. "Reducing the Industrial Base: Implications for Russian Submarine Construction." *FSRC Analytical Note*. A92-029/UL. Greenwood, CO: Science Applications International Corporation.

Kriwoken, L.K. 1991. "Antarctic Environmental Planning and Management: Conclusions from Casey, Australian Antarctic Territory." *Polar Record* 27:1–8 (January).

Kvint, V. 1990. "Eastern Siberia Could Become Another Saudi Arabia." *Forbes* 17:130–132 (September 17).

Kwaczek, A.S., W.A. Kerr, and S. Mooney. 1990. "Chernobyl Lessons in Nuclear Liability." *Forum for Applied Science and Technology* 5(2):21–27.

Kwiatkowska, B. 1991. "Creeping Jurisdiction beyond 200 Miles in the Light of the 1982 Law of the Sea Convention and State Practice." *Ocean Development and International Law* 22:153–188.

———. 1993. "The High Seas Fisheries Regime: At a Point of No Return?" *International Journal of Marine and Coastal Law* 8:327–355.

Laevastu, T., and F. Favorite. 1977. *Preliminary Report on Dynamic Numerical Marine Ecosystem Model (DYNUMES II) for Eastern Bering Sea*. Progress Report. Seattle: National Marine Fisheries Service, National Oceanic and Atmospheric Administration, US Department of Commerce.

Lander, R.H., and H. Kajimura. 1982. "Status of Northern Fur Seals." In: *Mammals in the Seas*, Volume 2. FAO Fishery Series 5. Rome: Food and Agriculture Organization of the United Nations.

Langseth, M., T. Delaca, G. Newton, B. Coakley, R. Colony, J. Gossett, C. May, P. McRoy, J. Morison, W. Smethie, D. Steele, and W. Tucker. 1994. "SCICEX-93: Arctic Cruise of the U.S. Navy Nuclear Powered Submarine USS Pargo." *Marine Technology Society Journal* 27(4):4–12.

Larkin, P. 1988. "Pacific Salmon." In: J.A. Gulland (ed.). *Fish Population Dynamics*. New York: John Wiley and Sons.

Larson, D.L. 1989. "United States Interests in the Arctic Region." *Ocean Development and International Law* 20:167–191.

Latukhov, S.V. 1990. "Transportnoie ispol'zovanie Severnogo Morskogo Puti i zagryaznenie Arktiki." *Geokologiya Morovogo Okeana*. Leningrad: Nauka (Science Press) (in Russian).

Laws, R.M. 1989. *Antarctica: The Last Frontier*. London: Boxtree.

———. 1991. "Unacceptable Threats to Antarctic Science." *New Scientist*, 30 March, p. 4.

Lefever, E.W. 1993. "Reining in the UN: Mistaking the Instrument for the Actor." *Foreign Affairs* 72(3):17–20.

Linnerooth, J. 1990. "The Danube River Basin: Negotiating Settlements to Transboundary Environmental Issues." *Natural Resources Journal* 30(3): 629–660.

Livingston, H.D. 1985. "Anthropogenic Radiotracer Evolution in the Central Greenland Sea." *Rit Fiskideildar* 9:43–54.

———. 1988. "The Use of Cs and Sr Isotopes as Tracers in the Arctic and Mediterranean Seas." *Philosophical Transactions of the Royal Society of London, Series A*, 325:161–176.

Livingston, H.D., and V.T. Bowen. 1977. "Windscale Effluent in the Waters and Sediments of the Minch." *Nature* 269:586–588.

———. 1982a. "Radionuclides from Windscale Discharges: I. Nonequilibrium Tracer Experiments in High Latitude Oceanography." *Journal of Marine Research* 40:253–272.

———. 1982b. "Radionuclides from Windscale Discharges: II. Their Dispersion in Scottish and Norwegian Coastal Circulation." *Journal of Marine Research* 40:1227–1258.

Livingston, H.D., S.L. Kupferman, V.T. Bowen, and R.M. Moore. 1984. "Vertical Profile of Artificial Radionuclide Concentrations in the Central Arctic Ocean." *Geochimist-Cosmochimist Acta* 48:2195–2293.

Lockwood, A. 1993. International Affairs, Office of the Assistant Secretary for Domestic and International Energy Policy, US Department of Energy. Telephone communication (July).

Loughlin, T.R., A.S. Pervlov, and V.A. Vladimirov. 1990. "Survey of Northern

Sea Lions in the Gulf of Alaska and Aleutian Islands during June 1989." NOAA Technical Memo NMFS-F/NWC-176. Washington, DC: National Marine Fisheries Service, National Oceanic and Atmospheric Administration, US Department of Commerce.

Lowry, L.F., and K.J. Frost. 1984. "Biological Interactions among Marine Mammals and Commercial Fisheries in the Bering Sea." In: J.R. Beddington, R.J.H. Beverton, and D.M. Lavigne (eds.). *Marine Mammals and Fisheries*. London: George Allen and Unwin.

Lowry, L.F., K.J. Frost, and T.R. Loughlin. 1989. "Importance of Walleye Pollock in the Diets of Marine Mammals in the Gulf of Alaska and Bering Sea and Implications for Fishery Management." In: *Proceedings of the International Symposium on the Biology and Management of Walleye Pollock*. University of Alaska Sea Grant Report 89-01. Fairbanks: University of Alaska.

Lundquist, T.R. 1977. "Iceberg Cometh? International Law Relating to Antarctic Iceberg Exploitation." *Natural Resources Journal* 17:1–41.

Lyster, S. 1985. *International Wildlife Law: An Analysis of International Treaties Concerned with the Conservation of Wildlife*. Cambridge, UK: Grotius Publications.

Lystsov, V.N. 1993. Russian Ministry of the Environment, Moscow. Personal telephone communication, 20 October.

Mackenzie, A.B., and R.D. Scott. 1982. "Radiocesium and Plutonium Intertidal Sediments from Southern Scotland." *Nature* 299:613–616.

MacKenzie, D. 1989. "Environmental Issues Surface at the Summit of the World." *New Scientist*, 25 February, p. 29.

MacKenzie, W.H., and R.S. Rootes. 1985. "Implementing the Convention on the Conservation of Antarctic Marine Living Resources: The Legislative Process." In: L.M. Alexander and L.C. Hansen (eds). *Antarctic Politics and Marine Resources: Critical Choices for the 1980s*. Kingston, RI: Center for Ocean Management Studies, University of Rhode Island.

Mackinder, H. 1904. "The Geographical Pivot of History." *Geographical Journal* 23:421–437.

Mahan, A.T. 1897. *The Influence of Seapower upon History*. Boston: Little, Brown.

"Major Hydrocarbon Structures in Barents Sea, Kara Regions." 1993. *Offshore/Oilman* 53:34–35 (January).

Malthus, T. [1798] 1976. *An Essay on the Principle of Population*. New York: W.W. Norton.

The Management of Marine Regions: The North Pacific. 1982. Berkeley and Los Angeles: University of California Press.

Marine Mammal Commission. 1994. "Annual Report to Congress: 1993." Washington, DC: Marine Mammal Commission.

Marine Policy Center. 1991. *Toward Improved World Fishery Statistics*. Final Report of an International Planning Group/Workshop Sponsored by the Ford Foundation. Woods Hole, MA: Marine Policy Center, Woods Hole Oceanographic Institution (27 August).

Marine Technology Society Journal. 1994. 27(4). (Winter 1993–94 issue, "Military Assets for Environmental Research.")

Mathews, J.T. 1989. "Redefining Security." *Foreign Affairs* 68(2):162–177.

Maxwell, J.B., and L.A. Barrie. 1989. "Atmospheric and Climatic Change in the Arctic and the Antarctic." *Ambio* 18(1):46–47.

McConnell, K.E. 1986. "The Damages to Recreational Activities from PCBs in New Bedford Harbor." Report prepared for the Ocean Assessment Division of NOAA. College Park, MD: University of Maryland-College Park.

McRoy, C.P., J.J. Goering, and W.S. Shields. 1972. "Studies of Primary Production in the Eastern Bering Sea." In: A.Y. Takenouti (ed.). *Biological Oceanography of the Northern Pacific Ocean.* Tokyo: Idemitsu Shoten.

Meadows, D.H., D.L. Meadows, J. Randers, and W.W. Behrens III. 1972. *The Limits to Growth.* New York: Universe Books.

Mendelsohn, R. 1986. "Assessment of Damages by PCB Contamination to New Bedford Harbor Amenities Using Residential Property Values." Draft report to the Ocean Assessment Division of NOAA. New Haven: Yale University.

Meng, Q-n. 1987. *Land-based Marine Pollution: International Law Development.* London: Graham and Trotman.

MEPC (Marine Environment Protection Committee). 1990. MEPC 29 (29th Session), 12–16 March. London: International Maritime Organization.

Metaxas, B.N. 1981. "Flags of Convenience." *Marine Policy* 5:65.

Miles, E.L., and W.T. Burke. 1989. "Pressures on the United Nations Convention on the Law of the Sea of 1982 Arising from New Fisheries Conflicts: The Problem of Straddling Stocks." *Ocean Development and International Law* 20:343–357.

Miles, E.L., and D.L. Fluharty. 1991. "US Interests in the North Pacific." *Ocean Development and International Law* 22:315–342.

Miller, D.G.M., and I. Hampton. 1989. "Biology and Ecology of the Antarctic Krill." *BIOMASS Scientific Series* 9.

Minerals Management Service. 1990. *5-Year Outer Continental Shelf Oil and Gas Leasing Program: Final Supplemental Environmental Impact Statement (Mid-1987–Mid-1992).* Washington, DC: US Department of the Interior.

Ministry of the Environment, Finland. 1989. "Background Information Prepared for the Consultative Meeting on the Protection of the Arctic Environment." 20–26 September, Rovaniemi, Finland.

Miovski, L. 1989. "Solutions in the Convention on the Law of the Sea to the Problem of Overfishing in the Central Bering Sea: Analysis of the Convention, Highlighting the Provisions Concerning Fisheries and Enclosed and Semi-Enclosed Seas." *San Diego Law Review* 26:525–574.

Mirovitskaya, N.S., and J.C. Haney. 1992. "Fisheries Exploitation as a Threat to Environmental Security: The North Pacific Ocean." *Marine Policy* 14(4):243–259.

Mitchell, R.C., and R.T. Carson. 1989. *Using Surveys to Value Public Goods: The Contingent Valuation Method.* Washington, DC: Resources for the Future.

Mitin, L.I. 1990. "Ecologischeskoye sostoyanie Chernogo i Azovskogo morey i puty ego ululchsheniya." *Eiev,* pp. 1–15.

Moore, J.N., F.S. Tipson, and R.F. Turner (eds.). 1990. *National Security Law.* Durham, NC: Carolina Academic Press.

Myers, N. 1986. "The Environmental Dimension to Security Issues." *Environmentalist* 6(4):251–257.

———. 1987a. *Not Far Afield: U.S. Interests and the Global Environment.* Washington, DC: World Resources Institute (June).

———. 1987b. "Population, Environment, and Conflict." *Environmental Conservation* 14(1):15–22.

———. 1989. "Environmental Security: The Case of South Asia." *International Environmental Affairs* 1(2):138–154.

Nance, J. 1993. National Marine Fisheries Service, Galveston Laboratory, Galveston, TX. Telephone communication (May).

National Research Council. 1991. *Tanker Spills: Prevention by Design.* Washington, DC: National Academy Press.

———. 1993a. *Alternative Technologies for the Destruction of Chemical Agents and Munitions.* Washington, DC: National Academy Press.

———. 1993b. *Science and Stewardship in the Antarctic.* Washington, DC: National Academy Press.

Neimeier, P. 1993. Organizations and Agreements Division, Office of International Affairs, National Marine Fisheries Service/NOAA. Personal telephone communication (April).

Nezhdanov, G. 1993. "Cs-137 Contamination of Sea Water Surrounding the *Komsomoletz* Nuclear Sumbarine." Paper presented at the Conference on Radioactivity and Environmental Security in the Oceans: New Research and Policy Priorities in the Arctic and North Atlantic, 7–9 June, Woods Hole Oceanographic Institution, Woods Hole, MA.

Nicol, S. 1989. "Who's Counting on Krill?" *New Scientist,* 11 November, pp. 38–41.

———. 1990. "The Age-Old Problem of Krill Longevity." *BioScience* 40:833–834 (December).

———. 1991a. "CCAMLR and Its Approaches to Management of the Krill Fishery." *Polar Record* 27:235.

———. 1991b. "Life after Death of Empty Shells." *New Scientist* 1755:46–48 (9 February).

Nies, H. 1988. "The Radioactive Contamination of the Baltic Sea during the Years 1973 to 1987 and Its Radiological Consequences." *Deutsche Hydrographische Zeitschrift* 41:39–44.

Nies, H., J. Herrmann, and I. Goroney. 1993. "Temporal Evolution of Artificial Radioactivity in the North and Baltic Sea after the Reduction of the Sellafield Discharges and the Chernobyl Accident." Paper presented at the Conference on Radioactivity and Environmental Security in the Oceans: New Research and Policy Priorities in the Arctic and North Atlantic, 7–9 June, Woods Hole Oceanographic Institution, Woods Hole, MA.

Nies, H., and C. Wedekind. 1988. "The Contamination of the North Sea and Baltic Sea by the Chernobyl Fallout." In: J.C. Guary et al. (eds.). *Radionuclides: A Tool for Oceanography.* London: Elsevier Applied Sciences.

NMFS (National Marine Fisheries Service). 1993a. *Fisheries of the United States, 1992.* Silver Spring, MD: National Marine Fisheries Service, National Oceanic and Atmospheric Administration, US Department of Commerce (May).

———. 1993b. "International Agreements Concerning Living Marine Resources of Interest to NOAA Fisheries." Report for internal distribution. Silver Spring, MD: National Marine Fisheries Service, National Oceanic and Atmospheric Administration, US Department of Commerce (February).

———. 1993c. "Proposed Regime to Govern Interactions between Marine Mammals and Commercial Operations: Final Legislative Environmental Impact Statement." National Marine Fisheries Service, National Oceanic and Atmospheric Administration, US Department of Commerce (March).

NOAA (National Oceanic and Atmospheric Administration). 1983. *Assessing the Social Costs of Oil Spills: The Amoco Cadiz Case Study.* Washington, DC: US Department of Commerce (July).

———. 1985. *Gulf of Mexico Coastal and Ocean Zones Strategic Assessment: Data Atlas.* National Ocean Service, National Oceanic and Atmospheric Administration (December).

———. 1990. *50 Years of Population Change along the Nation's Coasts: 1960–2010.* Report of the Office of Oceanography and Marine Assessment, Strategic Assessments Branch. Rockville, MD: National Oceanic and Atmospheric Administration, US Department of Commerce.

———. 1992. *National Status and Trends Program: Marine Environmental Quality.* Report of the National Ocean Service, Office of Ocean Research, Conservation, and Assessment. Rockville, MD: National Oceanic and Atmospheric Administration, US Department of Commerce.

Nollkaemper, A. 1991. "Deep Sea-Bed Mining and the Protection of the Marine Environment." *Marine Policy* 15(1):55–66.

———. 1993. "Agenda 21 and Prevention of Sea-based Marine Pollution: A Spurious Relationship?" *Marine Policy* 17:537–556 (November).

Northridge, S.P. 1984. *World Review of Interactions between Marine Mammals and Fisheries.* FAO Fishery Paper 251, Rome.

NPFMC (North Pacific Fishery Management Council) Newsletter. 1993. No. 3-93, 7 July, Anchorage, AK.

Ocean Science News. 1993. 35(16):1 (August 30).

———. 1993. "INSROP Startup—Why INSROP? The INSROP Newsletter 1(1):1–2.

Oceans Policy News. 1992. 9(11) (November).

———. 1993a. 10(5):2 (May).

———. 1993b. 10(6):4–5 (October).

———. 1993c. 10(8):4–5 (December).

"Oil from Russia." 1990. *Fortune*, 22 October, pp. 115–118.
Osherenko, G., and O.R. Young. 1989. *The Age of the Arctic: Hot Conflicts and Cold Realities*. Cambridge: Cambridge University Press.
Østreng, W. 1991. "The Northern Sea Route: A New Era in Soviet Policy?" *Ocean Development and International Law* 22(3):259–287.
Oswald, J. 1992. "Security Law Has New Meaning." *Naval Institute Proceedings/Naval Review 1992* 118(5):46–51 (May).
OTA (Office of Technology Assessment, United States Congress). 1985. *Oil and Gas Technologies for the Arctic and Deepwater.* OTA-O-270. Washington, DC: US Government Printing Office.
———. 1989. *Polar Prospects: A Minerals Treaty for Antarctica.* OTA-O-428. Washington, DC: US Government Printing Office (September).
———. 1990. *Coping with an Oiled Sea: An Analysis of Oil Spill Response Technologies*. Washington, DC: US Government Printing Office (March).
———. 1991. *Complex Cleanup: The Environmental Legacy of Nuclear Weapons Production*. OTA-O-484. Washington, DC: US Government Printing Office (February).
"Overfishing in Antarctica." 1984. *Marine Pollution Bulletin* 15(11):396.
Owens, D.W. 1992. "National Goals, State Flexibility, and Accountability in Coastal Zone Management." *Coastal Management* 20(2):143–165.
Oxman, B.H. 1991. "Environmental Warfare." *Ocean Development and International Law* 22:433–437.
Ozhan, E. 1993. "Saving the Black Sea." *Ocean and Coastal Management* 19(1):95–96.
Packman, G.A., and R.G. Shearer. 1989. "An Overview of Marine Environmental Quality in the Canadian Arctic." Background paper prepared for the Consultative Meeting on the Protection of the Arctic Environment, 20–26 September, Rovaniemi, Finland.
Paine, R.T. 1979. "Disaster, Catastrophe and Local Persistence of the Sea Palm *Postelsia palmaeformis*. *Science* 205:685-687.
———. 1980. "Food Webs: Linkage, Interaction Strength and Community Infrastructure." *Journal of Animal Ecology* 49:667–686.
Palmquist, R.B. 1990. "Hedonic Methods." In: J.B. Braden and C.D. Kolstad (eds.). *Measuring the Demand for Environmental Improvement*. Document ILENR/RE-EA-89/18, Illinois Department of Energy and Natural Resources.
Panel on the Law of Ocean Uses. 1993. *United States Interests in the Law of the Sea Convention*. Washington, DC: Council on Ocean Law (July).
Pease, W.M. 1989. "Program to Improve the Safety of Tankers and Other Hazardous Liquid Cargo Carriers Operating in U.S. Waters." Unpublished paper. Raytheon Company.
Pentreath, R.J., and M.B. Lovett. 1976. "Occurrence of Plutonium and Americium in Plaice from the Northeastern Irish Sea." *Nature* 262:814–816.
Perelet, R. 1993. "Environmental Security and Governance Issues in the Arctic Ocean." Paper presented at the Conference on Radioactivity and Environmental Security in the Oceans: New Research and Policy Priorities in

the Arctic and North Atlantic, 7–9 June, Woods Hole Oceanographic Institution, Woods Hole, MA.
Perkovich, G. 1993. "The Plutonium Genie." *Foreign Affairs* 72:153–165 (Summer).
Peterson, M.J. 1986. "Antarctica and the New Law of the Sea." *Ocean Development and International Law* 16:137–181.
Piatt, J.F., B.D. Roberts, and S.A. Hatch. 1990. "Colony Attendance and Population Monitoring of Least and Crested Auklets on St. Lawrence Island, Alaska." *Condor* 92:97–106.
PICES Fact Sheet. (n.d.) PICES Secretariat. Sidney, BC: Institute of Ocean Sciences.
Pimm, S.L. 1982. *Food Webs*. London: Chapman and Hall.
Pitt, D.E. 1993. "Fishing Countries Split on Harvests." *New York Times*, 1 August.
Pittenger, R. 1993. Director, Marine Operations, Woods Hole Oceanographic Institution, Woods Hole, MA. Internal WHOI memorandum, with attachment (28 September).
Plan Team for Groundfish Fisheries of the Bering Sea/Aleutian Islands of the North Pacific Fishery Management Council. 1992 (also 1989 ed.). *Stock Assessment and Fishery Evaluation Document for Groundfish Resources in the Bering Sea–Aleutian Islands Region as Projected for 1993*. Report submitted to the International North Pacific Fisheries Commission by the United States National Section. Seattle, WA: Alaska Fisheries Science Center, US Department of Commerce.
Plante, G. 1992. "Legal Aspects of Marine Pollution during the Gulf War." *International Journal of Estuarine and Coastal Law* 7(3):217–231.
Polar Record. 1972. 16:272–279.
———. 1988. 24:173–174.
———. 1989. 25:264–282.
———. 1990a. 26:161–162.
———. 1990b. 26:176–177.
———. 1990c. 26:179–180.
Pontecorvo, G. 1991. "Conference Report: Case of the Missing Data." *Marine Policy* 15(5):372–373 (September).
Portenko, L.A. 1972. *Birds of the Chukchi Peninsula and the Wrangel Islands, Part 1*. Leningrad: Nauka (Science Press) (in Russian).
———. 1973. *Birds of the Chukchi Peninsula and the Wrangel Islands, Part 2*. Leningrad: Nauka (Science Press) (in Russian).
Protsenko, A.N. (n.d.) "Accident at the South Kishtim on September 29, 1957." Unpublished lecture notes. Nuclear Safety Institute, Russian Academy of Sciences.
Puffer, D. 1993. Gulf of Mexico Program, Stennis Space Center, MS. Telephone communication (May).
Puissochet, J.-P. 1991. "CCAMLR—A Critical Assessment." In: A. Jorgensen-Dahl and W. Østreng (eds.). *The Antarctic Treaty System in World Politics*. London: Macmillan.

Quigg, P. 1983. *A Pole Apart: The Emerging Issue of Antarctica.* New York: McGraw-Hill.
Rabalais, N.N. 1993. Louisiana Universities Marine Consortium, Chauven. Telephone communication (7 July).
Rabalais, N.N., R.E. Turner, W.L. Wiseman, Jr., and D.F. Boesch. 1991. "A Brief Summary of Hypoxia on the Northern Gulf of Mexico Continental Shelf: 1985–1988." In: R.V. Tyson and T.H. Pearson (eds.). *Modern and Ancient Continental Shelf Anoxia.* Geological Society Special Publication No. 58. London: Geological Society.
"Ratifications of the Law of the Sea Convention." 1993. *Oceans Policy News* 10(4):1 (April).
Readman, J.W., S.W. Fowler, J.-P. Villeneuve, C. Cattini, B. Oregioni, and L.D. Mee. 1992. "Contamination of the Gulf Marine Environment Following the War." *Nature* 358:662–665.
Reed, R.K. 1990. "A Year-Long Observation of Water Exchange between the North Pacific and the Bering Sea." *Limnology and Oceanography* 35:1604–1609.
"Report of the First Meeting of the U.S.-Russian Joint Committee on Cooperation in Ocean Studies." 1993. (Held 14–23 July in Seattle.) Silver Spring, MD: National Oceanic and Atmospheric Administration, US Department of Commerce.
"Report of the 1988 Meeting to Review the Operation of the Convention for the Conservation of Antarctic Seals." 1991. In: W.M. Bush (ed.). *Antarctica and International Law.* Booklet AT5. New York: Oceana (February).
"Report of the Working Group on Risk Assessment and Remediation." 1993. Conference on Radioactivity and Security in the Oceans: New Research and Policy Priorities in the Arctic and North Atlantic, 7–9 June, Woods Hole Oceanographic Institution, Woods Hole, MA.
Richardson, L.F. 1938. "The Arms Race of 1909–13." *Nature* 142:792–793.
Roberts, B. 1992. *Chemical Disarmament and International Security.* International Institute for Strategic Studies, Adelphi Paper No. 267. London: Brassey's.
Roginko, A.Y., and M.J. LaMourie. 1992. "Emerging Marine Environmental Protection Strategies for the Arctic." *Marine Policy* 16(4):259–276.
Rother, D. 1980. "Ship Casualties—An Analysis of Cause and Circumstances." Lectures and Contributions No. 28, Institute of Shipping Economics, Bremen, Germany.
Røttingen, I. 1990. "Changes in the Distribution and Abundance of Norwegian Spring Spawning Herring and Barents Sea Capelin." *Polar Research* 8:33–42.
Rowland, J.R. 1988. "The Treaty Regime and the Politics of the Consultative Parties." In: C. Joyner and S. Chopra (eds.). *The Antarctic Legal Regime.* Dordrecht, Netherlands: Nijhoff.
"Russians, U.S. Differ on Arctic Sub Threat." 1993. *Science* 260:1881 (25 June).
Sahrhage, D. 1988. "CCAMLR: Its Practical Side—The Polarstern Expedition." In: R. Wolfrum (ed.). *Antarctic Challenge III: Conflicting Interests, Coop-*

eration, *Environmental Protection, Economic Development*. Berlin: Dencker and Humblot.
Sakshang, E., and E. Skjdal. 1989. "Life at the Ice Edge." *Ambio* 18:67.
Saliba, L.J. 1989. "Regional Measures for Marine Pollution Control in the Mediterranean." *Marine Pollution Bulletin* 20(1):12–17.
Sambrotto, R.H., J.J. Goering, and C.P. McRoy. 1984. "Large Yearly Production of Phytoplankton in Western Bering Strait." *Science* 225:1147–1150.
Sanger, D.E. 1993. "Nuclear Material Dumped Off Japan." *New York Times*, 19 October, pp. A1, A6.
SCAR (Scientific Committee on Antarctic Research). 1991. Bulletin No. 100. *Polar Record* 27 (January).
Scarff, J.E. 1977. "International Management of Whales, Porpoises and Dolphins." *Ecology Law Quarterly* 6:323–427, 571–688.
Schachte, W.L., Jr. 1991. "The Value of the 1982 UN Convention on the Law of the Sea—Preserving our Freedoms and Protecting the Environment." Speech delivered at the 25th Annual Conference of the Law of the Sea Institute, August, Malmø, Sweden. Printed in "Special Report, August 1991." Washington, DC: Council on Ocean Law.
Schaefer, M.B. 1957. "Some Considerations of Population Dynamics and Economics in Relation to Management of Marine Fisheries." *Journal of the Fisheries Research Board of Canada* 14:669–681.
Schneider, K. 1991. "Ozone Depletion Harming Sea Life." *New York Times*, 16 November, p. 6.
Schröder, P.C. 1992. "UNEP's Regional Seas Program and the UNCED Future: Après Rio." *Ocean and Coastal Management* 18(1):101–111.
"Search for Nuclear Wastes May Hold Up Sakhalin Program." 1993. *Offshore/Oilman* 53:42 (Asia/Pacific Report, November).
"Seismic Oil Searches: Are They Hurting Fishermen?" 1985. *National Fisherman*, 18 August, pp. 18–19.
Shelling, G.C. 1963. *The Strategy of Conflict*. New York: Oxford University Press.
Shepherd, D.S. 1993. "Environmental Liability: Now It's Personal." *Naval Institute Proceedings* 119(8):41–45 (August).
Shlyakhter, A., and R. Wilson. 1992. "Chernobyl and Glasnost: The Effects of Secrecy on Health and Safety." *Environment* 34:25–30.
Shubik, M. 1975. *Games for Society, Business and War*. New York: Elsevier.
Shuntov, V.P. 1972. *Seabirds and the Biological Structure of the Ocean*. Vladivostok: Far Eastern Publishing House. (Translated from Russian, NTIS, Springfield, VA, 1974.)
Shusterich, K.M. 1984. "International Jurisdictional Issues in the Arctic Ocean." In: W.E. Westermeyer and K.M. Shusterich (eds.). *United States Arctic Interests: The 1980s and 1990s*. New York: Springer-Verlag.
Simma, B. 1986. "The Antarctic Treaty as a Treaty Providing for an Objective Regime." *Cornell International Law Journal* 19:189–209.
"Sixtieth Ratification of the Law of the Sea Convention Deposited." 1993. *Oceans Policy News* 10(7):1–3.

Sjoeblom, K-J., and G. Linsley. 1993. "International Arctic Seas Assessment Project, IASAP (1993–1996)." Project summary presented at the Conference on Radioactivity and Environmental Security in the Oceans: New Research and Policy Priorities in the Arctic and North Atlantic, 7–9 June, Woods Hole Oceanographic Institution, Woods Hole, MA.

Skogan, J. 1991. "International Arctic Cooperation: Scope and Limitations." *Northern Perspectives* 19:2 (Summer).

Smith, R. 1986. "Oil and Gas Activities in the U.S. Arctic." Unpublished briefing report on the US-Canada Beaufort Sea Meeting, 13 May.

Smith, V.K., and Y. Kaoru. 1990. "Signals or Noise? Explaining the Variation in Recreation Benefit Estimates." *American Journal of Agricultural Economics* 72:419–433.

Soroos, M.S. 1992. "The Odyssey of Arctic Haze: Toward a Global Atmospheric Regime." *Environment* 34(10):6–11, 25–27 (December).

"Sostoyanie prirodnoi sredy i prirodookhrannaia deyatelnost v SSR v 1989: Gosudarstvennyi doklad." 1990. *Problemy okruzhayushchei sredy i prirodnykj resursov*, No. 11–12 (in Russian).

"Soviets' Secret Nuclear Dumping Raises Fears for Arctic Waters." 1992. *New York Times*, 4 May, pp. A1, A8–9.

Sowls, A.L., S.A. Hatch, and C.J. Lensink. 1978. *Catalog of Alaskan Seabird Colonies*. Anchorage: US Fish and Wildlife Service.

"Spec Seismic Survey Planned in Black Sea." 1994. *Interfax Offshore Report* 1(5):7–8 (March).

Spilhaus, A. 1991. *Atlas of the World with Geophysical Boundaries Showing Oceans, Continents, and Tectonic Plates in their Entirety*. Philadelphia: American Philosophical Society.

Springer, A.M., C.P. McRoy, and K.R. Turco. 1989. "The Paradox of Pelagic Food Webs in the Northern Bering Sea: II. Zooplankton Communities." *Continental Shelf Research* 9:359–386.

Springer, A.M., E.C. Murphy, D.G. Roseneau, C.P. McRoy, and B.A. Cooper. 1987. "The Paradox of Pelagic Food Webs in the Northern Bering Sea: I. Seabird Food Habits." *Continental Shelf Research* 7:895–911.

Staples, W. 1993. US Arms Control and Disarmament Agency. Personal communication (telephone), 1 March.

State of Alaska Office of the Governor. 1989. Enclosure 1:11 to official correspondence dated 11 November.

Stern, J.P. 1993. "Investment in Russian Oil and Gas: A Time for Realism." *Geopolitics of Energy* 15(12):1–3.

Stokke, O.S. 1990. "Environmental Threats in the Arctic." In: L. Heininen (ed.). *Arctic Environmental Problems*. Tampere Peace Research Institute Occasional Papers 41:22–41.

Suhrke, A. 1993. "Pressure Points: Environmental Degradation, Migration and Conflict." Occasional Paper Series of the Project on Environmental Change and Acute Conflict, No. 3 (March). University of Toronto and the American Academy of Arts and Sciences.

Sulden, D. 1989. "The Polar Environment: Illusion and Reality." *Ambio* 18:2–5.

Sullivan, W. 1992. "Soviet Nuclear Dumps Disclosed." *New York Times*, 24 November.

Swartzman, G.K., and R.T. Haar. 1984. "Interactions between Fur Seal Population and Fisheries in the Bering Sea." In: J.R. Beddington, R.J.H. Beverton, and D.M. Lavigne (eds.). *Marine Mammals and Fisheries*. London: George Allen and Unwin.

Takekawa, J.E., H.R. Carter, and T.E. Harvey. 1990. "Decline of the Common Murre in Central California, 1980–1986." In: S.G. Sealy (ed.). *Studies in Avian Biology*, No. 14. Lawrence, KS: Allen Press.

Texas Water Commission. 1990. *The State of Texas Water Quality Inventory*, 10th ed. Austin: Texas Water Commission.

Thorne-Miller, B., and J.G. Catena. 1991. *The Living Ocean: Understanding and Protecting Marine Biodiversity*. Washington, DC: Island Press.

Tikhii okean. 1982. Moscow: Mysl.

Tinker, C. 1990. "Environmental Planet Management by the United Nations: An Idea Whose Time Has Not Yet Come?" *New York University Journal of International Law and Politics*. 22(4):793–830.

———. 1992. "'Environmental Security' in the United Nations: Not a Matter for the Security Council." *Tennessee Law Review* 59(4):787–801.

———. 1993. "NGOs and Environmental Policy: Who Represents Global Civil Society?" Paper presented at the Annual Meeting of the International Studies Association, 25 March, Acapulco, Mexico.

Tischbin, M. 1993. Chief, Public Affairs Office, US Army Chemical Materiel Destruction Agency, Aberdeen Proving Ground, MD. Personal written communication (30 July).

Titz, M.A. 1989. "Port State Control versus Marine Environmental Pollution." *Maritime Policy and Management* 16(3):189–211.

Trukhin, A.M., and G.M. Kosygin. 1987. "Distribution of Seabirds in the Ice in the Western Portion of the Bering and Chukchi Seas." In: N.M. Litvinenji (ed.). *Distribution and Biology of Seabirds of the Far East*. Vladivostok: USSR Academy of Sciences. (Translated from Russian by Canadian Wildlife Service, Ottawa, Toronto, 1989.)

Tver, C.F. 1979. *Ocean and Marine Dictionary*. Centreville, MD: Cornell Maritime Press.

Twitchell, M. 1989. "Implementing the U.S.-Canada Pacific Salmon Treaty: The Struggle to Move from 'Fish Wars' to Cooperative Fishery Management." *Ocean Development and International Law* 20:409–428.

Ullman, R. 1990. "Enlarging the Zone of Peace." *Foreign Policy* No. 80:102–120.

UNCCORS Preparatory Committee Report. 1984. Composite text of Session I, January 3, 1984, and October 12, 1984, United Nations Conference on Conditions for Registration of Ships. TR/RS-CONF/3—TR/RS/CONF/PC/4.

"Underwater Risks." 1990. *Maclean's*, 12 February.

United Nations. 1972. UN DOC.A/CONF.48/14 and Corr. 1 (16 June). Reprinted in *International Legal Materials* 11:1416.
———. 1979. General Assembly Official Records, 34th session, Supplement 46.
———. 1982. UN DOC.A/CONF.62/122 (7 October).
———. 1984. UN DOC.A/39/583, part II (2 November).
———. 1990. General Assembly Official Records, 45th session, First Committee, 42nd Meeting (20 November 1990).
UNEP (United Nations Environment Programme). 1987. *State of the Marine Environment in the Black Sea Region* (Draft). Meeting of Rapporteurs of the Regional Task Teams for Review of the State of the Marine Environment, 14–18 December, Geneva. UNEP(OCA)/WG.1/11.
———. 1992. *Report of the Meeting of Experts on Land-based Sources of Pollution, 6–10 July, Veracruz, Mexico.* UNEP (OCA)/CAR WG.9/4.
United Nations General Assembly. 1992. *State of the Environment in Antarctica: Report of the Secretary-General.* UN DOC.A/47/624, 47th session, Agenda Item 66, Geneva (11 November).
United Nations Secretary General. 1989. "Law of the Sea: Protection and Preservation of the Marine Environment." Report of the Secretary General A/44/461 (18 September).
———. 1990. "Large Scale Pelagic Driftnet Fishing and Its Impact on the Living Marine Resources of the World's Ocean and Seas." Report of the Secretary General A/45-663 (26 October).
———. 1991. "Law of the Sea: Report of the Secretary General to the United Nations General Assembly." A/46/724 (5 December).
———. 1992. "Law of the Sea: Report of the Secretary-General." A/47/623 (24 November).
United Nations Treaty Series. 1953. 161:72–82.
———. 1977. 1046:120.
United States Army Program Manager for Chemical Demilitarization. 1988. *Chemical Stockpile Disposal Program Implementation Plan.* Aberdeen, MD: US Army Program Executive Officer-Program Manager for Chemical Demilitarization.
United States Navy. 1982. *Draft Environmental Impact Statement on the Disposal of Decommissioned, Defueled Naval Submarine Reactor Plants.* Annex to Appendix D: *Radiological Environmental Monitoring at Sites of Nuclear-powered Submarine Thresher and Scorpion Sinkings.* Washington, DC: US Department of the Navy.
———. 1984. *Final Environmental Impact Statement on the Disposal of Decommissioned, Defueled Naval Submarine Reactor Plants,* Volume 1. Washington, DC: US Department of the Navy.
US Secretary of Commerce. 1992. *Report of the Secretary of Commerce to the Congress of the United States Concerning U.S. Actions Taken on Foreign Large-Scale High Seas Driftnet Fishing Pursuant to Section 206 of the Magnuson Fishery Conservation and Management Act.* Washington, DC: US Department of Commerce.
Vader, W., R.T. Barrett, K.E. Erikstad, and D.B. Strann. 1990. "Differential Responses of Common and Thick-billed Murres *Uria* Spp. to a Crash in

the Capelin Stock in the Southern Barents Sea." In: S.G. Sealy (ed.). *Studies in Avian Biology*, No. 14. Lawrence, KS: Allen Press.
VanderZwaag, D., and D. Lamson. 1986. "Ocean Development and Management in the Arctic: Issues in American and Canadian Relations." *Arctic: Journal of the Arctic Institute of North America* 39(4):327–337.
Van Dyke, J.M. 1988. "Ocean Disposal of Nuclear Wastes." *Marine Policy* 12(2):82–95.
Vartanov, R. 1991. "Greening of the USSR." *Christian Science Monitor*, August 12.
———. 1992. "Commentary Article—The Arctic: One of Many Concerns of the New Russian Decision-Makers." *International Challenges* 12(3):40–46.
Vartanov, R., and A.Y. Roginko. 1990. "New Dimensions of Soviet Arctic Policy: Views from the Soviet Union." *Annals of the American Academy of Political and Social Science* 512:69–78 (November).
Veksler, M. 1990. "The USSR and USA: Possible Ways of Cooperation in Preventing World Ocean Radioactive Pollution." Unpublished translation of a research paper prepared for IMEMO. Moscow: Institute of International Relations and World Economy, USSR Academy of Sciences.
Vicuna, F.O. 1988. "The Law of the Sea and the Antarctic Treaty System." In: C. Joyner and S. Chopra (eds.). *The Antarctic Legal Regime*. Dordrecht, Netherlands: Nijhoff.
———. 1991. "The Effectiveness of the Decision-making Machinery of CCAMLR: An Assessment." In: A. Jorgensen-Dahl and W. Østreng (eds.). *The Antarctic Treaty System in World Politics*. London: Macmillan.
Von Neumann, J., and O. Morgenstern. [1944] 1953. *Theory of Games and Economic Behavior*. Princeton: Princeton University Press.
Voronina, N.M. 1990. "Anthropogeneous Evolution of the Pelagic Community of the Antarctic." *Vestnik*: USSR Academy of Sciences.
Wallace, R.L. (compiler). 1994. *The Marine Mammal Commission Compendium of Selected Treaties, International Agreements, and Other Relevant Documents on Marine Resources, Wildlife, and the Environment*. (3 vols.) Washington, DC: Government Printing Office.
Washington Letter of Oceanography. 1993. 27(13):1 (June 28).
WCED (World Commission on Environment and Development). 1987. *Our Common Future*. (The Brundtland Commission Report.) Oxford: Oxford University Press.
Weiss, E.B. 1989. *In Fairness to Future Generations: International Law, Common Patrimony and Intergenerational Equity*. Dobbs Ferry, NY: Transnational.
Weiss, T.G., and K.M. Campbell. 1991. "Military Humanism." *Survival* 33(5):451–465.
Westermeyer, W.E., and V. Goyal. 1986. "Jurisdiction and Management of Arctic Marine Transportation." *Arctic: Journal of the Arctic Institute of North America* 39(4):338–349 (December).
Westing, A. 1989. "The Environmental Component of Comprehensive Security." *Bulletin of Peace Proposals* 20:129–134.
———. 1992. "Environmental Security Dimensions of Maritime Security."

In: J. Goldblat (ed.). *Maritime Security: The Building of Confidence.* Geneva: UN Institute for Disarmament Research Doc. No. UNIDR/92/89.

Weston, B. (ed.). 1990. *Alternative Security: Living without Nuclear Deterrence.* Boulder: Westview Press.

WHOI (Woods Hole Oceanographic Institution). 1991. *U.S. Strategies for Cooperation with the Soviets on Ocean Science.* Report of a Workshop held 29–31 October at Woods Hole, MA. Woods Hole, MA: Woods Hole Oceanographic Institution.

―――. 1993. Proceedings of the Conference on Radioactivity and Environmental Security in the Oceans: New Research and Policy Priorities in the Arctic and North Atlantic, 7–9 June, Woods Hole Oceanographic Institution, Woods Hole, MA.

WHOI Conference Statement. 1993. Press Release of the Conference on Radioactivity and Environmental Security in the Oceans: New Research and Policy Priorities in the Arctic and North Atlantic, 7–9 June, Woods Hole Oceanographic Institution, Woods Hole, MA.

Wohl, K., and P. Gould. 1991. "Marine Birds and the High Seas Drift Net Fisheries in the North Pacific." Paper presented at Pacific Seabird Group Annual Meeting, 23–27 January, Monterey, CA.

Wolfers, A. 1952. "'National Security' as an Ambiguous Symbol." *Political Science Quarterly* 67:481–502.

Wright, N.A., and P.L. Williams. 1974. "Mineral Resources of Antarctica." US Geological Survey, Circular 705.

Yablokov, A.Y. 1993. "Facts and Problems Connected with Disposal of Radioactive Waste in the Littoral Ocean Space of the Russian Federation." Moscow: Governmental Commission for Nuclear Waste Disposal in the Sea.

Yancey, M. 1993. "Judge Blocks Free Trade Pact." *Cape Cod Times,* July 1, p. A-3.

Yatsu, A. 1990. "A Review of the Japanese Squid Driftnet Fishery." Working Paper for IWC Workshop on Mortality of Cetaceans in Passive Fishing Nets and Traps, 20–25 October, La Jolla, CA.

Yegang, L. 1993. "Chinese Cargo Ship 'Yinhe' Cleared of Suspicion." *Beijing Review,* 20–26 September, pp. 13–14.

Young, O.R. 1986. "The Age of the Arctic." *Foreign Policy* 61:169–183 (Winter).

―――. 1989. "The Arctic in World Affairs." Donald L. McKernan Lectures in Marine Affairs. Seattle: University of Washington (10 May).

Young, O.R., and G. Osherenko. 1991. "International Cooperation in the Arctic: Soviet Attitudes and Actions." In: L.W. Brigham (ed.). *The Soviet Maritime Arctic.* Annapolis: Naval Institute Press; London: Bellhaven Press.

Yue, T. 1993. "US Breaches International Law on the 'Yinhe' Incident." *Beijing Review,* 20–26 September, pp. 12–13.

Zilanov, V., et al. 1989. "International Problems of Conservation and Exploitation of Bering Sea Living Resources." *Mirovoe Rybolovstvo* (in Russian).

Zumberge, J.H. (ed.). 1979. *Possible Environmental Effects of Mineral Exploration and Exploitation in Antarctica.* Cambridge, UK: Scientific Committee on Antarctic Research.

———. 1982. "Potential Mineral Resource Availability and Possible Environmental Problems in Antarctica." In: J.I. Charney (ed.). *The New Nationalism and the Use of Common Spaces.* Totowa, NJ: Allanheld, Osmun.

List of Abbreviations

ACP	African, Carribean, and Pacific
AHL	allowable harvest level
AMAP	Arctic Monitoring and Assessment Program
ANILCA	Alaska National Interest Lands Conservation Act
ANWR	Arctic National Wildlife Refuge (US)
ANZUS	Australia-New Zealand-United States (defense pact)
ARPA	Advanced Research Projects Agency (US)
ASEAN	Association of South East Asian Nations
ASOC	Antarctic and Southern Ocean Coalition
BMP	best management practices
CCAMLR	Convention on the Conservation of Antarctic Marine Living Resources
CCAMRA	Convention on the Regulation of Antarctic Mineral Resources Activities
CCMP	Comprehensive Conservation and Management Plan (US Clean Water Act)
CEC	Commission of the European Communities
CEPPOL	Program for Assessment and Control of Marine Pollution (Action Plan for the Wider Caribbean, UNEP Regional Seas Programme)
CIS	Commonwealth of Independent States
CITES	Convention on International Trade in Endangered Species
CLC	International Convention on Civil Liability for Oil Pollution Damage
CRAMRA	Convention on Regulation of Antarctic Mineral Resource Activities
CSCE	Conference on Security and Cooperation in Europe
CV	contingent valuation
CWA	Clean Water Act (US)
CWC	Chemical Weapons Convention (formally, the Convention on the Prohibition of Development, Production, Stockpiling, and Use of Chemical Weapons and on their Destruction)

CZM	Coastal Zone Management
CZMA	Coastal Zone Management Act (US)
DOE	Department of Energy (US)
DWT	dry weight tonnage
EC	European Community
EDF	Environmental Defense Fund
EEZ	exclusive economic zone
EIS	environmental impact statement (US)
EMAP	Environmental Monitoring and Assessment Program (US EPA)
EPA	Environmental Protection Agency (US)
ESA	Endangered Species Act (US)
FAO	Food and Agriculture Organization (UN)
FBIS	Foreign Broadcast Information Service (US)
FOCI	Fisheries-Oceanography Coordinated Investigations (US Dept. of Commerce)
FUND	International Convention on the Establishment of an International Fund for Compensation for Oil Pollution Damage
GAO	General Accounting Office (US Congress)
GATT	General Agreement on Tariffs and Trade
GCEA	Global Commons Environmental Assessment
GEF	Global Environmental Facility
GESAMP	(Joint) Group of Experts on the Scientific Aspects of Marine Pollution (UN)
GIFA	governing international fisheries agreement
GIUK	Greenland-Iceland-United Kingdom (gap)
GOM	Gulf of Mexico (Program) (US EPA and NOAA)
GPS	Global Positioning System
GRT	gross tonnage
HNS	Draft Convention on the Carriage of Hazardous and Noxious Substances by Sea
IAEA	International Atomic Energy Agency
IASC	International Arctic Science Committee
ICES	International Council for the Exploration of the Sea
ICJ	International Court of Justice
ICRP	International Commission on Radiological Protection
ICRW	International Convention for the Regulation of Whaling
ICSU	International Council of Scientific Unions
IGPRAD	Intergovernmental Panel of Experts on Radioactive Waste Disposal
ILC	International Law Commission
IMCO	International Maritime Consultative Organization
IMDG	International Marine Dangerous Goods (Code) (IMO)
IMEMO	Institute for World Economy and International Rela-

	tions, Soviet/Russian Academy of Sciences
IMO	International Maritime Organization (formerly IMCO)
INPFC	International North Pacific Fisheries Commission
INQ	individual national quota
IOC	Intergovernmental Oceanographic Commission
ISA	International Seabed Authority
IUCN	International Union for the Conservation of Nature and Natural Resources
IWC	International Whaling Commission
JACADS	Johnston Atoll Chemical Agent Disposal System (US Army)
JGOFS	Joint Global Ocean Flux Study
LC	London Convention
LDC	London Dumping Convention (formally, the Convention on the Prevention of Marine Pollution by Dumping of Wastes and Other Matter; same as London Convention)
LLMC	Convention on Limitation of Liability for Maritime Claims
LNG	liquefied natural gas
MARPOL 73/78	International Convention for the Prevention of Pollution from Ships (1973) and the 1978 Protocol thereto
MEPC	Marine Environment Protection Committee (IMO)
MFCMA	Magnuson Fishery Conservation and Management Act (or Magnuson Act)
MMPA	Marine Mammal Protection Act (US)
MMS	Minerals Management Service (US Dept. of the Interior)
MSY	maximum sustainable yield
NAFTA	North American Free Trade Agreement
NATO	North Atlantic Treaty Organization
NCPDIP	National Coastal Pollutant Discharge Inventory Program (US EPA)
NEA	Nuclear Energy Agency (OECD)
NEP	National Estuary Program (US EPA)
NES	National Energy Strategy (US)
NGO	nongovernmental organization
NMFS	National Marine Fisheries Service (NOAA, US Dept. of Commerce)
NOAA	National Oceanic and Atmospheric Administration (US Dept. of Commerce)
NPAFC	North Pacific Anadromous Fish Commission
NPDES	National Pollution Discharge Elimination System (US EPA)
NPFMC	North Pacific Fishery Management Council (NMFS, US

	Dept. of Commerce)
NSR	Northern Sea Route
NS&T	National Status and Trends Program (NOAA, US Dept. of Commerce)
OCS	outer continental shelf
OECD	Organization for Economic Cooperation and Development (Europe)
OPRC	International Convention on Oil Pollution Preparedness, Response and Co-operation
OTA	Office of Technology Assessment (US Congress)
PBR	potential biological removal
PCB	polychlorinated biphenyl
PICES	North Pacific Marine Science Organization (informally, Pacific ICES)
POTW	publicly owned treatment works (US)
PSC	port state control (system)
SCAR	Scientific Committee on Antarctic Research
SES	Sanitary and Epidemiological Station (USSR/Russia)
SOI	State Oceanographic Institute (USSR/Russia)
SOLAS	International Convention for the Safety of Life at Sea
SPREP	South Pacific Regional Environmental Programme
TAC	total allowable catch
TAPS	Trans-Alaska Pipeline System
UNCCORS	UN Convention on Conditions for Registration of Ships
UNCED	UN Conference on Environment and Development
UNCTAD	UN Conference on Trade and Development
UNEP	UN Environment Programme
UNSCEAR	UN Scientific Committee on the Effects of Atomic Radiation
USFWS	US Fish and Wildlife Service (US Dept. of the Interior)
VTS	vessel traffic systems
WCED	World Commission on Environment and Development
WHO	World Health Organization
WHOI	Woods Hole Oceanographic Institution, Woods Hole, MA
WMO	World Meteorological Organization
WOCE	World Ocean Circulation Experiment

Index

Action Plan for the Human Environment, 172
Action Plan for the Wider Caribbean, 38
African, Caribbean, and Pacific (ACP) nations, hazardous waste transport and, 99
Agenda 21, 39, 66, 151, 223, 243, 248
Agreement on Arctic Cooperation (Canada and US), 172
Agreement on Mutual Fisheries Relations, US-USSR, 77
Airborne marine pollution, 230–31, 243
Alaska National Interest Lands Conservation Act, 169
Aleutian Islands, 54, 57, 59, 73
Algeria, 133
All-Union Service of Monitoring and Environmental Pollution Control, 45
Antarctic region, *see* Southern Ocean
Antarctic Treaty, 190–94, 198
 Consultative Meeting Recommendations, 192
 Protocol on Environmental Protection, 191, 194, 195, 199, 201, 207, 217–18
 annexes to, 194–95, 214–15
Antarctic Treaty Consultative Parties, 191, 194, 195, 196, 197, 199, 207, 214, 216, 217–18, 220, 222, 253
Antarctic Treaty system, 197, 199, 200, 206, 215, 222
ANZUS (Australia-New Zealand-United States defense pact), 143
Arctic Action Plan (proposed), 177–78, 179
Arctic Eight, 173, 175, 177
Arctic Environmental Protection Strategy, 175–76
Arctic National Wildlife Refuge (ANWR), 168–69
Arctic Ocean, environmental protection for the, 145, 163–89, 249
 cooperative agreements and emerging multilateral regime, 172–85
 enhancing the Arctic environmental regime, 176–78
 environmental impact assessment, information sharing, and research, 183–85
 MARPOL "special areas" regime, 178–80
 oil spill research and development, 182–83
 overview, 172–76
 shipping technology, transfer of, 181–82
 technology transfers, 180–85
economic development activities in the Arctic, 166–72
environmental vulnerabilities in the Arctic Basin, 163–66
hydrocarbon resource development, 167–71, 180–81
incentives for, 174
oil pollution, 165, 166
overview, 15, 163
polar projection of Arctic, 164
prospects and prerequisites for success, 187–89
radioactive contamination, 136, 137, 150, 164–65
research and development, 174, 179, 182–83

315

Arctic Ocean (*continued*)
  Russian political developments and, 185–87
  shipping, commercial, 170–71, 181–82
Argentina, 215, 221
Athens Protocol, 236, 239
Auklet, 59
Australia, 144, 216, 217, 218, 222
Austria, 216

Bahia Paraiso, grounding of, 205, 207, 215
Baltic Sea, 229, 230
Barcelona Convention for the Protection of the Mediterranean Sea Against Pollution, 231, 237
Basel Convention, hazardous wastes and, 89, 90, 91, 100
Belgium, 216, 217
Bering Sea, *see* Living resource problems in the North Pacific/Bering Sea
Bering Sea Fisheries-Oceanography Coordinated Investigations (FOCI), 81
Black Sea, 229
  land-based marine pollution affecting the, *see* Land-based marine pollution, the Black Sea
Black Sea Action Plan, 47
Black Sea Commission, 253
Borisov, Tengiz, 4
Brazil, 132, 144–45
British Nuclear Fuels, Ltd., 143
Brundtland, Gro Harlem, 149
Bucharest Convention on the Protection of the Black Sea Against Pollution, 47, 253
Bulgaria, 45, 47
Burke, W. T., 64
Bush, George, 3–4, 168, 217
Byelorussia, 45

Canada, 50
  Arctic cooperation, 172, 173, 180, 181, 182, 184
  driftnet regulation, 72

Canfield, J. L., 76
Cap de la Hague nuclear reprocessing plant, 128, 133
Cargo location devices, 118–19
Cartagena Convention, 89, 237
Chelyabinsk nuclear reprocessing center, 136, 138, 139
Chemical industry, radioactive waste disposal for, 125–26
Chemical weapons, transport of, 103–107
Chemical Weapons Convention (CWC), 105–106, 107
Chernobyl nuclear accident, 5, 8, 43, 46–47, 128–29, 146
Chile, 199, 215, 221
China, People's Republic of, 50, 60, 61, 106
Clean Water Act, 33–36
Coastal state authority:
  under Law of the Sea Convention, 101–102, 227, 229, 232
  in their EEZs, 101, 102, 232
Coastal Zone Management Act (CZMA), 36–37
Coast Guard, US, 73, 79, 91, 183
Commission on the European Community (CEC), 98
Commonwealth of Independent States (CIS), 49
Conclusions, 245–55
Conference on Security and Cooperation in Europe (CSCE), 253
Conoco, 180
Conservation and Management of Pollock Resources in the Central Bering Sea, draft agreement for, 66
Convention Concerning the Conservation of Migratory Birds and Their Environment, 1976 US-USSR, 79
Convention for the Conservation of Anadromous Stock in the North Pacific Ocean, 77, 248
Convention for the Conservation of Antarctic Seals, 192
Convention for the Prevention of Ma-

Index 317

rine Pollution by Dumping from Ships and Aircraft, 156
Convention for the Prevention of Marine Pollution from Land-based Sources in the North-East Atlantic, 156, 229, 231, 236
Convention for the Protection and Development of the Marine Environment of Wider Caribbean Region, 38, 229
1990 Protocol to the, 89
Convention for the Protection and Development of the Natural Resources and Environment of the South Pacific Region, 144
Convention for the Protection of the Marine Environment and Coastal Areas of the South-East Pacific, 156, 229
Convention on Limitation of Liability for Maritime Claims (LLMC), 100–101, 111
Convention on Regulation of Antarctic Mineral Resources Activities (CRAMRA), 194, 200, 201, 216–17, 218, 219
Convention on the Conservation of Antarctic Marine Living Resources (CCAMLR), 192–93, 200–201, 202, 209, 210, 218, 220
Commission of, 210–13
Scientific Committee, 210–11, 212, 213, 220
Convention on the Control of Transboundary Movements of Hazardous Wastes and Their Disposal, *see* Basel Convention
Convention on the International Trade in Endangered Species (CITES), 80
Convention on the Law of the Sea, *see* UN Convention on the Law of the Sea
Convention on the Prohibition of Development and Production, Stockpiling and Use of Chemical Weapons and on Their Destruction, 105–106, 107

Convention on the Protection of the Marine Environment of the Baltic Sea Area (1974 and 1992) (Helsinki Convention), 20, 155, 160, 177, 231, 236, 237
Convention on the Protection of the Marine Environment of the North-East Atlantic, 1992, 20, 156, 236
Convention on the Protection of the Mediterranean Sea against Pollution, 20
Criteria and standards for environmental policy, 251–52
Cross-cutting themes in ocean environmental security, 249–55
Cyprus, 108

Danube River, 40
Data, *see* Monitoring, data, and information management
Defense industry, radioactive waste disposal by, 125–26
Department of Defense, US, 104
Department of Energy, US, 138
Dispute settlement under the Law of the Sea Convention, 238–39, 240
Dolphin caught in driftnets, 69, 72
Double hull standard for tankers, 119, 182
Doughnut Hole, 60–61, 66, 83–84
Draft Convention on Liability and Compensation in Connection with the Carriage of Hazardous and Noxious Substances by Sea (HNS Convention), 91, 100–101, 111, 120
Driftnet fisheries, 77–80
incidental take and, 67–70
national legal initiatives on, 72–76
threats to environmental security and protection, 68
Duggins, D. O., 81–82
Dumping, pollution from:
Law of the Sea Convention, 228–29
London Dumping Convention, *see* London Dumping Convention (LDC)

Dumping, pollution from (*continued*)
  radioactive substances, *see* Radioactivity in the oceans

Economic development activities in the Arctic, 166–72
  commercial shipping, 170–71
  hydrocarbon resource development, 166–70
  mineral extraction, 171–72
Economic motivations and environmental protection, 246–47
Economic perspectives on hazardous waste transport, *see* Hazardous materials transport, economic perspectives on
Ecosystem protection under Law of the Sea Convention, 234–35, 240
Electronic chart technology, 118
Endangered species, 197
Endangered Species Act (ESA), 75
Environmental Canada, 182
Environmental impact assessment:
  under Law of the Sea Convention, 236–37
  Protocol on Environmental Protection provisions for, 194
  transfer of technology on, 183–85
Environmental impact statement (EIS), 104
Environmental Protection Agency (EPA), 21, 32
  Environmental Monitoring and Assessment Program (EMAP), 32–33
  hazardous waste transport and, 92
Environmental Protection Protocol, *see* Protocol to Antarctic Treaty on Environmental Protection
Environmental security:
  defined, 6
  emergence of concept of, 7–13
  major threats to ocean, 13–16, 247–49
  precursors of, 7
Estes, J. A., 81–82
Estonia, 73

European Community (EC), 73, 193
  hazardous waste transport and, 99
European Port State Control (PSC) system, 109, 117
Exclusive economic zones (EEZs), 99
  coastal state authority in their, 101, 102
  ecosystem protection in, 235
  fishing rights and, 60, 61, 63, 64, 68, 72–73, 83, 84
  introduction of concept of, 225, 231
Exxon Valdez oil spill, 5, 181, 183, 205

Finland, 173, 175
  technology transfer, 180, 181, 182, 184
Finned fish, 192, 203, 208, 212–13
Finnish Initiative, 175, 176, 177, 189
First Ministerial Conference on the Protection of the Arctic Environment, 175
Fisheries and fishing:
  in the North Pacific, *see* Living resource problems in the North Pacific/Bering Sea
  in the Southern Ocean, 192–93, 196–97, 208–14, 216
Fishermen's Protective Act of 1967, 74
Fishery Conservation Amendments of 1990, 73, 74
France, 193, 211, 214, 216, 217
  nuclear weapons testing by, 143
Frost, K. J., 59
Fur seals, 58–59, 68, 69
Fur Seal Treaty of 1911, 76

Gas exploration, *see* Hydrocarbon resources development
Geneva Convention, 105, 106
Geneva Convention on Long Range Transboundary Air Pollution, 231
Georgia, 46, 47
Germany, 217, 221
Global Commons Environmental Assessment (GCEA), 104
Global Environmental Facility (GEF), 47

Global perspective on environmental security, 250–51
Global Positioning System (GPS), 117
Global warming, 202
Gorbachev, Mikhail, 8, 174
Governing international fisheries agreements (GIFAs), 73, 74
Greece, 108, 216
Greenland, 173
Greenpeace, 131, 134, 143, 214
Griffiths, F., 179
Group of Experts on the Scientific Aspects of Marine Pollution (GESAMP), 94, 128
Gulf of Mexico, land-based marine pollution affecting the, *see* Land-based marine pollution, the Gulf of Mexico
Gulf of Mexico (GOM) Program, 38–39

Haiti, 89
Hanford, Washington atomic installation, 128, 133
Hatanaka, T., 82
Hayase, S., 82
Hazardous materials transport, 86–121
 accidents, 86, 87–88
 damage assessment, 114–16, 121
 records of, 91–92, 117
 conclusions, 120–21
 data inadequacies and casualty records, 91–92
 defining a material as hazardous, 89–91
 economic perspectives, 111–16
 damage assessment, 114–16, 121
 insurance markets, 113–14
 liability regimes and economically optimal levels of precaution, 111–13
 existing legal and regulatory structure, 92–99
 for transboundary movement, 97–99
 overview, 14, 86–89
 regulatory inadequacies and their implications for marine environmental protection, 99–111
 chemical weapons, 103–107
 coastal state authority to enforce preventative regulation, 101–103
 disparate standards, 107
 institutional responses to problem of substandard ships, 108–11
 open registers, 107–108, 110
 unratified and uncompleted conventions, 100–101
 technology for, 117–20
Health Effects Institute, 32
Helsinki Convention, 20, 155, 160, 177, 236, 237, 239
Hollister, Charles, 129–30
Hydrocarbon resources development:
 in the Arctic, 167–70, 180–81
 in the Southern Ocean, 193–94, 203–206, 216, 219
Hypoxia, 29

Iceland, 73, 145–46, 173, 247
Incidental take:
 driftnet fisheries and, 67–70
 policy implications of, 70–72
Information management, *see* Monitoring, data, and information management
Insurance, marine, 113–14
Intergovernmental Oceanographic Commission (IOC), 11
Intergovernmental Panel of Experts on Radioactive Waste Disposal (IGPRAD), 154
International Arctic Council, 176
International Arctic Science Committee (IASC), 174, 177, 185, 186
International Atomic Energy Agency (IAEA), 11, 46
 radioactive contamination of the oceans and, 126, 128, 134, 136, 153, 155, 158, 159, 160
International Commission on Radiological Protection (ICRP), 142, 158

International Conference of Arctic and Nordic countries, 1988, 174
International Convention for the Prevention of Pollution from Ships (1973) and its 1978 Protocol, see MARPOL 73/78
International Convention for the Regulation of Whaling (ICRW), 192
International Convention for the Safety of Life at Sea, see SOLAS Convention
International Convention on Oil Preparedness and Response, 101, 120
International Convention on Salvage, 96–97, 156
International Convention on the Protection and Preservation of the Marine Environment of the Black Sea, 47
International Convention Relating to Intervention on the High Seas in Cases of Oil Pollution Casualties (Intervention Convention), 95
1973 Protocol to, 95–96
International cooperation and coordination, 252
International Council for the Exploration of the Sea (ICES), 80
International Council of Scientific Unions (ICSU), 174
International Court of Justice (IJC), 11, 239
International Geosphere Biosphere Programme, 174
International law:
functions of, 223
sources of environmental, 224–26
see also names of specific laws and conventions
International Maritime Organization (IMO), 11, 89, 90, 99, 119–20, 182, 231
International Maritime Dangerous Goods (IMDG) Code, 90–91
Legal Committee, 101
Marine Environmental Protection Committee (MEPC), 94
reporting on hazardous wastes incidents and, 91
International North Pacific Fisheries Commission (INPFC), 76–77, 82, 83
International organizations, 252–53
see also specific organizations
International Seabed Authority (ISA), 200
International security, 6
critical problems of environmental, 247–49
International Tribunal for the Law of the Sea, 239
International Union for the Conservation of Nature and Natural Resources, 172, 197
International Whaling Commission (IWC), 192, 197
Intervention Convention, 95
1973 Protocol to, 95–96
Intervention on the High Seas in Cases of Marine Pollution by Substances other than Oil, 1973 Protocol relating to, 95–96
Inuit Circumpolar Conference, 177, 179
Iran, 106
Iraq, 4
Irish Sea, radioactivity in the, 142, 143, 160
Italy, 216, 217

Japan, 5, 170, 247
civilian nuclear power generation, 131–32
driftnet fisheries, 68–69, 78
fishing in the Southern Ocean, 210
fishing rights and, 50, 53, 60, 61, 65, 77, 83–84
future of the Southern Ocean and, 217, 218, 221
transport of spent nuclear fuel, 133–34, 141–42
Johnston Atoll Chemical Agent Disposal System (JACADS), 103–107
Johnston, D. M., 77–78

Joint Global Ocean Flux Study
  (JGOFS), 185

Keating, Paul, 218
King George Island, 208, 214
Kiss, A., 233
Kohl, Helmut, 103
Komsomolets, 146–49, 185
Krill, 192, 202–203, 205, 207–13
Kuwait Convention, 237

Land-based marine pollution, 17–49
  the Black Sea, 39–47
    coastal development, effects of, 41
    facts about, 26–27, 39–41
    industrial and municipal sources, 42–43
    information gathering, 44–45
    international regional cooperation, 46–47
    legislation and regulation, 45–46
    management approach of the Soviet Union, 42–46
    nutrient enrichment, effects of, 41
    public health, effects on, 41–42
  Convention on the Law of the Sea, 229–30, 243
  emission "export" through capital mobility, 19
  export goods, damage to, 18
  global legal guidelines, 19
  the Gulf of Mexico, 24–39
    coastal development, effects of, 25–29
    facts about, 24–25, 26–27
    industrial and municipal sources, 31
    information gathering, 32–33
    international regional cooperation, 38–39
    legislation and regulation, 33–38
    map, 25
    nutrient enrichment, effects of, 29
    public health, effects on, 29–30
    US management approach, 31–38
  non-use values and, 19
  overview, 13, 17
  regional cooperation, 21–23
  regional regimes for, 19–20
  shared resource stocks, 18
  in the Southern Ocean, 206–207
  tourism and, 18
  transboundary pollution, 18
Land-based Marine Pollution Sources Protocol, 38–39
Law of the Sea Convention, *see* UN Convention on the Law of the Sea
Liability limits on hazardous materials transport, 111–13
Liberia, 108
Lima Convention, 156, 229
Lithuania, 73
Living resource problems in the North Pacific/Bering Sea, 50–85, 248
  Doughnut Hole and the problem of straddling stocks, 60–61
  driftnet fisheries, 77–80
  incidental take and, 67–70
  national legal initiatives on, 72–76
  threats to environmental security and protection, 68
  economic considerations for straddling stocks, 61–63
  ecosystem perturbation from fisheries depletion, 58–59
  ecosystems of the Bering Sea-North Pacific region, 51–53
  fisheries depletion, 53–58
  gaps in scientific information and future research projects, 81–82
  incidental take, 67–72
  international legal responses to, 76–80
  legal options for straddling stocks, 63–67
  overview, 14
  principal issues, 50
  regional regime for environmental security, 82–85
  straddling stocks, 60–61
    Doughnut Hole and, 60–61
    economic considerations for, 61–63
    legal options for, 63–67
  total allowance catch (TAC), 57

Living resource problems in the Southern Ocean, 192, 197, 203, 208–14
Lomé IV Convention on Trade and Aid, 99
London Convention (LC), *see* London Dumping Convention (LDC)
London Dumping Convention (LDC), 99, 207, 229, 230
   radioactive substances and, 126, 134, 135, 149, 151, 153–54, 159
LORAN-C, 117
Louisiana Wetland Conservation and Restoration Act, 37
Lowry, L. F., 59
Lystsov, Vitaliy, 135–36

Magnuson Fishery Conservation and Management Act (MFOMA), 54, 57, 69, 72, 73
Marine insurance, 113–14
Marine Mammal Protection Act (MMPA), 73, 74–75
Marine Plastic Pollution Research and Control Act, 254
MARPOL 73/78, 91, 94–95, 120, 195, 254
   "special areas" regimes;
      for Arctic Ocean, 178–80
      for Southern Ocean, 207
Military conversion efforts, 254
Military nuclear disposal practices, 134–38
Mineral extraction:
   in Arctic region, 171–72
   in Southern Ocean, 193–94, 196, 206, 216, 219
      CRAMRA, 194, 200, 201, 216–17, 218
Mineral Management Service (MMS), 169
Miovski, L., 64
Moldova, 46
Monitoring, data, and information management, 252
   on hazardous materials transport, 91–92

on land-based marine pollution, 32–33, 44–45
for Law of the Sea Convention, 235–37
on living resource problems in the North Pacific/Bering Sea, 81–82
on radioactivity in the oceans, 159–61
Montreal Guidelines for the Protection of the Marine Environment against Pollution from Land-based Sources, 20
Municipal sources of land-based marine pollution, 31, 42
Murmansk initiatives, 174, 185–86

National Coastal Pollutant Discharge Inventory Program (NCPDIP), 32
National energy strategy (NES), 168–69
National Estuary Program (NEP), 35–36
National Institute of Standards and Technology, 32
National Marine Fisheries Service (NMFS), 73, 75, 79, 81, 82
National Oceanic and Atmospheric Administration (NOAA), 36, 252
National Status and Trends Program (NS&T), 32
   Sea Grant programs, 37
National Petroleum Reserve, 168
National Pollution Discharge Elimination System (NPDES), 33–34
National security, 6
NATO, 140, 145
New Zealand, 143, 144, 198–99, 216–17, 222
NGOs (nongovernmental organizations), 253
   Antarctic region, monitoring the, 220
   land-based marine pollution, agenda for managing, 31–32
North American Free Trade Agreement (NAFTA), 19, 32

North Atlantic Ocean, radioactive contamination in the, 136, 137
North Atlantic Treaty Organization (NATO), 140, 145
Northern Forum, 176, 189, 253
Northern Sea Route (NSR), 170–71, 181–82
North Korea, 50
North Pacific, living resource problems in the, *see* Living resource problems in the North Pacific/Bering Sea
North Pacific Anadromous Fish Commission (NPAFC), 77, 82, 83, 248, 253
North Pacific Driftnet Proclamation, 72
North Pacific Fishery Council, 63
North Pacific Fishery Management Council (NPFMC), 73
North Pacific Marine Science Organization, 248
Norway, 146, 173, 181, 184, 217, 222, 247
Novaya Zemlya, 134, 135, 138, 165
Nuclear accidents and unintended releases, 128–31, 132
potential for future, 140, 165
see also specific accidents, e.g. Chernobyl nuclear accident
Nuclear fuel reprocessing, 133–34
Nuclear power and reprocessing plants, operational effluent discharges from, 126–28
Nuclear weapons, 122
decommissioning of, 138–40, 164
lost at sea, 5, 123, 129
proliferation of, 123
testing of, 124–25, 143, 164–65
see also Radioactive wastes
Nunn, Sam, 4
Nutrient overenrichment, effects of, 29, 41

Oil exploration:
in the Arctic, 167–70, 180–81
in the Southern Ocean, 193–94, 203–206, 216, 219
Oil pollution:
in the arctic seas, 165, 166
in the Southern Ocean, 205–206, 207
Oil Pollution Act, 182
Oil spill research and development, 182–83
"On Strengthening Nature Protection in the Areas of the Far North and the Marine Areas Adjacent to the Northern Coast of the USSR," 181–82
Open registries, 107–108, 110
Organization for Economic Cooperation and Development (OECD), 97–98
Oslo Convention, 156
Ozone depletion, 201–202

Pacific Marine Science Organization (PICES), 80
Pacific white-sided dolphin, 69
Pakistan, 133
Panama, 108
Paris Convention, 156, 229, 231, 236
Penguins, 202, 205, 207, 208, 209, 214
Perintis, 86
Persian Gulf, 4, 229
Persian Gulf war, 4, 168
Poland, 50, 60, 61, 73
Polar Bear Conservation Agreement, 172
Pollock, 51, 53–54, 57–66, 68, 71, 77, 83–84, 248
Population growth, influence of, 251
Port and Tanker Safety Act, 182
Port state control (PSC) system, European, 109, 117
Precautionary principle, 12–13, 151
Program for Assessment and Control of Marine Pollution (CEPPOL), 39

324 Index

Project for Environmental Management and Protection of the Black Sea, 47
Protocol for the Prevention of Pollution of the Mediterranean Sea by Dumping from Ships and Aircraft, 155
Protocol to the Antarctic Treaty on Environmental Protection, 191, 194, 196, 199, 201, 207, 217–18
  annexes to, 194–95, 214–15
Prudhoe Bay, 168
Public health, land-based marine pollution's effects on, 29–30, 41–42

Radioactivity in the oceans, 5, 122–62
  conclusions, 157–62
  legal rights, obligations, and controls, 152–56
  overview, 14–15, 122–23
  potential sources of, 123–31
    accidents and unintended releases, 128–31, 132
    naturally occurring radionuclides, 124, 125
    nuclear weapons testing, 124–25, 164–65
    operational effluent discharges and nuclear power and reprocessing plants, 126–28
    radioactive waste disposal: industrial, defense, and medical inputs, 125–26, 127
    uranium mining and milling, 124
  projections of future sources of contamination, 131–40
    accidents and unintended releases, 140, 165
    civilian nuclear power generation, 131–33
    decommissioning military assets, 138–40, 164
    fuel reprocessing issues, 133–34
    military nuclear disposal practices, 134–38
  public perceptions and resulting countermeasures, 142–51

  Russia and, see Russia, radioactive wastes and
  Soviet Union and, former, see Soviet Union, former, radioactive wastes and
  transport of nuclear material, threat of, 133–34, 140–42
Radionuclides, naturally occurring, 124, 125
Rainbow Warrior, 143
Rathbun, G. B., 81–82
Reagan, Ronald, 103
Recreational activities in Antarctica, 215–16
Red tides, 29–30, 42
Reprocessing of nuclear fuel, 133–34
Research and development:
  in the Arctic, 174, 179, 246
  oil spill, 182–83
  in the Southern Ocean, 214–15
Responsibility, liability, and compensation, absence of criteria for assignment of, 249–50
Rio Declaration of 1992 UN Conference on Environment and Development, 39, 192, 237, 247
  precautionary principle, 12–13, 151
Romania, 47
Russia, 4, 5, 41, 47, 61, 221, 246
  Arctic agreements and, 173, 183
  arctic hydrocarbon resources, 169–70, 181
  environmental regulations, 181–82
  fisheries regulation, 76
  fishing in the North Pacific, 50–85 *passim*
  fishing in the Southern Ocean, 208, 209–210
  Governmental Commission for Nuclear Waste Disposal in the Sea, 150
  Navy, 139
  political developments in, and Arctic policy, 185–87
  radioactive wastes and, 122, 126, 129–30, 134–36, 139, 146–50, 160–61, 247–48, 254–55

shipping in the Arctic region, 170–71
technology transfer in the Arctic environment, 180, 181, 182

Safety standards for marine radioactivity, 157–59
Salmon, 60, 70–71, 72, 73, 77, 78
Salvage Convention, 96–97, 156
Savannah River nuclear reprocessing installation, 133, 136–38
Scientific and technical assistance provisions of the Law of the Sea Convention, 237–38, 240
Scientific Committee on Antarctic Research (SCAR), 174, 214
Seabirds, 59
  driftnet fisheries and catches of, 69, 71–72
  protective treaties, 79
Seals, 192, 203, 208, 209
Sea of Japan, 5, 9, 135, 254
Sea of Okhostsk, 246
"Seas of the USSR" project: 44–45
Sea turtles, 69
Second World Conference on National Parks, 197
Security, concept of, 6–7
Sellafield nuclear reprocessing plant, 128, 133, 142–43, 160, 252
Sewage treatment, 31, 42–43
Shared resource stocks, pollution affecting, 18
Shelton, D., 233
Shipping:
  Arctic commercial, 170–71, 181–82
  of hazardous materials, see Hazardous materials transport
  Law of the Sea Convention on vessel-source pollution, 231–34, 243
  radioactive materials transport, 133–34, 140–42
  in the Southern Ocean, hazards of, 207–208
  transfer of arctic technology through, 181–82
Short-tailed albatross, 75

Singapore, 108
Socialist International, 8
SOLAS Convention, 90, 93
Southern Ocean, 190–222, 249
  competing management principles, 195–201
  conservationist regimes, 195, 200–201
  preservationist regimes, 195, 196–99
  territorial division, 201, 221–22
  as three-sided, 195–96
  environmental hazards in the, 201–216
  deep seabed mining, 193–94, 196, 206
  exploitation of living marine resources, 192, 197, 203, 208–14
  land-based marine pollution, 206–207
  navigation, 207–208
  offshore oil and gas exploration, 193–94, 203–206
  ozone hole, 201–202
  recreation, 215–16
  research activity, 214–15
  tourism, 215–16
  waste disposal, 207–208
  existing legal rules, 190–95
  future management of, 216–22
  map, 191
  overview, 15–16, 190
  UN management of nature preserve, possibilities for, 197–99, 200, 215, 216, 219, 221, 222
  UN Trusteeship for Antarctica, proposed, 198
South Korea, 50, 60, 61, 73, 84, 170
  driftnet fisheries, 69, 70, 79
South Pacific, 156
  antinuclear sentiment in the, 143–44
South Pacific Forum, 144, 253
South Pacific Nuclear Free Zone Treaty, 156
South Pacific Regional Environmental Programme (SPREP), 144

Soviet-American Agreement on Mutual Assistance in the Event of a Large-Scale Oil Spill in the Bering and Chukchi Seas area, 172
Soviet Union, former, 53, 246
American Agreement on mutual fisheries regulations, 1988, 65
Arctic policy of, and Russian Arctic policy, 185–87
chemical weapons stockpile, 104–105
EEZs of, 61, 68
fishing in the Southern Ocean, 208, 210
management approach to land-based marine pollution, 43–47
Ministry of Fisheries, 75
Ministry of Marine Transportation, 183
radioactive wastes and, 122, 126, 129–30, 134–36, 146–50, 160–61, 165, 247
Spain, 217
State Oceanographic Institute (SOI), 45
Stellar's sea cow, 58, 59, 75
Stockholm Declaration on the Human Environment, 1972, 11, 192, 226
Straddling stocks, *see* Living resource problems in the North Pacific/Bering Sea
Strategic interdependence, 7
Swartzman, G. K., 58
Sweden, 146, 173, 181, 184, 217

Taiwan, 50, 53, 60
driftnet fisheries, 68, 69, 79
Technical assistance provisions of Law of the Sea Convention, 237–38, 240
Technology for hazardous materials transport, 117–20
enabling technologies, 117–19
gaps among developing countries, 119–20
Technology transfer, 253–54
in Arctic region, 180–85

on environmental impact assessment, 183–85
Titz, M. A., 108, 117
Tokai nuclear reprocessing plant, 128
Tourism, pollution and, 18, 215–16
"Toward Environmental Security: A Strategy for Long-Term Survival," 8
Trans-Alaska Pipeline System (TAPS), 170, 171
Tuna, 72
Turkey, 45, 47

Ukraine, 40–41, 45, 46, 181, 210
UNESCO Man and Biosphere Programme, 198
United Kingdom, 217, 221
United Nations (UN), 252–53
agencies of, role in environmental security of, 9–11
Antarctic nature preserve to be established under the, 197–99, 200, 215, 216, 219, 221, 222
UN Conference on Environment and Development (UNCED), 20, 66, 145, 198
Agenda 21, 39, 66, 151, 223, 243, 248
UN Conference on Straddling Fish Stock and Highly Migratory Fish Stocks, 67
UN Conference on the Human Environment, 153, 172, 226
UN Conference on Trade and Development (UNCTAD) Code for Liner Conduct, 107
UN Convention on Conditions for Registration of Ships (UNCCORS), 107, 108–109
International Seabed Authority, 233
overview, 16
Part XII of, 226–28
pollution control under, 192, 228–35
radioactivity in the oceans and, 154–55, 159, 160, 161

Index 327

ratifications and concessions, 240, 241–42
UN Convention on the Law of the Sea, 13, 63–64, 67, 201, 223–44, 247, 249
Arctic environment and, 172–73, 178, 179
Article 66, 60
coastal state authority under, 101–102, 227, 229, 232
conclusions, 239–44
drawbacks of, 227, 240–44
driftnet fisheries and, 67–68, 78
ecosystem protection, 234–35, 240
EEZ concept introduced by, 225, 231
as framework for assurance building, 235–39
hazardous waste transport and, 92–93, 111, 120
UN Environmental Programme (UNEP), 11, 39, 47–48, 98, 120, 189
Governing Council, 20
Regional Seas Programme, 38, 47, 230, 237
UN General Council, 11, 67, 197, 219
driftnetting, resolution on, 78, 79, 248
resolutions and declarations of, as source of international law, 225
UN International Law Commission, 248
UN Joint Group of Experts on the Scientific Aspects of Marine Pollution (GESAMP), 17–18, 126
UN Security Council, 11, 253
United States, 181, 182
Arctic agreements, participation in, 172, 173, 183–84
driftnet regulation, 72
EEZs of, 61, 68, 72
fishing industry coalition, 63
fishing in the North Pacific, 50–85 *passim*
future of the Southern Ocean and, 217, 221, 222

shipping standards, 119, 182
Soviet Agreement on mutual fisheries relations, 1988, 64–65
US Advanced Research Projects Agency (ARPA) Technology Reinvestment Project, 254
US Army, 103–107
US Coast Guard, 73, 79, 91, 183
US Department of . . . , *see* Department of . . .
US Driftnet Impact Monitoring Assessment and Control Act, 72
US Fish and Wildlife Service (USFWS), 82
US Geological Survey, 206
US Navy, 254
nuclear submarines, decommissioning of, 138, 139–40
US Palmer Station, 207, 208
US-USSR Agreement on Cooperation in the Field of Environmental Protection, 183
Uranium milling and mining, 124
USSR, *see* Soviet Union, former
USSR State Committee on Hydrometeorology, 45–46

Vaidagubuski, 183
Vessel Traffic Systems (VTS), 118
Vodnyi Transport, 145

Walruses, 205
Waste disposal:
protection of the Southern Ocean, 207–208, 214–15
radioactive materials, *see* Radioactivity in the oceans
Waste management:
hazardous materials transport, *see* Hazardous materials transport
Protocol on Environmental Protection on, 194, 195
Watanabe, Y., 82
Whales, 192, 197, 203, 208–209
White Sea Biological Station, 173
Woods Hole Oceanographic Institute (WHOI), 129–30

Woods Hole Oceanographic Institute (*continued*)
  1993 Conference, 149
  Marine Policy Center, 21
World Commission on Environment and Development, 223
World Conservation Strategy, 172
World Maritime University, 120, 178
World Meteorological Organization (WMO), 11
World Ocean Circulation Experiment (WOCE), 185

Yeltsin, Boris, 9, 150
Yinhe, 106
Young, O. R., 179

ISLAND PRESS BOARD OF DIRECTORS

SUSAN E. SECHLER, CHAIR, Executive Director, Pew Global Stewardship Initiative

HENRY REATH, VICE-CHAIR, President, Collector's Reprints, Inc.

DRUMMOND PIKE, SECRETARY, President, The Tides Foundation

ROBERT E. BAENSCH, Consulting Partner, Scovill, Paterson Inc.

PETER R. BORRELLI, Executive Vice President, Open Space Institute

CATHERINE M. CONOVER

LINDY HESS, Director, Radcliffe Publishing Program

GENE E. LIKENS, Director, Institute of Ecosystem Studies

JEAN RICHARDSON, Director, Environmental Programs in Communities (EPIC), University of Vermont

CHARLES C. SAVITT, President, Center for Resource Economics/Island Press

PETER R. STEIN, Managing Partner, Lyme Timber Company

RICHARD TRUDELL, Executive Director, American Indian Resources Institute